THE SLAIN WOOD

Studies in Industry and Society

Philip B. Scranton, Series Editor

The Slain Wood

Papermaking and Its Environmental Consequences in the American South

WILLIAM BOYD

Johns Hopkins University Press
Baltimore

The publication of this book was supported by a gift from Darlene Bookoff.

Printed in the United States of America on acid-free paper
2 4 6 8 9 7 5 3 1

Johns Hopkins University Press
2715 North Charles Street
Baltimore, Maryland 21218-4363
www.press.jhu.edu

Library of Congress Cataloging-in-Publication Data

Boyd, William, 1967–
The slain wood : papermaking and its environmental consequences in the American South / William Boyd.
pages cm
Includes bibliographical references and index.
ISBN 978-1-4214-1878-0 (hardcover : alk. paper) — ISBN 978-1-4214-1331-0 (electronic) — ISBN 1-4214-1878-9 (hardcover : alk. paper) — ISBN 1-4214-1331-0 (electronic) 1. Papermaking—Southern States—History. 2. Paper industry—Environmental aspects—Southern States. 3. Clearing of land—Environmental aspects—Southern States. 4. Southern pines. 5. Wood-pulp. I. Title.
TS1095.U6B69 2015
676.0973—dc23 2014047755

A catalog record for this book is available from the British Library.

Special discounts are available for bulk purchases of this book. For more information, please contact Special Sales at 410-516-6936 or specialsales@press.jhu.edu.

Johns Hopkins University Press uses environmentally friendly book materials, including recycled text paper that is composed of at least 30 percent post-consumer waste, whenever possible.

In this actual world there is then not much point in counterposing or restating the great abstractions of Man and Nature. We have mixed our labour with the earth, our forces with its forces too deeply to be able to draw back and separate either out. Except that if we mentally draw back, if we go on with the singular abstractions, we are spared the effort of looking, in any active way, at the whole complex of social and natural relationships which is at once our product and our activity. . . . But it is even harder than that. If we say only that we have mixed our labour with the earth, our forces with its forces, we are stopping short of the truth that we have done this unequally; that for the miner and the writer, this mixing is different, though in both cases real; and that for the labourer and the man who manages his labour, the producer and the dealer in his products, the difference is wider again. Out of the ways in which we have interacted with the physical world we have made not only human nature and an altered physical world; we have also made societies.

—Raymond Williams, 1980

CONTENTS

PREFACE

The pine trees seemed to go on forever, sweeping by in endless rows as we drove along the two-lane road to the beach. There was a monotony to it all—the kind that can only come from something artificial, something planned and stamped hard against the sandy soils of the South Carolina coastal plain. Neat, even-aged stands of loblolly pine stretching in all directions—interrupted here and there by fields and farms, old barns, trailers, a small crossroads, and occasionally a brief dip through some shady bottomland with a blackwater creek and a glimpse of the diverse, chaotic, more natural forests that used to thrive on this land. And every now and then, more so as we got closer to the coast, we saw the giant live oaks standing over an old farmhouse or some other structure in the middle of a field—witnesses to a world that long ago surrendered to something new.

Driving every summer with my family from our home in Columbia to Pawleys Island on the South Carolina coast provided an unmatched visual record, even to the eyes of a child, of the indelible imprint that the modern pulp and paper industry was making on the southern landscape. Pine trees by then had long been the region's number one "cash crop"—planted across millions of acres of previously cutover and degraded lands as if to cover and repair, even if only slightly, a harsher time when some southerners (and their partners from the North) took far too much from the land.

We saw people too, black and white, as we drove through the same hollowed out towns: Wedgefield, Pineville, Paxville, Greeleyville, Andrews—towns that had been swept up in one way or another by the great agrarian transition brought on by the New Deal that changed the regional economy as much as the Civil War; towns that now seemed emptied out of all but a few. Where had they all gone? What happened to these main streets? The New South was somewhere else.

As we drove toward the coast, trucks hauling pulpwood logs to the mills would sometimes pass us—a menace to everyone on the road. We had heard stories of fatal accidents; some people even stopped driving the old way because of the trucks. But we never paused to ask how these pulpwood haulers lived—who they were and what they did. They were feeding the vast appetites of the big pulp and paper mills (there were three within a hundred miles or so), hauling logs cut from somebody's land down some side road in some forgotten corner of the rural South. Most of these men (and they were all men) were independent contractors, at least in the eyes of the law, with limited formal education and few resources. And back in the woods there were more men, many of them black, cutting and loading that wood in the relentless heat (it may be hotter now; but it was plenty hot then) and the bugs and the heavy machinery, hot metal blades and booms, gasoline and diesel, intense noise. Timber harvesting, even the mechanized operations of today, has always had a certain violence to it that seems missing in agriculture.

These men too were almost certainly independent contractors, rounded up by the so-called wood dealers—local powerbrokers who were good at hustling wood and tapping into the informal labor markets that marked so much of this region—playing on deep vulnerabilities to dispatch the "producers" as they were known to cut wood for the mills in one of the most dangerous occupations in America. These woods workers, some of them barely one step away from sharecropping, lived on the extreme margins, excluded from basic protections, unable to organize because of their status as independent contractors, and exempt for more than twenty-five years from the wage and hours provisions of the Fair Labor Standards Act as a result of intense lobbying by industry leaders. Even though they provided the basis for a flexible labor system that worked to secure the competitive advantage of the southern pulp and paper industry, they were always on the outside looking in.

Sometimes we would pass a cutover field, with the stumps and trash trees twisted up in the dirt and left behind in snarls. If it was fresh and the windows were down, you could smell the mix of churned up earth and broken bits of root and tree. It was clear that something big and powerful had been there, taken value from the land, and left the rest behind.

There were replanted fields too—small green pine trees neatly planted in rows with the improved stock from the company tree nurseries that were so busy breeding the new, faster-growing varieties that would start the cycle all over again. Some of this land was company land, but most of it wasn't. Most of it was under some kind of management deal with the mills and the forestry consultants, with shiny green tree farm signs posted on the edge as a testament to their role in the complex network of actors who serviced the industry.

And as we got closer to Georgetown, depending on the wind, we could sometimes smell the big International Paper mill. We always knew we were close then. People have long described it as a rotten egg smell. It was sulfurous to be sure, leaving a vague sense of some kind of dangerous off-gassing from the big cook that was going on. Chemicals and heat and water, dissolving all of those pine trees, all of that tough lignin that makes wood hard, to liberate the long fibers and make the brown bags, the cardboard, the clean bleached white paper that so much of modern life took for granted.

Driving out of Georgetown on Highway 17 over the bridges to the barrier islands on the coast, we could see back behind us off to the right the blocky structure and tall stacks of the paper mill—a vast industrial complex, something much bigger and more modern than anything that had come before—a tightly bound cluster of fixed capital that embodied many hundreds of millions of dollars in investment—always in motion, with an appetite for raw materials—water, fiber, energy—that was difficult to comprehend.

And there were people working in that mill, though you would never know it from the outside, at least from afar. We never got close enough to see, and of course we never had any reason to see these people. They were different, separated from us by the deep lingering divisions of class and race that marked so much of the New South. But these were good jobs, high-paying jobs, union jobs—much better than those of the woods workers. And because of that, in the kind of fight that divided the working class in the South for so long, many of the best jobs were reserved for whites during the decades before civil rights. As I learned later, the unions did much to make this happen, creating segregated locals and working with management to construct seniority lines that made it all but impossible for black workers to get access to the better, higher-paying jobs in the mills. It took courageous activism on the part of black workers and their advocates together with the federal government, acting under the power of the new civil rights laws, to straighten that out. During the 1960s, the pulp and paper industry was the most litigated industry in the South.

And finally the signs, yellow and red with black writing—maybe a skull and cross bones—it's hard to remember—warning against eating the shellfish from Winyah Bay and the surrounding tidal complex because of some kind of contamination. Why the signs? They had not always been there. What had happened here? Was this somehow related to the mill? These connections were hard to grasp as a child, overshadowed by a half-shaped feeling that something was broken, out of balance, not right. The massive discharges from the mill had used up most of the oxygen in the water, suffocating much of the life that was there. And later, when they found dioxin contamination downstream from some of the pulp and paper mills, it was clear that there was also a different kind of violence at work here—silent, invisible, latent—something that could initiate a cascade of harms that might one day end up as cancer or some other dreadful disease. When they tested these waters in the 1990s, the fish downstream from the International Paper mill in Georgetown had some of the highest concentrations of dioxin in the country. The government issued fish advisories, but it's hard to know how much of a difference that made. Some people depended on the fish they caught from the Sampit River for protein. Some of them probably couldn't read the signs anyway.

These are some of the memories that I have of an industry that reached deep into the rural South to mobilize the productive capacities of southern land and labor on a previously unimaginable scale and build a regional industrial complex that dominated national and international production for much of the twentieth century. In many ways, this industry represented the very best of the New South. But in others, it reinforced and fed on a legacy of inequality, discrimination, and disregard for people and place.

This book is about that industry. It traces the connections that I was only vaguely aware of (and in some cases completely unaware of) on those many drives to the coast. It shows how the pulp and paper companies, in concert with other actors, constructed a highly successful regional industrial system out of the distinctive institutional and biophysical landscape of the post–New Deal South.

Although I now teach at a law school, and while there is plenty of law in this story, this is not a legal history. Most of the research and a fair amount of the writing were done before I went to law school. Had I set out to write this book after that experience, it would almost certainly be more narrow with less attention to the many people and places that made this industry such an important part of the New South. While I believe very strongly in the constitutive role of law in economic and social relations, I also believe, as Willard Hurst observed, that any understanding of the actual role of law in society must be situated within larger social, political, and economic processes.

If anything, this book represents an effort to integrate southern economic and environmental history through a careful examination of one of the region's most important industries. As such, it is a modest attempt to come to terms with the burdens of southern history in our own time, finding inspiration in the powerful writing of C. Vann Woodward and other historians of the New South, the remarkable work of regional social scientists such as Howard Odum and Rupert Vance, and the incomparable portraits of the region present in the works of writers such as William Faulkner. In the end, though, if this book has any value, it will largely be because of the stories and perspectives that I collected in the many interviews conducted with people working in the industry and living in various communities touched by it in one way or another. More than anything else, it was through the experiences they so generously shared with me that I was able to see the industry and its connections to the region in a more direct and palpable way. For this I am thankful.

ACKNOWLEDGMENTS

This book has been a long time coming, and it has benefited immensely from the generous support and guidance I have received from many different people and institutions. Chief among them, as mentioned in the preface, are the people who allowed me to interview them during my fieldwork. Needless to say, this book would not have happened without them.

As a graduate student at the University of California, Berkeley, where this project was born, I was fortunate enough to be in the company of a remarkable group of faculty and fellow students. Despite the fact that I have lost touch with some of them over the years, they have always been with me throughout the long and winding course of this project, and I am happy to have the chance to thank them here. Michael Watts has served as a source of insight and inspiration since I first met him in the early 1990s. He showed me, and so many others, how to study agrarian transitions and what good, critical social research looks like. Annalee Saxenian taught me how to think about regional economic development and industrial systems and the great value of getting out and talking to people. Gavin Wright, even though he was down the road at Stanford, provided an unparalleled source of knowledge and expertise on southern economic history and, more importantly, invaluable guidance in the planning and execution of the research that became this book. My fellow graduate students at Berkeley deserve special mention as intellectual colleagues, friends, and fellow travelers in field-based social research. In particular, I want to thank Scott Prudham, Dara O'Rourke, Navroz Dubash, James McCarthy, Sharad Chari, Paul Sabin, Thomas Sikor, Sid Dietz, and Denny Kelso. Finally, the Energy and Resources Group at Berkeley, affectionately known as ERG, provided a wonderful setting for interdisciplinary research on problems of resource use and environmental change. There should be more places like ERG in the American academy. I am thankful to have found it.

At the University of Colorado, both in the Law School and across campus, I have had the good fortune of working with a wonderful group of colleagues. Several were kind enough to read all or part of this manuscript. Charles Wilkinson, Ahmed White, and Paul Sutter gave me important comments on various parts of the manuscript as it took final form. Cynthia Carter, Anshul Bagga, and

Diana Avelis deserve a great deal of thanks for helping to get the manuscript in shape. Librarians Jane Thompson and Matt Zafiratos were always available to track down obscure and hard-to-find sources. And, even though this project is largely unfamiliar to them, my law faculty colleagues Fred Bloom, Pierre Schlag, Kristen Carpenter, and Sarah Krakoff, among others, have been an important source of intellectual engagement and support over the last six years.

Beyond Colorado, I have been fortunate to work with and learn from a stellar group of environmental law colleagues. This book is surely better because of my interactions over the years on various topics with Ann Carlson, Jed Purdy, Doug Kysar, Jim Salzman, and Buzz Thompson, among others. I have also benefited immensely over the past decade from collaborations with an amazing group of people working in tropical forest regions all over the world. Dan Nepstad, Tony Brunello, Heather Wright, Claudia Stickler, Toby McGrath, Steve Schwartzman, John Carter, Rosa Maria Vidal, Romeo Dominguez, Mariano Cenamo, Monica Julissa De Los Rios de Leal, Avi Mahaningtyas, and many others have help me understand the challenges of agrarian change and rural development in the often unruly tropical frontiers of Brazil, Indonesia, Mexico, and Peru, providing an important comparative perspective for my work on the American South.

In conducting the research for this book, I benefitted greatly from assistance provided by librarians and archivists at several institutions. Specifically, librarians and staff at the Forest History Society in Durham, North Carolina; the Georgia Historical Society in Savannah; the University of Georgia Library; the Southern Historical Collection at the University of North Carolina; the National Archives; and the Bioscience and Natural Resources Library at UC Berkeley were all extremely helpful.

My editor at Johns Hopkins University Press, Robert J. Brugger, and his assistant, Kathryn Marguy, have been very patient (heroically so!) and supportive during the course of this project. They deserve great thanks for helping me get all of the pieces in place, for editing and fine-tuning the manuscript, and for seeing it through to completion. The book also benefitted from the astute comments and suggestions provided by an anonymous reviewer for the Press and from extensive comments and encouragement from Phil Scranton at the beginning of the process.

I was fortunate enough to be able to present parts of this work at conferences, colloquia, and workshops at Berkeley, Johns Hopkins University, Georgia Tech, and the Yale Agrarian Studies Program. An earlier version of chapter 1 was published in *The Second Wave: Southern Industrialization from the 1940s to the*

1970s, Philip Scranton, ed. (Athens: University of Georgia Press, 2001), and portions of chapter 4 were published in the *Stanford Environmental Law Journal* 21, no. 2 (2002). Thanks are also due to the Forest History Society for assisting with some of the photographs included in the book, to Gabriel Benzur for permission to use two of his father's photographs, and to Bill Nelson for help with the maps.

The research on which this book is based was generously supported by several grants and fellowships. A three-year U.S. EPA STAR Fellowship provided critical support for my fieldwork and much of the early writing. The MacArthur Foundation and the University of California also provided much-needed support during the research and writing phases.

Finally, I want to thank my family. My father, Bill, my mother, Eve, who sadly died way too young, my sister, Margaret, and my brother, John, have always been there for me. They were with me on those many summer drives to the coast, helping me to begin making sense of the complex social and physical landscapes that make up the rural South. Together with our larger extended family (and now families), they have taught me so much about the great joys of belonging. Most importantly, I owe more thanks than can be expressed here to my wife, Norrie, who also grew up with me driving those same roads to the coast, and to our three wonderful daughters, Eve, Mary Wallace, and Norrie (wee). They have provided unconditional love and support throughout, always reminding me of the truly important things in life.

THE SLAIN WOOD

Introduction

> The biggest single thing that has stimulated the South has been the coming of the pulp and paper industry on a very large scale. It wasn't alone, but the things that it brought with it—namely the chemical industry, and the land they bought and paid taxes for, coupled with the stimulus that it gave to the railroads, truck transportation, and everything else—were a real massive shot in the arm . . . There are hardly a half dozen counties in this region that haven't had their annual income boosted tremendously through wages from pulp companies, or selling wood or land to pulp companies to put money into circulation—outside money, Yankee money, real money, not the same old soiled bills that we used to pass around taking in each other's washing. —*Inman F. Eldredge, 1959*

Yankee money. Nothing stirred the hearts and minds of southern boosters more. Few things, outside of race, captured the predicament of postbellum southern economic development better. Without that money, the South seemed doomed to economic backwardness. With it, anything seemed possible. Of course, capital alone would never solve the problems of underdevelopment that plagued the region after the collapse of Reconstruction. For that to happen, institutions would have to change—a new industrial order would have to be constructed. But outside investment was clearly a key ingredient in the building of a new, industrialized South. Despite the persistent cries of dependency and colonial exploitation voiced by southern intellectuals and populist political leaders during the late nineteenth and early twentieth centuries,[1] most local politicians and businessmen were only too happy to have a new factory or a new rail line built in their area. Any investment was better than none at all.

To be sure, much of the outside capital that flowed into the region during the decades after Reconstruction concentrated on exploiting the region's plentiful natural resources and cheap labor with an all too often brutal disregard for the South's long-term economic prospects. Seeking to turn a quick profit, business ventures in extractive sectors such as lumber and mining plundered the region's natural wealth without generating any significant economic linkages. Other manufacturing concerns, notably textiles, built their comparative advantage on the region's low-cost workforce.

Any effort to spur real and lasting industrial development in the region also had to contend with the distinctive agrarian class structure and institutions that emerged after the Civil War. As regional sociologist Rupert Vance remarked in 1932, agriculture "sets the background for all other industries in the South."[2] Notwithstanding their waning fortunes, southern agrarians continued to exert a disproportionate influence on regional politics well into the twentieth century.[3] And the southern economy itself was largely hostage to the dominant institutions of sharecropping and farm tenancy, which together stifled labor mobility and stunted the growth of normally functioning labor markets. Bound through ties of debt and dependency, many southerners found it almost impossible to escape farming. As a result, the South remained a predominantly rural and agricultural region well into the twentieth century, trailing far behind the rest of the country in major indicators of economic development.[4]

Indeed, by 1930, more than half a century after Reconstruction, the region found itself in the midst of a protracted economic crisis. Prices for cotton—the staple of the regional economy—had been in significant decline since the early 1920s. Per capita income had fallen to around half of the national average, back to where it had been in 1880. Other major indicators of development—nutrition, literacy, access to health care and housing—were shockingly deficient when compared to the rest of the nation.[5]

The region's biophysical environment hardly fared better. Driven by the perverse incentives embedded in the farm tenancy system, tenants and sharecroppers found themselves trapped in the diminishing returns of farming worn-out soil, creating a vicious cycle of deepening poverty and accelerating environmental degradation.[6] Because merchants and landowners would only extend credit for a commitment to plant cotton—a row crop that was notoriously hard on the soil—tenants and sharecroppers had neither the ability nor the incentive to maintain the productivity of the land. Merchants squeezed the tenants, who in turn squeezed the land. By the early 1930s, more than 32 percent of the region's *total* land area had been severely eroded. In the hilly areas of the southern piedmont, land degradation affected more than half of the total land area (figure I.1).[7] For the famous southern regionalist Howard Odum, "the destroyed land" carried with it a "long chain of consequences" for the region. Images from Farm Security Administration photographers such as Dorothea Lange, Walker Evans, and Marion Post Wolcott revealed a land that was seemingly beyond repair. In Odum's words, "vast gullies and gulches, wagon wide and tree deep, spotty hillsides and great stretches of fields marred like some battle field—each year destroyed more and more, each decade added ugliness and havoc to the landscape."[8]

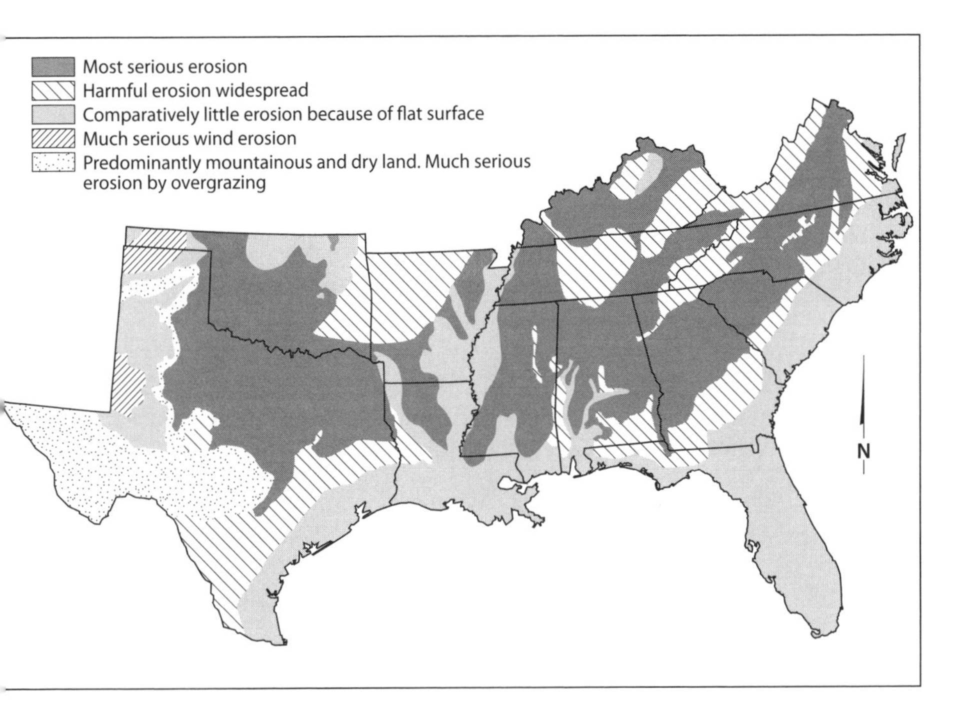

Figure I.1. The South suffered widespread erosion during the 1920s and 1930s. This map shows the extent of erosion circa 1933 based on surveys conducted by the U.S. Department of Agriculture. Adapted from Howard Odum, *Southern Regions of the United States* (Chapel Hill: University of North Carolina Press, 1936), 38.

The severe degradation visited on the agricultural landscape was matched by a relentless assault on the region's forests. During the half century after Reconstruction, the lumber and naval stores industries marched virtually uninhibited through the South's old-growth forests, leaving massive areas of denuded land. Much of the destruction was fueled by the rapid transfer of public lands to private ownership at rock-bottom prices. During the twelve years after the 1876 repeal of the Southern Homestead Act, which permitted unrestricted cash entry into the public lands states of the region,[9] some 5.7 million acres of federal lands in the southern public lands states were sold—the vast majority to speculators and lumbermen from the Lake States for cash at the minimum price of $1.25 per acre.[10] By 1888, when Congress moved to restrict purchases, the best of the southern public lands were in private hands.[11] In the words of one historian, "The South had invited the investment of northern capital but instead had gotten mere speculation and retarded development. When it awoke to its error it reverted to the policy of 1866 but the change came too late. Northerners controlled the best stands of yellow pine and cypress lands and were to reap the benefit by taking the cream of the profits from the rising lumber industry."[12] In comparison to the federal disposal, state land sales were even more generous. Florida, for example, sold some four million acres at twenty-five cents an acre to a group of Philadelphia capitalists in 1881, while the government of Texas granted twelve railroad companies more than 32 million acres—an area larger than the state of Indiana.[13]

The overall result was an "enormous loss of wealth to the public and its monopolization by a few interests."[14] Shortly after the Civil War, at least three-fourths of the standing timber in the United States was publicly owned. By the early 1910s, about four-fifths was privately owned.[15] Although there still were vast public lands in parts of the West, very little remained in the South. Meanwhile, those who acquired the best southern timberlands reaped extraordinary profits. Much of the southern pine that had been sold for $1.25 an acre in the 1880s fetched more than $60 an acre by the 1910s.[16]

Widespread cutting of the region's forests ensued. Motivated by the prospect of spectacular returns, lumbermen cut virtually all of the virgin forests in the region, with severe overcutting in the southern Appalachians and in the vast pineries of the coastal plain. Given the distinctive insecurities and constraints that accompanied timberland ownership—threat of fire and disease, an unfavorable tax system, liquidity constraints, faulty price expectations, and inadequate knowledge of forestry—most landowners operated on the basis of high discount rates and short time horizons. The basic incentive was to cut and get out. The destruction reached a crescendo during World War I, declining rapidly as the bottom fell out of the market in the wake of the war. But the same insecurities and constraints that drove excessive cutting also stifled attempts to regenerate cutover timberlands. Instead of replanting, the owners most often sold cutover lands to desperate farmers or left them idle as the lumber barons moved on in search of new opportunities.

By 1920, the cutover area in thirteen southern states exceeded 156 million acres. Most of this land, perhaps as much as 100 million acres, was in the yellow pine belt stretching from South Carolina to Texas.[17] In that region, less than 24 million acres of old-growth pine remained in 1920. Seven years later, the figure stood at less than 13 million acres. In the state of Georgia, which had once boasted as much as 18 million acres of old-growth pine, only 700,000 acres remained in 1920. By 1927, this figure had been reduced by half.[18]

As the industry began to "cut out" in the 1920s, countless rural towns and villages that had grown up to support it suddenly found their existence threat-

Figure I.2. Extensive cutting by the lumber and naval stores industries during the late nineteenth and early twentieth centuries left behind large areas of cutover lands. This photograph, taken by Arthur Rothstein of the Farm Security Administration, shows a "submarginal farm on cutover land" in Hernando County, Florida (1937). Library of Congress, Prints & Photographs Division, FSA/OWI Collection, LC-DIG-fsa-8b35759.

ened. Many simply "curled up and died."[19] Throughout the South, the devaluation associated with this ghost town phenomenon became a major public policy issue (figure I.2). In 1922, Georgia's state board of forestry reported to the general assembly that "dismantled mill plants and deserted communities throughout the lumber regions of the state are forbidding reminders of the migration of an industry which, under wise and proper management of our forestlands, should be a permanent and leading industry in the state."[20] Leading intellectuals such as William Faulkner and Thomas Wolfe saw the destruction of the southern landscape as part of a larger regional tragedy.[21] The South had sold off a large part of its natural endowment and had very little to show for it.

Well before the depression of the 1930s, therefore, the South found itself in the grip of a protracted economic and environmental crisis. Much of the region's soil and forest resources had been ruined. Millions of southerners struggled to scratch out a living in farming or some other low-wage occupation, trapped in the "worn grooves of a tributary economy."[22] Racial animosities deepened as southern political leaders worked to divide the working class and consolidate white rule.[23] As the rest of the nation sank into depression, the situation went from bad to worse. For many, the region became synonymous with intractable poverty and economic backwardness—summed up in President Roosevelt's famous 1938 characterization of the South as "the nation's number one economic problem."[24]

It was against this backdrop that pulp and paper firms began investing in the South. Moving into the region in the wake of the destruction wrought by the lumber industry, these companies promised a new kind of development. The large investments in fixed capital embodied in a new pulp and paper mill and the substantial demand for land and timber represented something the region had never seen before. As Inman F. Eldredge, a prominent forester and director of the Southern Forest Survey, pointed out, this was an industry that reached deep into the rural South, creating the kinds of economic linkages considered so important to balanced industrial development.[25]

In fact, despite the devastation of the region's land and timber resources, the South offered significant opportunities for pulp and paper firms. Once it became clear that the "Kraft" or sulfate process could be used to make commercial grades of paper from southern pine and that southern timberlands were capable of rapid regeneration, companies raced to establish integrated pulp and paper complexes throughout the region.[26] Facing a declining raw material base in the Northeast and lured by the promise of dramatically lower production costs, leading firms began to set up pulp and paper operations in the South during the 1920s and 1930s.[27]

By the latter half of the 1930s, the "grand march South" had turned into a stampede.[28] Companies such as International Paper and Union Bag built new mills on a gigantic scale and integrated backward into raw materials supply. Profitability in the industry soon came to depend on a southern production base. With the opening of its Savannah, Georgia, pulp and paper complex in 1936, for example, Union Bag transformed itself from "a dopey, mismanaged corporation," which "even managed to show a deficit straight through the golden boom years" of the late 1920s, into a profitable company. Net profits from its new Savannah mill were estimated to be around 30 percent.[29] For the region as a whole, investments in pulp and paper mills totaled more than $100 million between 1935 and 1940, doubling regional pulping capacity. At the beginning of 1940, some forty-seven mills were operating or under construction in the region,

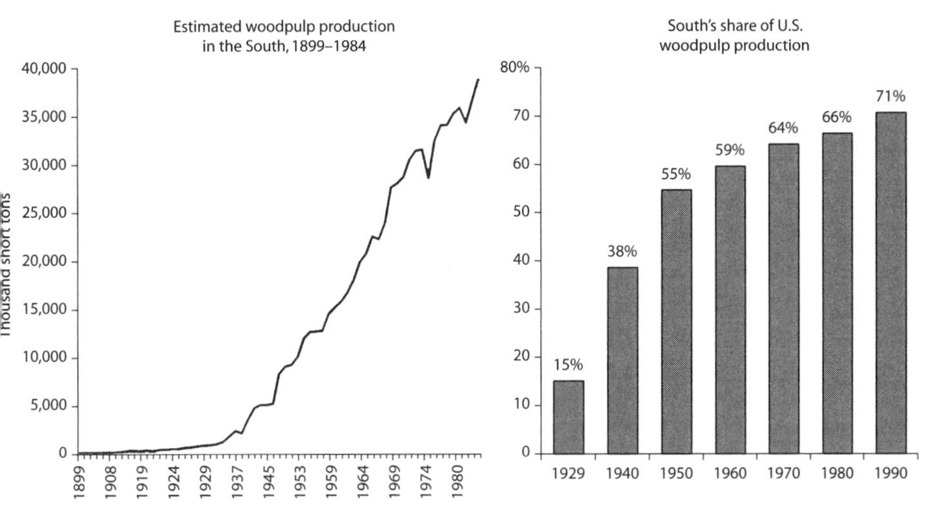

Figure I.3. Woodpulp production grew dramatically in the South after 1930. Data from U.S. Forest Service, *The South's Fourth Forest: Alternatives for the Future*, Forest Resources Report no. 24 (Washington, DC: Government Printing Office, 1988), 329, table 2.27; U.S. Forest Service, *An Analysis of the Timber Situation in the United States: 1952–2030*, Forest Resource Report no. 23 (Washington, DC, 1982), 298–99; American Forest & Paper Association, *Paper, Paperboard, and Woodpulp Statistics* (Washington, DC, 1995), 49. The South includes Delaware, Maryland, Virginia, West Virginia, North Carolina, South Carolina, Georgia, Florida, Tennessee, Kentucky, Alabama, Mississippi, Texas, Oklahoma, Arkansas, and Louisiana.

representing a total investment of over $200 million.[30] After a brief interruption due to war, the procession resumed at an even faster pace.

At midcentury, the South's share of domestic woodpulp production had grown to 55 percent, up from a mere 15 percent in 1929. Two decades later, the South enjoyed undisputed national leadership in the production of pulp and paper, accounting for almost two-thirds of domestic woodpulp production and one-half of paper and paperboard production (figure I.3).[31] By the 1990s, the region was home to roughly three-fourths of domestic pulpwood production, more than two-thirds of woodpulp capacity, and over half of the nation's paper and paperboard production (figure I.4).[32]

National dominance translated into global leadership as well. By the mid-1990s, the southern pulp and paper industry was far larger than that of all other countries (figure I.5).[33] The South also accounted for about half of the world's total acreage of industrial timber plantations, despite containing less than

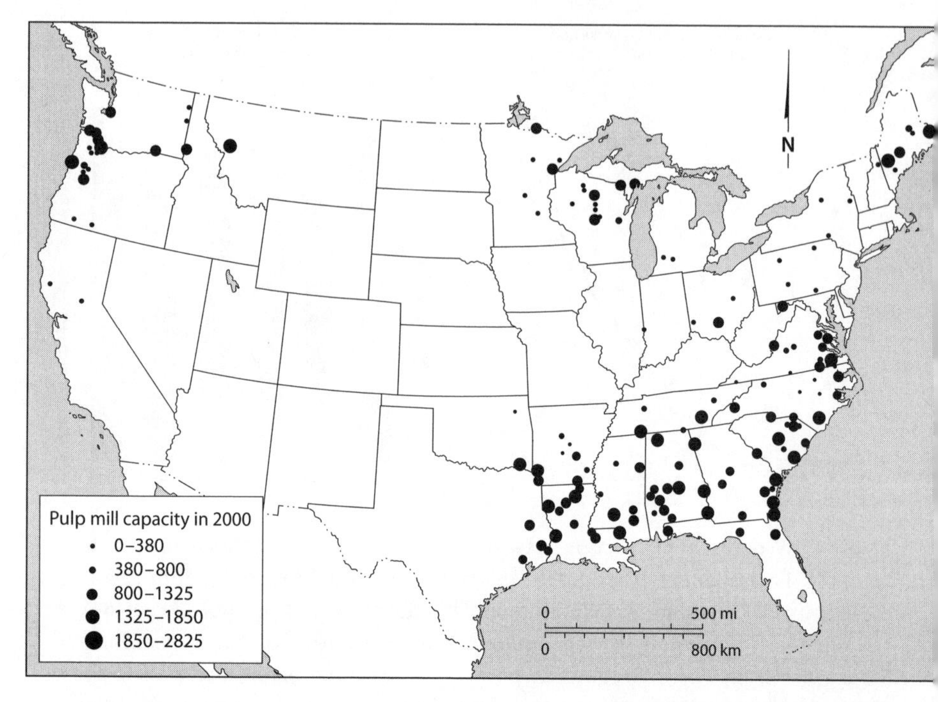

Figure I.4. Over the course of the twentieth century, the South emerged as the dominant pulp and paper producing region in the country. This map shows the location and capacity of pulp mills in the United States circa 2000. Adapted from data compiled by Jeffrey Prestemon et al. at the UDSA Forest Service Southern Research Station, Asheville, NC.

3 percent of world forest area (figure I.5).[34] As the global leader in the production of pulp and paper based on the intensive cultivation of timber, the southern pulp and paper industry clearly represented one of the region's more impressive industrial success stories.

Viewed in the aggregate, the dramatic concentration of pulp and paper production in the South during the post–New Deal period might be explained simply as the result of regional comparative advantage in key factors of production and the successful attraction of branch plants controlled largely by northern corporations. Once the southern economy was liberated from the chokehold of cotton tenancy—largely through New Deal agricultural and labor policies—labor and capital could move freely in and out of the region according to relative factor endowments.[35] Key resources needed for paper production—land, water, and energy—were in plentiful supply. Meanwhile, state and federal policies aimed at promoting forest conservation and fire control created incentives for

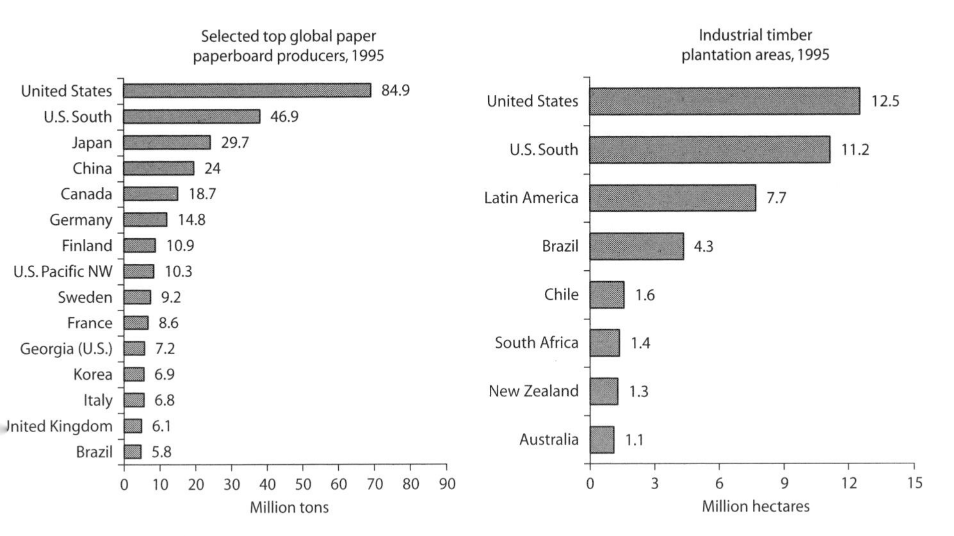

Figure I.5. By the 1990s, the South was the global leader in pulp and paper production and industrial timberland area. Paper/Paperboard Production Figures: *Pulp and Paper International*, January 1997; American Forest and Paper Association, *Paper, Paperboard, and Wood Pulp, 1995 Statistics*. Industrial Timber Plantation Areas F. Cubbage, R. Abt, W. Dvorak, and G. Pacheco, "World Timber Supply and Prospects: Models, Projections, Plantations and Implications" (presentation at Central American and Mexican Coniferous Resource Cooperative Meeting, Bali, Indonesia, October 21–23, 1996). Figures for the United States and the South are based on linear projections of U.S. Forest Service data and represent more conservative estimates than the Forest Service's own projections. Figures for other regions and countries are taken from Bob Flynn, "Latin America: The Future of Fiber Exports," and Mike Edwards, "The South African Forestry and Forest Products Industry: A Synopsis," both in *Proceedings of the International Woodfiber Conference, May 13–14, 1996, Omni Hotel at CNN Center, Atlanta, Georgia* (San Francisco: Miller Freeman, 1996).

firms to invest in timberlands and promote industrial forestry among southern landowners. With few pollution control laws on the books and a strong desire to recruit industry, southern states provided a hospitable climate for big polluting industries such as pulp and paper. Finally, the region's cheap labor supplies and limited protections for workers proved attractive to firms investing in the region. Together these various factors promised substantial opportunities for firms seeking to develop new low-cost production complexes in the region.

While such an explanation is fine as far as it goes, it fails to grasp the ways pulp and paper firms, in concert with other actors, mobilized and organized the productive capacities of southern land and labor and integrated them into a highly competitive industrial system. It fails, in other words, to account for the actual process of regional industrialization and, specifically, the social and environmental changes that underwrote the rise of the southern pulp and paper industry. This book investigates these processes of regional industrialization and environmental change, explaining how a new and highly competitive industrial system was fashioned out of the distinctive mix of regional institutions and resources that marked the postbellum South. Rather than focusing on traditional notions of comparative advantage, the perspective taken here views regional industrialization as a historically situated process of social, political, and ecological construction. The goal is to understand how, out of the wreckage of the interwar decades, various social actors constructed a new industrial order that came to dominate national and global production within a couple of decades.[36]

In place of top-heavy narratives that privilege national patterns of industrialization, therefore, the story told here looks from the ground up at the construction of new organizational forms and strategies necessary to deal with the various institutional and biophysical challenges specific to the region.[37] The southern pulp and paper industry, in this view, did not represent a regional version of the large vertically integrated, mass production paradigm that marked so much of the American economy during this time. Nor did this process of regional industrialization derive solely from the South's comparative advantage in key factors of production and the successful attraction of outside investment. This was a distinctive industrial system deeply inflected with the character of regional institutions.

The book is organized around a series of problems that confronted firms as they expanded their operations in the region. Each of the four chapters deals with one of these problems. Put most succinctly, they consisted of the making of a highly productive "industrial forest" in the South (chapter 1), the social organization of logging and wood procurement (chapter 2), the management of race and class in the context of mill labor and the distinctive capital requirements in the industry (chapter 3), and the politics of environmental pollution associated with pulp and paper production (chapter 4). Because each of these problems embodied a distinctive aspect of regionalism, solving them required deep engagement with both the biophysical and the social landscape of the South. In the process, the history and geography of the region came to be intimately bound up with the construction of a new industrial system that dramatically reshaped not only the structure of the industry but also the land and its people.

Chapter 1 begins with one of the immediate problems that confronted firms as they moved into the region: ensuring a dependable long-term supply of timber. Because the new southern mills represented such huge investments in fixed capital, they were incapable of migrating throughout the region in search of timber. As a result, pulp and paper firms faced an imperative of establishing a regime of forest management unlike anything previously experienced in the lumber industry. Solving the timber supply problem, in other words, required an entirely new approach to southern timberlands—one that treated the forest as crop rather than as mine. Given the long time lags involved in timber growth and the fact that small landowners controlled two-thirds or more of the region's timberlands, this was not an easy task. Small southern landowners had to be persuaded through education and incentives of the virtues of viewing timber as a crop.

Such an effort entailed extensive public-private cooperation and institution building in areas such as fire control, tax reform, reforestation, and tree improvement, all situated against the backdrop of deep structural changes in southern agriculture facilitated by federal agricultural policies during the New Deal and after.[38] Together these policies and programs opened up substantial acreages of cropland for conversion to other uses and created a more rational landscape for longer-term investments in timberlands. In the process, the southern forest was regenerated and subordinated to the dictates of industrial production. Between 1940 and 1990, more than 38 million acres were artificially regenerated in the South. During this time, southern tree nurseries produced over 23 billion tree seedlings, the vast majority of which, since the 1980s, were genetically improved. Timber soon replaced cotton as the region's number one cash crop—one that was subject to an increasingly sophisticated effort to accelerate growth rates and maximize desirable characteristics.[39]

This rather dramatic transformation of the southern forest into a highly productive "organic machine"[40] derived in part from a process of regional collective learning—an interpretation that raises important questions about the supposed lack of an indigenous technological community in the postbellum South and the region's capacity for innovation. That said, the remaking of the southern forest also entailed a far-reaching process of ecological simplification. While replacing worn-out cotton lands with industrial timber plantations clearly represented a more ecologically rational form of land use, the conversion of mixed, diverse forests to intensively managed pine monocultures also brought with it a significant loss of biodiversity and habitat. Not surprisingly, by the end of the twentieth century, southern timberlands had become an increasingly contested terrain as environmentalists and others challenged the ongoing industrialization of the landscape.

Building a new regime of forest management, of course, only solved part of the supply problem facing firms as they established operations in the region. Securing access to the wood and delivering it to the mills proved to be an equally challenging problem—the subject of chapter 2. Because of regional land tenure patterns, new mills depended on the thousands of small landowners scattered throughout the region for three-fourths or more of their timber supply. In order to access such timber on favorable terms, pulp and paper firms carefully constructed a series of highly localized timber markets in which they or their agents, known as wood dealers, typically had the upper hand in price negotiations. By acquiring large tracts of timberlands around the mills, these firms also created a very effective hedge against price pressures. Most small timberland owners seeking to sell timber to the mills found themselves in a take-it-or-leave-it situation. Not surprisingly, more than a few expressed disappointment with the prices they received for their timber—an issue that became the subject of congressional investigations during the 1940s.

Firms also faced the considerable challenge of organizing a labor force capable of harvesting timber from the many small tracts dispersed throughout the region in a manner that delivered wood to the mills in a timely and dependable fashion. Because southern pine deteriorated rapidly in storage, moreover, stockpiling timber was not an option. Most mills also varied their production runs on a frequent basis, leading to weekly fluctuations in timber demand by as much as 30 percent. Given the variable nature of the work and the many risks and liabilities associated with logging, company managers understandably hesitated to employ company logging operations, finding it much more desirable to leave the recruitment and organization of loggers to local wood dealers and others who were familiar with the complexities of local labor markets and comfortable with the racial norms of the postbellum South. As chapter 2 illustrates, the procurement system that emerged exhibited principles common to flexible "just-in-time" production systems. While this system typically worked well for the mills, it translated into deep insecurities for those who worked in the woods. As independent contractors operating at "arm's length" from the pulp and paper companies, pulpwood loggers ended up shouldering virtually all of the risks and liabilities associated with one of the most dangerous occupations in the country.

Far from a simple response to various transactions costs, the southern wood procurement system was very much a product of active political and legal maneuvering on the part of the pulp and paper industry. Beginning in the 1940s, for example, advocates for the pulp and paper companies lobbied successfully to exempt southern loggers from the wages and hours provisions of the Fair Labor Standards Act—a situation that lasted for more than twenty-five years. And, with few exceptions, the companies almost always prevailed in the courts against efforts to reach past the "arm's length" relationship and impose liability on them for accidents and other mishaps associated with logging operations. In a region marked by callous indifference and, at times, outright hostility to labor, the southern logger was one of the least visible and most vulnerable workers in the region.

Chapter 3 takes up the challenges of managing fixed capital in the industry and the imperative of maintaining order and stability in the mill labor force. When pulp and paper firms moved into the region during the 1920s and 1930s, most brought their existing union contracts with them. Unlike conditions prevailing in many other New South industries, and in sharp contrast to the harsh realities of woods work, pulp and paper mill workers typically enjoyed high wages and the benefits of union representation. Given the tremendous investments in fixed capital embodied in a mill and the imperative of maintaining throughput, industry managers and executives preferred the industrial stability that went with long-term union contracts. Moving into the South, however, required both the paper companies and the unions to confront the harsh realities of southern race relations. Having little experience with black workers outside the South, pulp and paper firms acquiesced while racist union locals erected systematic barriers to deny black workers any chance of moving up the job hierarchy. Confined to segregated union locals with little or no bargaining power, black workers found themselves locked into the dirtier, lower-paying jobs. The racial divisions of the Jim Crow South thus came to be deeply embedded within the mill labor force. Only after courageous activism by black workers and their advocates together with massive federal intervention did the mills finally impose a new desegregated racial order on the workforce where everyone enjoyed equal opportunity. This was not one of the bright spots in the history of the southern pulp and paper industry, and it challenges the view that outside business interests and organized labor acted as forces for progressive racial change in the South.

The last set of problems that confronted pulp and paper firms as they expanded in the region involved environmental pollution—the subject of chapter 4. As a consequence of its vast appetite for timber, water, energy, and chemicals, the industry was one of the nation's largest polluters throughout the second half of the twentieth century, generating unimaginably large pollution loads that severely disrupted local and regional ecological systems. The International

Paper mill in Bastrop, Louisiana, for example, was discharging 10,000 gallons of wastewater *per minute* into the surrounding bayou during the 1940s, wiping out local fish populations and leaving a permanently blackened landscape. In North Carolina, the Champion International mill in Canton diverted the entire 45-million-gallon flow of the Pigeon River for its operations, returning most of the water to the river in a severely degraded state.

Initially, most communities were probably more than happy to trade reduced environmental quality for the jobs and income that accompanied a new pulp and paper mill. Alabama governor George Wallace's famous characterization of the rotten-egg odor of a nearby paper mill as the "smell of prosperity" captured the prevailing sentiment. As the industry expanded in the region, however, people began to voice concern about the magnitude of air and water pollution associated with pulp and paper production. In the early years of the industry's expansion, when no environmental laws were on the books, adjacent landowners sometimes sought recourse in the courts under nuisance and trespass laws. And while these actions occasionally resulted in damages paid to landowners and others for the harms done to their property and livelihoods, the courts refused to order any sort of injunctive relief given the substantial economic value (investment, jobs, taxes) that came with the new mills.

Early state laws directed at the growing pollution problem were not much better, with most legislation heavily influenced by industry and expressly based on the need to subordinate environmental concerns to the priority of full industrial development. During the 1960s, as the push for a stronger federal role in controlling air and water pollution gained force, pulp and paper executives joined with state political and business leaders in an effort to maintain local and state autonomy over environmental regulation. States' rights became their rallying cry as they fought against the federalization of pollution control. Ultimately, however, they failed in their bid to maintain state control, a sign that their influence was waning and increasingly subject to larger national trends. Starting in the 1970s, pulp and paper firms spent a great deal of money reducing their pollution loads in order to comply with new federal environmental laws. The overall result in many cases was a dramatic reduction in gross insults to air and water and a marked improvement in environmental quality.

In the meantime, other environmental problems emerged. Dioxin contamination from pulp and paper bleaching, for example, became a major source of concern in the mid-1980s, putting the industry in the national spotlight and subjecting it to a rash of tort cases and new regulations. In the 1990s, local citizens groups in various parts of the South began to agitate against ongoing releases of toxic air and water pollution from the mills, pushing for additional controls and improved practices. In both cases, it was clear that the industry no longer wielded the power and influence it once enjoyed. With the growing prominence of environmental justice concerns during this time, moreover, renewed attention was directed to the fact, long obvious to people living close to the mills, that many of the environmental harms associated with pulp and paper production all too often fell disproportionately on poor and minority communities. As in other parts of the region and throughout the country, prevailing class structures, racial discrimination, and the distribution and exercise of political power all worked to limit the capacity of some people to avoid the environmental hazards associated with industrial development.

Thus, while it would be hard to overestimate the overall contribution of the pulp and paper industry to the economic development of the South (whether measured in jobs, wages, taxes, or value-added), not everyone participated equally in this industrial success story.[41] Small contract loggers, white and black, faced considerable danger and deep economic insecurity as they worked to deliver wood to the mills. Up until the late 1960s, black mill workers found themselves segregated into black union locals and excluded from higher-paying jobs. Poor and minority communities all too often bore the brunt of the pollution problems associated with the industry's expansion in the region. Such unevenness, of course, is part and parcel of any process of industrial development, but it is all too often left out of standard economic accounts. This book shows how that unevenness, and how the many people who lived and worked in some of those "uneven places," to use George Tindall's phrase,[42] were so central to the success of the southern pulp and paper industry.

The story told here also illustrates the central role that the region's natural resources and local environments played in underwriting and sustaining the industry's success. In building what has surely been one of the most important industries in the modern South, pulp and paper firms, together with other actors, substantially remade the biophysical landscape. In some cases, this remaking brought dramatic improvements to a scarred and degraded land. In others, it resulted in massive disruptions of regional and local ecosystems. In all cases, the ecological transformations that accompanied this process of industrialization were as profound as anything the region had ever witnessed. Along with the New South came a new nature—one that deserves far more attention in future studies of the region and its industrial history.

Industrializing the Southern Forest

> About five years ago I came to the conclusion that the Union Bag and Paper Corporation—without a Southern plant—would make about as much progress on the road to fame and success as Colonel Lindbergh would have made without an aeroplane. —*Alexander Calder, 1936*

The 1930s proved to be an auspicious time for pulp and paper firms to establish operations in the South. Though there were already a dozen or so mills operating in the region, lingering concerns over the destruction of the southern timber resource and the appropriateness of immature southern pine as a furnish for papermaking continued to prompt questions in the board rooms of major northern paper manufacturers into the 1920s. By the 1930s, however, the so-called sap problem[1] had been dispensed with, and the preliminary results of the first forest surveys indicated that the southern forest, which had been under virtually continuous assault since the collapse of Reconstruction, was capable of rapid regeneration. Indeed, despite a series of government reports and hearings, as well as rising public concern over the possibility of a national timber famine during the 1920s,[2] the "second" forest that was growing up in the wake of the destruction of the South's old-growth forest testified both to the resilience of the timber resource and, more importantly, to the rapid growth rate of southern pine.

At the same time, of course, southern states and towns, still reeling from the depression, were eager for any sort of industrial development and provided generous incentives to pulp and paper mills seeking to locate in their areas.[3] Key raw materials—wood, water, and energy—were in abundant supply. Land and labor were plentiful and cheap. Unions were virtually nonexistent, and the big eastern markets lay close at hand. By the end of the decade, the South had radically reshaped the industry's competitive landscape.[4] Without a southern mill, as Alexander Calder, president of Union Bag and Paper, suggested, it was unlikely that one would be able to remain in the business for long.

With its 1936 decision to build what would become the world's largest pulp and paper complex in Savannah, Georgia, the Union Bag and Paper Company launched a race to find and develop the most suitable sites for pulp and paper mills in the South. Because the new mills were massive by the standards of the day, reflecting the huge economies of scale available in Kraft pulp and paper manufacture, their investments in fixed capital dwarfed anything previously seen in the forest products sector.[5] Unlike the small "peckerwood" sawmills that could be moved in search of new timber supplies, these new mills weren't going anywhere. Operating under an imperative to keep their fixed capital in motion, many of them ran twenty-four hours a day, seven days a week. Their appetites for water, energy, and timber were almost inconceivable.[6] As a result, the mills depended for their timber on what could be grown in their immediate procurement areas (a 100–150 mile radius). The earlier extractive logic employed by the lumber industry no longer sufficed. A new regime in forest management would be necessary.[7]

As pulp and paper expanded in the South, industrial advocates throughout the region heralded the dawn of a new age. Echoing the boosterism of Henry Grady, the *Savannah Morning News* proclaimed that the Union Bag plant signaled a "New Industrial Epoch" for Southeast Georgia, noting that the "lowly pine tree . . . has taken on a new aura of grandeur and significance as a symbol of vast potential wealth."[8] Speaking at a banquet celebrating the formal opening of the Union Bag plant, Mayor Thomas Gamble pointed to the vast opportunities awaiting his city and the South: "No one can safely limit the possibilities which center in the pine tree and its various possible products . . . We apparently are on the threshold of discoveries which will incalculably broaden the use of products common to our section. The South, especially our own immediate territory, is recognized as a coming industrial empire. A new world seems to be opening before us. Savannah must be prepared to enter and possess it."[9] Savannah's own Dr. Charles Holmes Herty, who played an important role in promoting the virtues of southern pine during the 1920s and 1930s, declared in 1933 that "some day, in the not far distant future, King Cotton is going to be replaced by King Pulp."[10]

So it came to be. By the end of 1947, slightly more than a decade after Union Bag opened its mill, Savannah's last cotton compress had been dismantled. King Cotton, which had shaped the fortunes of the city for so long, had been replaced by the pine tree as the staple of regional economic development. As a *New York Times* reporter put it: "The industrializing of the New South is being dramatically illustrated in Savannah, as cotton is no longer brought here and the old river and wharves support a new business that still draws upon the countryside and the farmer. The cotton farmer has been succeeded by the tree farmer in a large section of the South. The pine tree has become the new cash crop."[11]

Such proclamations, of course, had to be tempered with the massive logistical and institutional challenges involved in providing these mills with a continuous source of fiber. This supply problem had two essential components. First, a new forest management regime would have to be constructed to ensure that timber was treated as a crop rather than as a mine. Moreover, because small farmers and other small private landowners owned two-thirds or more of the timberland in the South (which meant that mills depended on such owners for more than half of their overall fiber supply), it was not enough to establish such a regime solely on industry-owned lands. The small southern landowner also had to be persuaded, through education and incentives, of the virtues of viewing timber as a crop. Given the long time lags associated with growing timber, the risks and uncertainties involved, and the perverse incentives embedded within prevailing tax and credit systems, this was no easy task. The second component of the supply problem, perhaps even more challenging, involved securing access to the small landowners' timber and getting the wood to the mills. Accomplishing this, of course, meant that the mills would have to engage the institutional realities of the rural South, particularly those involving land and labor. Again, no easy task.

This chapter focuses explicitly on the first of these two aspects of the timber supply problem, while the following chapter takes up the second. After a brief discussion of the state of the southern forest in the 1920s and 1930s, it will examine the transition to industrial forestry that transpired in the South during the interwar and postwar years and transformed timber into the region's number one cash crop.[12] This transition, which went into high gear during the 1950s, had its origins in the Progressive conservation era and the New Deal, and it represented a remarkable example of public-private cooperation aimed at rationalizing the use of a renewable resource in the interest of efficiency and industrial development. Three major phases marked the transition: (1) *rationalization*: the creation of a more stable environment for investment in timber growing, particularly through improved fire protection and management; (2) *regeneration*: the reforestation and afforestation of cutover and marginal agricultural lands; and (3) *intensification*: the acceleration of biological productivity through advances in forest genetics and tree improvement. Within these three phases, which were overlapping and complementary, a complex and changing division of labor between federal, state, and private actors emerged. New institutions would have to be constructed. Extensive cooperation would be required. The whole process, moreover, was part of the larger agrarian transition taking place in the post–New Deal South—a transition marked by the decline of cotton tenancy, the industrialization of agriculture, the opening up of rural labor markets, and the transformation of southern land use. By the 1960s, the combined effect of such processes was manifest in the establishment of the South's "third forest," much of which was the result of "artificial regeneration."[13] The South was on its way to becoming the woodbasket of the world.

The emphasis here is thus decidedly more upstream than most studies of the pulp and paper industry, not to mention those of southern industrialization. There are two main reasons for such a focus. First, most general studies of twentieth-century southern industrialization and southern economic history have paid little attention to the forest products sector and almost none whatsoever to the attendant rise of industrial forestry in the region during the interwar and postwar decades. Second, to divorce a study of the pulp and paper industry from the timber supply issue would miss some of the more important and distinctive elements associated with resource-based industrialization. Indeed, one could argue that the complex set of challenges involved in subordinating biological systems to the dictates of industrial production, in the context of a particular regional institutional matrix, represented one of the more difficult tests facing firms attempting to construct a viable and competitive industrial form in the South. The outcome depended heavily on the construction of new institutions for public-private cooperation and new patterns of state intervention.

Given the importance of such interventions and the overall instrumentalization of nature that resulted, one could view the entire process as an example of "state simplifications," that is, a state-directed project aimed at achieving legibility and control over nature in pursuit of economic development.[14] Such an emphasis, however, would miss important elements of a story that is also very much about industrialization as a process of regional collective learning, whereby the institutional arrangements that facilitated the "making" of industrial timberlands function as vital elements of a larger industrial system. Viewed in this way, firms' evolving organizational capabilities are embedded in a larger process of building networks for collective learning and knowledge accumulation.[15]

That these networks were primarily regional in scope raises some interesting questions regarding southern industrialization. In contrast to arguments about the lack of an "indigenous technological community," which served to reinforce the South's status as late developer, it is quite clear that in the case of industrial forestry just such a community was starting to emerge in the region during the interwar years.[16] Moreover, because this community played such an important role in the rise of the South as a national and international leader in the forest products sector, one has to question the utility of applying traditional comparative advantage approaches to southern industrialization. As noted in the introduction, the emergence of a distinctively southern pulp and paper industry cannot be explained simply as the result of regional comparative advantage based on key factors of production and the successful attraction of branch plants controlled largely by northern corporations. One must also attend to the ways these firms, in concert with other actors, mobilized and organized the productive capacities of southern land and labor and integrated them into a highly competitive industrial form. Regional advantage, from this perspective, did not simply derive from factor endowments but was instead a historically situated process of social and ecological construction.

The Slain Wood

> The plain truth of the matter is that in county after county, in state after state of the South, the piney woods are not passing but *have passed*. Their villages are Nameless Towns, their monuments huge piles of sawdust, their unwritten epitaph: "The mill cut out!" Locally the catastrophe has already arrived of a vanished industry, unreplaced by any new industries remotely adequate to redeem the situation. —*R. D. Forbes, 1923*[17]

From the end of Reconstruction to the close of World War I, the southern forest suffered unmitigated destruction. Spurred by a growing domestic market for lumber and the prospect of timber depletion in the heavily cutover Great Lakes states, speculators and lumbermen moved into the South during the last decades of the nineteenth century to capture the spectacular profits available in extensive stands of old-growth timber being sold at nominal prices. As they had done in the Lake States, the lumbermen cut the southern forest with unprecedented haste. They operated according to a logic of extraction, their imperative: "cut and get out." By 1920, their work was virtually complete, with the cutover area in thirteen southern states estimated at more than 156 million acres. Most of this cutover land, perhaps as much as 100 million acres, was in the yellow pine belt stretching from South Carolina to Texas (figure 1.1). In that area, estimates indicated that less than 24 million acres of old-growth pine timber remained in 1920.[18] One government forester, who surveyed the region at the height of the cutting, assessed the situation as "probably the most rapid and reckless destruction of forests known to history."[19] William Faulkner called it simply "the slain wood."[20]

The forces driving such rapid exploitation are fairly easy to discern. By the early 1880s, concern over timber depletion in the Lake States was mounting. In

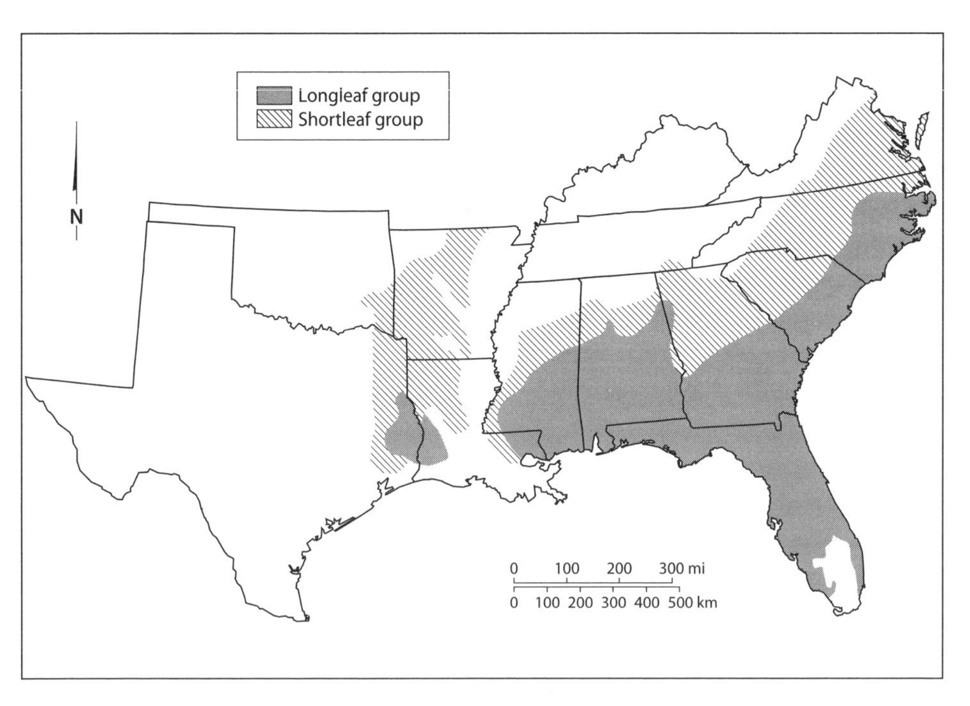

Figure 1.1. The South's yellow pine region suffered massive cutting between 1880 and 1920. This map shows the natural range of longleaf and shortleaf pines that together made up the yellow pine belt. Adapted from Howard Odum, *Southern Regions of the United States* (Chapel Hill: University of North Carolina Press, 1936), 80.

1884, Harvard professor Charles S. Sargent issued his famous *Report on the Forests of North America*, arguing that current rates of timber extraction in the Lake States were unsustainable while also highlighting the vast expanse of old-growth timber available in the states of the South and the Pacific Northwest.[21] Meanwhile, the South was emerging from war and Reconstruction, with the newly ensconced "Redeemer" governments intent on opening the region to outside capital. Land policies, such as the numerous state sales to railroads and other interests and the 1876 repeal of the Southern Homestead Act of 1866, which permitted unrestricted cash entry into the public land states of the region, provided substantial opportunities for acquisition of vast tracts of timberland. The availability of a mobile, "flexible" workforce made up largely of itinerant freedmen ensured that labor costs would be well below those prevailing in other lumber-producing regions.[22] Finally, the growing national consensus that yellow pine made good merchantable lumber combined with the proximity to eastern markets and the expansion of railroads in the region during the post-Reconstruction decade, gave the South considerable comparative advantage vis-à-vis other lumber-producing regions.

Although early conservationists had voiced concern over timber depletion and the lack of conservation in the South since the turn of the century,[23] it was not until the 1910s and 1920s that the issue began to receive serious attention at the national level. During this time, the voice heard most often on the subject was that of Gifford Pinchot, the man credited with introducing the term "timber famine." In 1919, Pinchot authored a report for the Society of American Foresters that predicted an imminent timber shortage. This prompted Kansas senator Arthur Capper to request a U.S. Forest Service study of the timber supply situation. The resulting document, known as the Capper report, fueled the debate over timber supply and the role of federal policy in regulating forest practices.[24] Based on an assemblage of previously collected information, the report provided data on trends in depletion, prices, trade, and concentration of timberland ownership in all of the major producing regions. The authors identified timber depletion as the "fundamental problem" underlying the decline of interregional competition, increased scarcity of timber products, rising prices, and the growing concentration of timberland ownership in the United States.[25]

In the southern pine region, the Capper report estimated that four-fifths of the original yellow pine forests had been cut since 1870, and classified more than 30 million acres of cutover lands (almost a third of the original pine forest) as "not restocking." With an annual cut that exceeded annual growth by a factor of three, the report predicted that the southern yellow pine industry would soon cease to be a national factor and that by the early 1930s the South would be importing lumber from the Pacific Northwest to meet its own consumption needs.[26] In light of the systematic neglect of conservation, state and federal foresters feared that the second-growth timber in the region would be unable to support a substantial forest products industry in the future. Major George P. Ahern of the Forest Service captured such fears when he concluded in 1928, "The end of virgin timber in the South is definitely in sight, while already the younger timber is being cut as fast as it grows."[27]

Had the South traded the possibility of a stable, long-term industry for short-term profit, much of which was captured by interests from outside the region? Would there be enough timber to support a lumber industry in the future, not to mention the nascent pulp and paper industry? No one really knew, for there had been no systematic survey of the forest resources of the South. Many simply assumed the worst. Added to such uncertainty were the substantial obstacles facing investments in conservation, forest regeneration, and the imposition of a new forest management regime. Forest fires, for example, burned large areas of the southern woods well into the 1940s, undermining efforts to promote the long-term investments required for timber growing. Tax policies, particularly at the county level, also created uncertainty as to the long-term tax obligations associated with timberland ownership. Insects and disease threatened the viability of young trees, especially those already weakened by turpentining and wildfire. Finally, the unpredictability of timber markets, exacerbated both by endemic overproduction and the growing concern over the future of the industry in the region, added to a highly unstable and shifting investment environment that offered few incentives to invest systematically in forest protection and regeneration. In short, before the forest could be regenerated, before prudent investments in tree improvement could be made, the landscape of timberland investment would have to be made amenable to rational economic calculation.

Rationalization

> The chief obstacle to forestry in the South is the destruction caused by forest fires. Fires run riot in the pineries, lay waste the cut-over lands, and constantly endanger, often seriously damage[,] the virgin timber. They set a premium upon premature, hasty, and destructive logging, and effectually discourage the holding of forest land for a continuous succession of crops, even for a single additional crop. As long as they go unchecked they will furnish lumbermen and forest owners with an unanswerable argument in support of their present destructive methods. Unless fires are checked, forestry in the Southern pineries will never appeal to men of good business sense. —*Gifford Pinchot, 1910*[28]

Fire was, by virtually all accounts, the "chief obstacle" standing in the way of rational, scientific forestry in the South during the early twentieth century. Yet the fire problem was compounded by the almost complete lack of reliable information on the state of the southern timber resource. Any successful effort in fire control, as part of a larger program of scientific forestry, thus required an accurate assessment of the volume and distribution of the timber resource, its condition and quality, the various threats facing it, and the ratio of annual growth to annual drain.[29] In the absence of such information, systematic forest land use planning would be impossible, and the large investments in fixed capital and forest management necessary for a successful pulp and paper operation would be very difficult to justify. Indeed, despite all the reports and hearings about timber depletion and forest destruction during the 1920s, no one could speak with certainty about the actual state of the southern forest resource. The task of accomplishing a regional forest survey (there were almost 200 million acres of timberland in the twelve states of the South) was mind-boggling. Even in a single state, the financial and logistical requirements of developing an accurate assessment of the timber resource greatly exceeded the capacities of state governments and private firms. Such an undertaking clearly required federal assistance.

In 1928, the McSweeney-McNary Forest Research Act authorized the first nationwide forest survey. Major objectives included a field inventory of timber supplies, an accurate estimate of annual growth and annual drain, and an assessment of management options and public policies necessary for "the most effective and rational use of land suitable for forest production."[30] For the southern states, the survey began in 1931 under the direction of the Southern Forest Experiment Station in New Orleans.[31] Professional foresters carried out field-level inventories and assessments for individual states, which were divided into clusters of counties known as forest survey units. It was, at one level, a gigantic timber cruise. Nothing on such a scale had ever been attempted before.

Given the magnitude of the task, it took the better part of a decade before the final results were available. As the preliminary findings of early subregional surveys were released in the mid-to-late 1930s, however, they revealed a somewhat surprising recovery of the southern pine forest.[32] Despite decades of extensive logging, an almost complete neglect of conservation, and widespread incidence of forest fires, young second-growth pine had naturally regenerated over a wide area, including large acreages of abandoned or idled cropland.[33] Timber, which was already reclaiming many abandoned fields, was arguably the highest and best use for such lands.

Although the surveys revealed a larger volume of timber than previously thought to exist, they also pointed to considerable pressure on the forest resource. In roughly half of the area surveyed, timber drain exceeded timber growth. The vast majority of southern timberlands exhibited widespread damage due to fire, adding to the general lack of positive incentives for investment in scientific forestry. In Georgia, for example, the forest survey found that fires had occurred in the previous few years on more than three-fourths of the state's total forest area—a reflection not only of the natural role that fire had long played in the longleaf and other pine forests of the state but also of the concerted effort to domesticate the forest landscape to other economic imperatives. Half of the burned area lay in south Georgia, where cattlemen and turpentine operators used fire extensively to improve grazing conditions and to protect turpentine orchards from uncontrolled fires. Overall, fire represented the direct or indirect cause of three-fourths of the state's pine timber mortality. For 1937 alone, some 3.75 million acres (18% of the total forest area) had been burned over. Echoing Pinchot and others, the report concluded that "frequent and indiscriminate forest burning has long been the most important factor militating against the development of well-stocked timber stands" and that the "first and most important step in the rebuilding of Georgia's forests is the control of the fire situation."[34] In short, without systematic investment in forest fire protection and forest management, southern timberlands would never reach their economic potential. Industrializing the southern forest required a solution to the "fire problem" that would mark a fundamental break with older, more natural, rhythms of fire in the southern woods.

The forest survey was thus far more than a gigantic timber cruise. In addition to providing southern states with the first accurate assessment of their timber resources and the ratio of growth to drain, the survey came at a time when many pulp and paper firms were considering whether to establish operations in the South. By providing a realistic portrait of the southern timber resource as well as an indication of its potential, it undoubtedly stimulated an already growing interest in the South and facilitated the industry's further expansion in the region. Yet the survey also raised concerns about the future of the southern timber resource, particularly if large pulp and paper mills continued to establish themselves in the region, consuming an ever-larger share of the young second-growth timber in the area. In this respect, the survey represented an important intervention in the emerging discourse on forestry practices in the South, particularly in the context of fire control.

Forest fires had long been a vital part of the landscape of the rural South. Annual woodsburning, a common practice throughout the region since the arrival of the first human inhabitants, was practiced extensively during the late nineteenth and early twentieth centuries, both in order to control the "rough" or understory that grew so vigorously in the forest and as a means of improving livestock grazing and eliminating pests. Although fires sometimes assumed massive proportions, particularly in times of drought, most burned low, marching slowly through the woods, damaging grown trees, and destroying younger trees and seedlings. As statistics on wildland fire began to be collected in the early twentieth century, it became clear that the South accounted for the vast majority of wildland fire in the United States, in terms of both frequency and acreage burned.[35] According to the Regional Committee on Southern Forest Resources, during the 1920s and 1930s, the South accounted for some 85 percent of all forest fires in the country, and more than 90 percent of the burned-over acres, even though it contained less than one-third of the nation's total forest area. Roughly 40 percent of these fires were suspected to be of incendiary origin, a consequence of the practice of annual woodsburning. Fire, in short, was part of the fabric of rural life in the South.[36]

With the advent of industrialized logging—railroads, steam skidders, and trucks—fire had alternative vectors along which to travel. As rail and road opened up new tracts of timber, the scope of fire increased. The widespread and growing incidence of fire, however, was hardly compatible with industrial forestry. Not only did fire prevent natural regeneration of certain pine species on cutover lands, but it also provided a major disincentive to invest in artificial regeneration. Investments could all too easily go up in smoke. For early conservationists such as Gifford Pinchot, this represented the single greatest challenge facing southern forestry. Echoing these sentiments, W. W. Ashe, the secretary for the National Forest Reservation Committee, wrote in 1925, "Without adequate protection [from fire], especially for cut-over lands, all methods of management on private lands are futile."[37] By the early twentieth century, conservation and forest regeneration had become intimately bound up with the political economy of fire in the rural South.

Put simply, for industrial forestry to succeed, a new "cycle of fire" would have to be constructed, one subordinated to the needs of industrialism and scientific management rather than to the rhythms of rural life.[38] Given that the majority of the region's timberland was owned by small farmers and private landowners, many of whom practiced annual woodsburning, the institutional challenges of systematic fire protection were large indeed. People had to be persuaded of the value of the forest and the detrimental impacts of woodsburning. The physical and institutional infrastructure for fire protection had to be established. Huge sums of money would be needed.

Institutionally, such a program would have to be built on cooperation between state and federal agencies, the forest products industry, and private landowners. Because fire did not respect political or administrative boundaries, moreover, a successful strategy required a regional focus. Fire control, to put it crudely, represented a collective action problem that demanded new forms of coordination. Thus, although organized fire protection had been practiced piecemeal in the South since the turn of the century—primarily on the lands of several large lumber companies and through the experimental activities of some state agencies and cooperative associations—it was not until the 1910s and 1920s that the first systematic efforts at coordination were made. Congress took its first statutory steps in this direction in 1911, when it enacted the

Weeks Act, which provided matching funds for state agencies engaged in fire protection. Although progress under the law proved to be slow in the South, due to limited funding and a general reluctance among state governments to embrace the new regime of fire protection, the Weeks law did establish fire protection as a legitimate area of cooperation between federal and state agencies. Picking up where Weeks left off, the Clarke-McNary Act of 1924 gave fire protection efforts a significant boost. By 1930, federal cooperative funding for fire control in the South had grown in nominal terms from nothing to about $400,000, while state funding had increased to $750,000.[39] Almost 70 million acres of southern timberland (out of roughly 230 million) received some form of fire protection. Although this was a modest share of what ultimately would be required to realize the goal of regionwide fire protection, it was a start.

Beyond the challenge of establishing an institutional infrastructure and providing adequate funding for fire protection, perhaps the most difficult task associated with fire control was educating rural southerners and persuading them to abandon woodsburning. Federal and state agencies, conservation groups, and forestry associations all worked throughout the 1920s and 1930s to spread the gospel of fire protection through publications, films, rallies, conferences, informal meetings, and technical reports. Calling for "strict accountability" for "the man who burns the woods," one U.S. government pamphlet compared the southern woodsburner to the "boll weevil, the malaria germ, and the cattle tick . . . Because of him land values have suffered, industries and population have moved out, and idle acres have multiplied. Because of him every year millions of young forest seedlings, which in short time would have constituted a valuable asset to landowners, have been licked up by flames . . . The South cannot afford to let the woods burner block economic progress . . . The irresponsible burner must be banished from the woods, and the well-intentioned burner must squarely face the responsibility incurred when he starts fire on his own land." The report concluded with a plea for collective action: "No one agency alone can cope with this situation. All private landowners, all Southern States, and the Federal Government must agree on a common course of action and work together to put an end to forest destruction."[40]

As state forestry associations developed throughout the South, new rules and regulations stiffened the penalties for arson.[41] Woodsburning was criminalized, and fire control was transformed into a moral crusade. Starting in 1928, the American Forestry Association launched a three-year tour of the South to educate rural folks about the dangers of woodsburning. The "Dixie Crusaders," as the members of the fire prevention caravan were known, traveled 300,000 miles, holding rallies, distributing pamphlets, and showing self-produced movies to some 3 million people. One state forester described the fire prevention caravan as one of the most important events in the history of southern forest protection, "the first mass effort toward the solution of the South's woodland fire problem."[42]

Fire control and fire protection assumed an unprecedented sense of moral urgency during the early 1930s.[43] The establishment of the Civilian Conservation Corps (CCC) in 1933 created a virtual army of firefighters. Though most often remembered for their efforts in planting trees, the young men of the CCC also fought fires and constructed the roads, bridges, trails, fire towers, and telephone lines that provided the physical infrastructure needed for fire control in timber-producing regions. By 1942, when the CCC disbanded, the amount of southern land under fire protection had increased to almost 90 million acres.[44]

Through its fire protection work, the CCC also bestowed a certain legitimacy on the use of prescribed burning as a tool for fire control in the South, an issue that provoked considerable debate among foresters well into the 1940s.[45] Early proponents of prescribed burning, such as Austin Carey and H. H. Chapman, argued eloquently for "proper" burning as a means for controlling wildfires and, more importantly, as a tool for stimulating forest regeneration in the South. In a 1912 article Chapman argued that "to keep fire entirely out of southern pine lands might finally result in complete destruction of the forests."[46] In 1926, he issued his famous Yale School of Forestry Bulletin number 16, which caused a great stir throughout the southern forestry profession, arguing directly for the use of fire in longleaf pine regeneration.[47] Six years later, the U.S. Forest Service adopted a policy statement confirming that fire could play a constructive role in longleaf pine culture. The following year, the Southern Forest Experiment Station issued its own twenty-three point "Fire Statement," allowing for the judicious use of fire for both silvicultural and forest management practices.[48] Not until 1935, however, did the prohibition on controlled burning under the Clarke-McNary Act formally end. Moreover, only after another series of disastrous fires between 1941 and 1943, fueled by a combination of drought and the accumulation of debris on the forest floor, did professional opinion actually begin to shift in favor of prescribed burning. In the wake of these fires, Lyle Watts, chief of the Forest Service, reversed policy and initiated a program of prescribed burning in southern national forests.[49]

It would take another decade, however, before southern states developed formal institutions for multistate cooperation in fire control and forest fire research. In 1956, state governments signed two regional fire protection compacts (southeastern and south central) to coordinate efforts and resources among the various agencies involved in fire control.[50] Three years later, the Forest Service established the nation's first forest fire laboratory in Macon, Georgia. The Southern Fire Lab, as it was known, conducted research on fire ecology and hosted a series of seminars on prescribed burning for the fire control community. Beginning in 1962, under the leadership of fire ecology pioneers Herbert Stoddard, Leon Neel, and Ed Komarek, the Tall Timbers Research Station in Tallahassee, Florida, also began hosting a series of annual fire ecology conferences.[51] In the view of one southern fire expert, the Tall Timbers organization "became almost overnight the outstanding force for exposition and promotion of fire ecology and controlled burning in the entire world."[52] By the 1960s, the southern forestry community had firmly established itself at the vanguard of industrial fire control.[53]

For the twelve southern states,[54] fire protection expenditures increased consistently throughout the post–New Deal period, peaking in 1970, while the amount of protected acreage grew until the early 1980s, at which time roughly all of the timberland in the region was under some form of fire protection.[55] In Georgia, a leading state, fire control expenditures increased by 150 percent between 1950 and 1957, while protection expanded to roughly 21 million acres (close to 90% of the state's timberland).[56]

As a result of the increased investment in fire protection, between 1925 (a peak year for fire damage) and 1965, timberland burned annually in the southern states fell from over 21 million acres to slightly less than 600,000 acres. With increased fire protection, investments in timberland ownership and forest regeneration could proceed with much less risk.[57] Fire control, in short, brought a much-needed calculability to the landscape of timberland investments. By 1980, "the great fire problem of the South," as one 1940 government report had put it, was little more than a memory.[58]

The high incidence of forest fires in the South during the 1920s and 1930s however, was simply the most spectacular obstacle to systematic investment in forest regeneration and sound forest management. Other major problems stood in the way of industrial forestry as well. Like the fire problem, insects and diseases threatened long-term investments in forest management and represented a similar sort of collective action problem. The Forest Pest Control Act of 1947 provided for federal-state cooperation in insect and disease control in much the same way that Clarke-McNary provided for cooperative fire protection. Yet, pest control has historically involved much more public controversy than fire protection efforts, primarily because of the heavy use of chemicals. At the regional level, the forest products industry assumed primary responsibility for dealing with insect and disease issues and formed the Southern Forest Disease and

Insect Research Council as a mechanism for coordinating research and allocating funds to southern universities for investigation of pest problems.[59]

Other obstacles were more institutional in nature. The forest survey, for example, noted the need "to remove those causes that threaten the stabilized land ownership necessary for long-time forest management, such as unfair tax treatment, discriminating freight rates, hard credit terms, and unfavorable legislation."[60] In particular, the credit system that prevailed in the rural South meant that both small woodland owners and small sawmill owner-operators had limited access to the long-term financing necessary to invest in sustained-yield operations. In part, this situation derived from the vulnerability of timberland to forest fires and other destructive agents, which made it a risky investment for financial institutions. More importantly, it reflected the relative lack of capital market institutions in the rural South. Major credit providers, primarily local merchants and local banks, focused almost exclusively on agriculture, with its yearly cycles of planting and harvesting. This created a preference for short-term financing which, when combined with the high debt burdens and various risks associated with forest investments, reinforced incentives for small timberland owners to liquidate their holdings as quickly as possible. Indeed, not until the 1950s did long-term credit become available for the majority of southern nonindustrial timberland owners.[61] Based on a recognition of the improvements in fire protection and forest management, the U.S. Federal Reserve Board amended its regulations in 1953 to allow financial institutions to lend money on timberlands.[62] As a result, life insurance companies, southern banks, the federal land bank, and, a bit later, the Farmer's Home Administration began making long-term credit (from five to forty years) available for forest loans.[63] Timberlands had finally become creditworthy investments in the eyes of the financial community.

Closely related to the credit issue, and commanding far more attention, was the question of forest taxation—a major public policy concern throughout the first three decades of the twentieth century. For some observers, such as Rupert Vance, much of the destruction of the southern forest during the half century after Reconstruction resulted from a perverted tax structure in which timber "was forced to pay revenues as though it were an annual crop." Vance referred to this treatment of timber under the ad valorem property tax as "merciless taxation," arguing along with many others for adoption of a severance tax to be levied only when the timber was cut.[64] In Vance's view, the disadvantage of the prevailing property tax system lay in the long time lags associated with timber growth. While timberland owners paid property taxes on a yearly basis, they often had to wait years, even decades, before they received any timber-related income. According to the proponents of tax reform, the loss of interest on the investment in long-term timberland ownership, combined with the possibility of tax increases and the various risks associated with holding timberlands for extended periods of time (damage from fire, insects, and disease), encouraged the premature cutting of timber while discouraging investment in forest regeneration. Seen in this context, clear-cutting followed by abandonment made more fiscal sense than holding timber as a permanent investment.[65] This compounded the related problem of tax delinquency. As local timber supplies were liquidated and sawmills cut out, many timber-dependent counties suddenly found themselves facing substantial reductions in their tax base. Short of revenue, these local governments responded by increasing their tax rates on those lands not in default, which simply added to the incentive for timber owners to liquidate their holdings.[66] It was a fiscal catch-22.

Even as early as the 1910s, however, a number of states, including several in the South, amended their property tax systems in an effort to promote better forestry practices. During the 1920s and 1930s, in the face of mounting concern over forest destruction and depressed economic conditions throughout rural America, the tax problem began to garner national attention.[67] The 1920 Capper report, for example, called for both federal- and state-level investigations into forest taxation.[68] Four years later, the Clarke-McNary Act vested the secretary of agriculture with the authority to investigate "the effects of tax laws, methods, and practices, upon forest perpetuation."[69] Pursuant to these provisions, in April 1926, the Forest Service established the Forest Taxation Inquiry under the leadership of Professor Fred Rogers Fairchild of Yale University.[70] Based on a nine-year investigation, the 1935 Fairchild report, which ran to almost 700 pages, provided exhaustive detail on state and local taxation practices and their effects on forest management. It focused, not surprisingly, on "the principal instrument of local taxation—the property tax," arguing that state and local methods of property taxation "subject[ed] the forest business to an influence directly opposed to conservation."[71] In short, the committee found the property tax systems of many state and county governments inherently unfavorable to deferred-yield property such as timberlands. Moreover, it was not so much the actual tax burden but the *uncertainty* of future tax obligations that discouraged sound forest management. This presented a problem because it promoted rapid cutting of old-growth timber and provided a disincentive to invest in reforestation of cutover lands and proper management of immature timber.

Nonetheless, the committee did not find the property tax system to be the unitary force responsible for forest depletion and the general lack of conservation, as so many believed. According to the report, taxation figured as only one of several factors shaping the landscape of forest investment and management. The threat of fire, vulnerability to insect damage, the vicissitudes of the lumber market, and the long time lag needed for timber to reach maturity also discouraged investment in conservation.[72] "Taxation," the committee concluded, "is only one of the carrying charges that tend to bring about the rapid cutting of virgin timber and only one of the reasons why private capital is not embarking in timber-growing enterprises. There is no magic in forest-tax reform."[73]

More than anything else, then, the Fairchild report debunked the notion that there was a single, easy solution to the timber tax problem. Writing in the *Journal of Forestry* in 1938, R. Clifford Hall, a Forest Service economist, noted that with the Fairchild report, "the idea of a simple tax panacea, or even a model law that will fill the need in any and every state, ought to be dead. Anyone who examines the findings of this study can hardly fail to be convinced that no single and simple solution is possible."[74] Because the taxation of timberlands under the property tax was essentially a local and state issue, it required action at these levels. By the time Fairchild released his report, moreover, the depression was in full swing, and forest taxation no longer had the urgency it held during the prior decade. Tax delinquency and declining revenues still concerned local governments, but legislators and government officials effectively shelved the issue of systematic property tax reform in the face of larger and more immediate concerns.

Eventually, all southern states adopted some sort of special tax code provisions recognizing the distinctive nature of timberland investments and providing incentives for industrial forestry.[75] As state and county governments throughout the South realized that the forest products industry constituted a vital part of their economic development prospects, they recognized that a low and stable tax rate provided an important incentive for future investments. Finally, as cheap, long-term credit for timberland investments became available in the 1950s, much of the "tax problem" of previous years disappeared, as annual timber taxes could be paid more easily.[76]

By midcentury, forest taxation no longer elicited the concerns that it had during the 1920s and 1930s.[77] Substantial progress had been made in rendering forest taxation more hospitable for timberland investments. Such progress, combined with advances in fire control, the increased availability of long-term credit, and the development of a systematic and ongoing forest survey, meant that forest management could now proceed as a rational business enterprise rather than as a speculative venture.

Regeneration

> Here are two big birds of ill omen to be killed by one stone. We can put our unplowed acres to work growing a profitable crop for which there is no glutted market; repopulate our deserted forest regions and abandoned farm districts; give both the earth and the people something to do; and meet the impending shortage of forest products—by growing wood, east, west, north and south as part of a rational scheme of land use, with somewhat the same intelligence and skill that we put into the growing of cereals and fruit. National reforestation should command the interest and support of every thinking American citizen.
>
> —*Secretary of Agriculture Henry C. Wallace, 1923*[78]

Regenerating cutover lands in the South occupied a prominent place in national and regional policy debates during the first two decades of the twentieth century. Government officials decried the destruction and waste evident in vast areas of cutover lands and called for government intervention to deal with the problem.[79] The 1908 Conference of Governors on the Conservation of Natural Resources and the Weeks Act of 1911 both stemmed in part from concerns over how to utilize cutover forest lands, though neither had much impact. The first Southern Forestry Congress, held in 1916, also devoted considerable attention to the challenge of regenerating southern forests.[80] By this time, the problem had become so acute that in August 1917 the Southern Cut-Over Land Conference was convened in New Orleans, under the auspices of the Southern Pine Association and the Southern Settlement and Development Organization, to address the problem directly and to develop a program for restoring barren lands. Rather than focusing on reforestation, however, most of the discussion at the meeting evaluated the possibilities for conversion to pasture and cropland. Of the 340 people who attended the conference, only four had trained as foresters. Few considered forest regeneration a viable option, even though the vast majority of cutover lands could not support commercial agriculture. Southern agriculture, moreover, would soon be having its own problems.[81] By the early 1920s, as the farm sector sank into depression, the prospects of converting barren lands to cropland had dimmed considerably. With fears of a national timber famine on the rise, the attitude toward forest regeneration on cutover and marginal agricultural lands began to change. Timber was finally beginning to be seen as an alternative land use—as crop rather than as mine.

The first major commercial reforestation effort in the South was initiated by Henry Hardtner, president of the Urania Lumber Company in Louisiana, who developed an extensive program for "natural" regeneration on his company lands during the early twentieth century. Though well publicized, his efforts were not widely emulated by his contemporaries, and they offered little solace to those seeking a solution to the cutover land problem.[82] In 1920, however, shortly after Hardtner began his venture, the Great Southern Lumber Company of Bogalusa, Louisiana, initiated the South's first large-scale artificial reforestation program. Under the direction of F. O. Bateman, Great Southern converted thousands of acres of cutover longleaf pine lands to loblolly pine plantations to provide fiber for the company's lumber and pulp mills. To support these efforts, Bateman also established the first industry nursery at Bogalusa and pioneered the use of nursery seedlings as planting stock, a practice that became the standard for future southern pine regeneration efforts. By the early 1930s, the company had regenerated approximately 30,000 acres.[83]

Several years after the Bogalusa program commenced, the Clarke-McNary Act of 1924 authorized limited funding for state efforts to cultivate seedlings for planting on private lands. This led to the creation of a network of state nurseries throughout the South during the 1920s and 1930s, which in turn functioned as the biological foundation for southern regeneration efforts after 1930. Equally important, Clarke-McNary also established the institutional precedent for federal-state cooperation in assisting private timberland owners in forest regeneration and management.[84]

In both of these areas, as in fire control, Clarke-McNary ushered in an era of cooperation between the federal government, state governments, and private actors on matters of forest policy and management. The statute also represented the first tangible legislative result of a debate over public regulation of private forest practices that began at the close of World War I and lasted into the 1950s. This debate, which centered on whether the federal government should regulate private forestry directly or assist state governments and industry through cooperative institutions and programs, stemmed from the growing concern among professional foresters and political leaders over the extent of forest destruction in the United States during the 1910s and 1920s. Although the proponents of regulation included prominent foresters such as Gifford Pinchot and congressional leaders such as Senator Arthur Capper, those favoring the cooperative approach, including industry leaders and Forest Service chief William B. Greeley, carried the day.[85]

The onset of depression and the beginning of the New Deal, however, rekindled the debate over forest regulation. By lending an increased legitimacy to those voicing concern over the exploitative practices of natural resource industries, the Roosevelt administration created an administrative space for realizing some of the principles of earlier progressive conservationists.[86] Rational and efficient use of natural resources based on government planning became a mantra for many New Dealers. An early example of such convictions, the 1933 Copeland report, known formally as *A National Plan for American Forestry*, contained the most extensive account of forest conditions and practices in the United States ever completed and demonstrated convincingly that private timberland owners, even with the assistance of federal and state programs, had so far failed to initiate a significant program of rational, scientific forestry. Of the many recommendations made in the report, the most controversial was a massive program of public acquisition of timberland—totaling some 224 million acres, with 177 million acres in the East—as a way of ensuring that the nation's timberlands would be properly managed.[87] Although this particular proposal was never acted on, the Copeland report boosted efforts aimed at establishing forest conservation and scientific management as both civic duty and sound business practice.

Such efforts found their most prominent expression in the National Industrial Recovery Act (NRA) of 1933. Article X of the Lumber Code (the "conservation article") legally committed the lumber industry to principles of conservation and sustained yield. Even though the Supreme Court declared the NIRA unconstitutional two years later, article X provided a basis for the forestry policies and practices adopted by forest products companies during the late 1930s and 1940s.[88] The most immediate and tangible impact of the New Deal on southern forest regeneration, however, involved the work of the Civilian Conservation Corps in forest protection and forest regeneration. As with fire protection, the CCC helped transform reforestation into a moral crusade. By 1942, when the CCC closed shop, more than a million acres of timberland had been planted in the South (see figure 1.2).[89]

Nonetheless, by the end of the 1930s, despite the best efforts of the CCC and the Roosevelt administration, it was far from certain that the South's timber resource would be able to support a growing forest products industry. During this period, as firms in the pulp and paper industry accelerated their efforts to move south (fifteen new mills were built between 1934 and 1940, and regional pulping capacity more than doubled), forestry professionals along with representatives of other wood-using industries began to voice concern that the huge new mills would quickly strip the region of its young second-growth timber without making any attempt to regenerate it. Although the pulp and paper industry accounted for a small portion of the region's timber consumption (approximately 7% by 1939), the industry's growth potential combined with its preference for small trees raised concerns. Lumbermen and naval stores operators in particular opposed the new mills, fearful that younger trees would not be allowed to mature enough to provide raw materials for their own operations. Federal policy makers also voiced apprehension. In his annual report of 1937, the chief forester of the United States, F. A. Silcox, noted that pulp and paper firms' recent purchases of timber rights on large acreages of land from small farmers and landowners in the South had not involved any commitment to regeneration. In Silcox's view, this set the stage for "re-exploitation": "If such a practice continues, the land, the farmer, and the whole social and economic set-up must inevitably suffer. The South stands now at the crossroads."[90]

The following year, in March 1938, President Roosevelt entered the fray with a letter to Congress requesting an investigation of the "national forest problem." "The forest problem," he wrote, "is a matter of vital national concern, and some way must be found to make forest lands and forest resources contribute their full share to the social and economic structures of this country, and to the security and stability of all our people." Evoking images of "denuded" watersheds and "crippled" forest communities "still being left desolate and forlorn," Roosevelt urged the Congress to study the problem and propose legislation that would include "such public regulatory controls as will adequately protect private as well as the broad public interests in all forest lands." He concluded, "The fact remains that . . . most of the States, communities, and private companies have, on the whole, accomplished little to retard or check the continuing process of using up our forest resources without replacement. This being so, it seems obviously necessary to fall back on the last defensive line—Federal leadership and Federal Action."[91] Two months later, the Joint Congressional Committee on Forestry, under the direction of Alabama senator John H. Bankhead, launched a multiyear investigation of the nation's "forest problem." Based on extensive hearings in every major forest-producing region, the committee concluded that the management of commercial forest land under private ownership represented the crux of the so-called forest problem.[92] For pulp and paper firms moving into the South during the late 1930s, the threat of federal regulation proved quite serious indeed.

Many of these firms, however, had already begun to adopt practices of systematic forest conservation and regeneration. Because the new southern mills required a continuous supply of pulpwood at a reasonable cost, which effectively constrained mill procurement to the surrounding area, commitment to forest protection and forest regeneration on company-owned lands, as well as on the lands of the small owners who provided more than two-thirds of their total timber, was an economic imperative. From the mills' perspective, the overall objective was to generate a steady supply of timber to meet their needs without depleting the growing stock of timber in the immediate procurement area—a 100–150 mile radius from the mill site. Though most firms moving into the region during this period quickly acquired enough land to supply one-half to two-thirds of the wood necessary to support their mill operations on a continuous basis, they continued to depend significantly on nonindustrial timberland owners for the bulk of their timber, using their own timberlands to stabilize local pulpwood markets. By the late 1930s, for example, Union Bag already owned or controlled "considerably more" than the 400,000 acres it had estimated would be needed to supply its operations. With such a land base, the Savannah mill could effectively set the price for pulpwood in the region. Much of this land, however, was cut or burned over and thus required systematic investment in regeneration in order to reach its potential. To accomplish this, and in response to the mounting threat of regulation, Union Bag initiated its Forest Conservation Program in 1937 (one of the first in the industry) committing the company to establishing and maintaining high-productivity forests on company-owned land through fire protection, regeneration, and timber stand improvement.[93]

As important as such efforts were, however, they were not enough in themselves to stem the threat of federal regulation. Only a credible, industry-wide commitment would successfully preempt government intervention. Seeking to avoid such an outcome, representatives of the pulp and paper industry met in New Orleans in May 1937 to formulate a program for forest utilization and conservation in the South. The resulting "Statement of Conservation Policy of the Southern Pine Pulpwood Industry" committed firms in the industry to promote selective cutting practices, forest regeneration, and fire protection on company and noncompany lands. Two years later, in 1939, the major firms with southern operations formed the Southern Pulpwood Conservation Association (SPCA). Financed by the industry and headquartered in Atlanta, the SPCA provided technical forestry assistance and education to nonindustrial private timberland owners. Its motto—"Cut wisely, prevent fires, and grow more trees to build a better South"—symbolized the extent to which forest protection and forest regeneration were being framed in the language of moral duty. Over the course of its existence, the SPCA provided assistance to thousands of timberland owners. Although such a program was undeniably in the long-term interest of the industry, perhaps its most important impact was in silencing critics who claimed that firms were doing nothing to promote sound forest management among small private landowners.[94] Thus, even though the push for regulatory controls over private forest practices during the New Deal did not produce any long-lasting legislative results, the very threat of such regulation added to industry leaders' incentives to adopt minimum standards of forest protection and management on their own lands and to promote such standards among private landowners.

Notwithstanding such efforts to regulate private forestry directly, the most dramatic impact of the New Deal on southern forestry came not in the area of industrial policy but through agricultural policy and the profound restructuring of postwar land use that resulted. By initiating an agrarian transition in the South that radically reshaped the institutions that had governed land and labor since the end of Reconstruction, New Deal agricultural policies (and their successors) created unparalleled opportunities for forest regeneration, opening the door for the rise of timber as the South's number one cash crop. The era of farm forestry had arrived.

In response to these opportunities and incentives, small timberland owners developed representative organizations to pursue their collective interests. The Forest Farmers Association, for example, organized in Valdosta, Georgia, in 1941, provided a greater voice for small timberland owners in regional and national matters, such as credit and tax policy, and offered technical assistance. That same year, the American Tree Farm System began offering education and certification to private tree farmers (most of whom were large private landowners under industry sponsorship) throughout the United States and particularly in the South.[95]

By the early 1940s, then, the institutional foundations for southern forest regeneration were in place. After a short hiatus during World War II, forest planting and seeding accelerated significantly (see figure 1.2). Drawing on the network of state-supported nurseries established in the 1920s and 1930s under the Clarke-McNary program, seedling production moved into high gear.[96] It was the Soil Bank Act of 1956, however, that provided the most substantial incentives for forest regeneration.

In many respects, the Soil Bank Program simply extended the system of agricultural regulation established during the New Deal. As in the 1930s, the basic problem was overproduction, and the overall policy objective was to provide incentives for farmers to take acreage out of production—to bank soil. What was different about this program, however, was that it provided explicit provisions and financial incentives for forest regeneration as an alternative use for idled cropland. Thus, the 1956 Soil Bank Act had two primary components—an acreage reserve program designed to take cropland out of production temporarily in

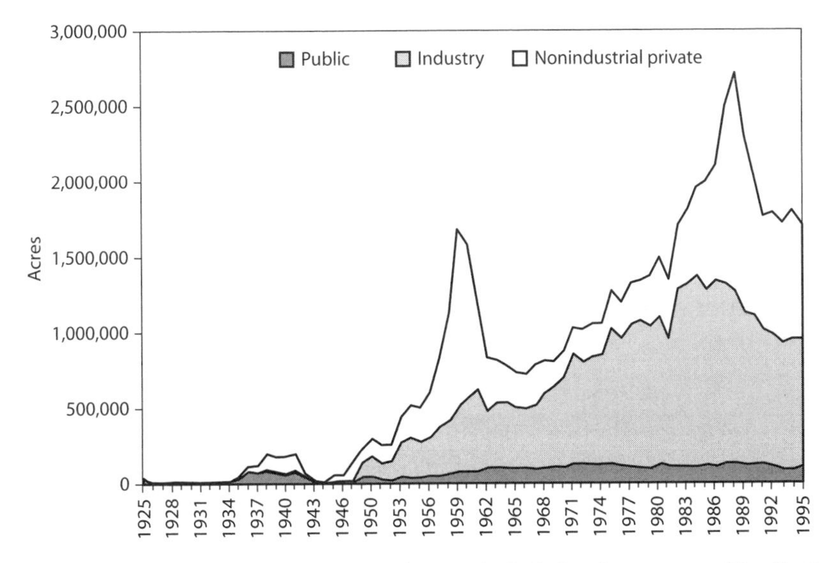

Figure 1.2. Annual forest planting or seeding in the U.S. South, 1925–1995. Hamlin L. Williston, *A Statistical History of Tree Planting in the South, 1925–1985*, Misc. Report SA-MR 8 (Atlanta, GA: USDA Forest Service, 1980); U.S. Forest Service, *U.S. Forest Planting Report* (1986–1989); U.S. State and Private Forestry (Cooperative Forestry), *Tree Planting in the United States* (1990–1995).

order to reduce overproduction problems for key commodities, and a conservation reserve program to provide incentives (cost-share arrangements) for landowners to convert their idled cropland to other uses such as timber production. Under the cost-share arrangement, the government reimbursed landowners who planted trees for 80 percent of their costs and provided annual payments determined by the estimated value of their land for the following ten years.[97] Many southern farmers obviously found the offer attractive. Over the course of the program (1956–1960), they converted more than 2 million acres of cropland to timber plantations. During the peak year of 1959, southern nurseries (public and private) produced more than a billion pine seedlings for planting on some 1.7 million acres of industry and nonindustry land. Overall, more than 70 percent of the total area planted and seeded in the region during this period was on nonindustrial private lands (figure 1.2).[98]

As an extension of New Deal agricultural regulation, the Soil Bank Act illustrated the power of government incentives to alter land use patterns and transform economic institutions. Without government assistance, landowners would find it difficult to embark on a substantial program of artificial forest regeneration. Indeed, significant acreages of nonindustrial private timberland were not

Figure 1.3. Union Bag and other paper companies actively promoted tree farming among private landowners in the areas surrounding their mills. Photograph, circa 1950. Courtesy of the Forest History Society, Durham, NC.

artificially regenerated in the South again until the advent of another government assistance program, the Conservation Reserve Program (CRP) of the mid-to-late 1980s (see figure 1.2).[99] From the perspective of the forest products industry, the Soil Bank Program was thus a phenomenal success. Indeed, the vast majority of cropland converted to timber plantations under the program remained in timber well after the annual cost-share payments ceased. A 1980 survey of Soil Bank lands in the South, for example, found that over 90 percent of the original plantations were still growing pines.[100] The program, though, was only a success for those whose interests it served. By creating incentives for established, primarily white, farmers to bank soil and by channeling the program through county political machines, the federal government gave a significant boost to the ongoing capitalization of southern agriculture—forcing marginal farmers and tenants, many of whom were African American, to leave the land.[101] The conversion of cropland to timberland, from this perspective, turned out to be a highly effective form of enclosure.

In the absence of government programs, many landowners simply allowed their lands to restock naturally, often with management assistance from forestry consultants and professionals working in industry-sponsored landowner-assistance programs (figure 1.3). The overall result was a substantial transformation of land use. Between 1948 and 1968, some 10 million acres of agricultural land in the South were converted to timberland through natural and artificial regeneration.[102] In contrast to earlier years, by the 1970s, most private timberland owners no longer actively engaged in farming. Many simply saw timberland as part of a larger portfolio of investments.[103]

As for industry lands, forest regeneration efforts grew steadily throughout the post–World War II period (see figure 1.2). By the early 1980s, the forest industry was regenerating more than a million acres per year in the South.[104] As land values increased, the premium on intensive forest management grew. For firms that owned large acreages, maximizing biological productivity became an economic imperative. Tree improvement represented the next frontier for industrial forestry.

Intensification

> Why . . . should the forester be satisfied to gather his seeds from the wild, unimproved forms that are only partially adapted to his needs? Should he not take steps to develop better strains of trees, and especially those capable of more rapid growth? —*Lloyd Austin, 1927*[105]

> Eventually, through the application of breeding methods, we may expect to produce high yielding and otherwise desirable genotypes at will.
>
> —*Scott S. Pauley, 1954*[106]

Though few foresters would likely have challenged the basic maxim of early inheritance studies that "like begets like," the science and practice of forestry was relatively slow in appreciating and incorporating the role of genetic variation and "parentage" in influencing the viability and success of trees. Despite the widespread use of improved seeds in agriculture, most American foresters in the 1930s and 1940s did not think twice about using unimproved seeds in their reforestation efforts. Unlike farmers, foresters typically had little appreciation for genetics and made almost no effort to match seeds to particular sites. In effect, the rediscovery of Mendel's work on inheritance at the turn of the century took almost fifty years to find its way into forestry practice. For some reason, foresters saw trees as different—beyond the scale of practical human manipulation.[107] Most operated on the assumption that trees of a given species were genetically uniform and that intraspecies variation was entirely the result of environmental influences.

Of course, there were those who did not accept such assumptions. Writing in 1929, Aldo Leopold chastised the forestry community for what he called "the highly improbable assumption that 'all trees are born free and equal.'" Leopold, along with Lloyd Austin and a few other pioneers, felt that forestry was missing an important opportunity and that both genetics and environment had to be taken seriously in forestry practice. By controlling both genetic and environmental variability, timber growing could achieve much-needed calculability. In Leopold's view, "what we are trying to create is not timber, but confidence that valuable timber can be made to grow; not dividends, but confidence that dividends can be made to accrue."[108] Part of the difficulty, of course, lay in the long biological time lags involved in tree growth. As Austin noted, "It is, at the present time, hard to interest people in the planting of trees that they know will not reach merchantable size in their generation."[109]

For those interested in applying classical breeding techniques to tree improvement, several obstacles stood in the way.[110] Compared to crop breeders, tree breeders had to deal with very large, immobile organisms with long nonreproductive periods and generation cycles that often exceeded human life expectancy. In contrast to most commercial crops, basic knowledge of forest and tree biology was quite thin, particularly regarding the relationship between the genetic structures of trees and tree characteristics.[111] Given the large heterozygosity of forest trees, finding superior trees and seeds, matching them to particular sites, and controlling the selection process in the context of open-pollination posed logistical challenges not found in crop breeding. Furthermore, the relative difficulty of controlling the biotic and abiotic environment of a particular stand of trees over the long time frame needed to judge breeding successes strongly limited the application of conventional breeding techniques.[112] Finally, from an operational standpoint, any successful tree improvement program would also have to produce improved seeds in commercial quantities sufficient to meet the needs of forest regeneration—a daunting task in a region such as the South where almost 2 million acres per year were being regenerated in the late 1950s.

Of all the obstacles facing tree improvement, though, it was the time constraint—time to achieve phenotypic stability, time to reach reproductive maturity, and time to harvest—that presented the greatest challenge. Maximizing the rate of return on investments in tree improvement thus would depend fundamentally on optimizing the amount of genetic enhancement per unit of time.[113] As gains accrued, the incentives to invest in tree improvement would increase. The problem, however, was that even under the best conditions, such gains would not be apparent until a decade or more after the initiation of a formal tree improvement program. Given the risks associated with such long time lags, combined with the considerable logistical and institutional challenges involved in such an effort, systematic investment in tree improvement greatly exceeded the scope of any individual firm's investment horizon. Developing a tree improvement program in the South would thus have to be based on cooperation among firms, as well as between private industry and the state—all of which would take time. As Harvard professor Scott Pauley indicated in the mid-1950s, directed manipulation of forest genetics on a commercial scale was still a rather distant prospect.

Yet, even as Pauley wrote, the institutional foundations of industrial tree improvement were being established in the South. Although early breeding efforts, particularly in the development of hybrid pines, had taken place on an *experimental* scale at the Eddy Tree Breeding Station in Placerville, California,[114] there had been no systematic effort to develop a tree improvement program on an *industrial* scale in the United States until the early 1950s. During this time, a constellation of factors in the South—the growth of the pulp and paper industry and its appetite for wood, the rising value of timberlands, the increased demand for seedlings to furnish regeneration efforts, and the growing appreciation and acceptance of population genetics as a tool for tree breeding—created the opportunity for those interested in tree improvement to move toward operationalizing an applied tree breeding program for the region.[115] One of the first formal steps came in early 1951, when a group of foresters convened in Atlanta to develop a tree improvement strategy. This meeting, which led to the formation of the Southern Forest Tree Improvement Committee (SFTIC) later that year, underscored the growing concern among some in the forestry profession that southern regeneration efforts then under way were missing important opportunities by giving only passing attention to forest genetics. Going forward, the SFTIC, which operated under the auspices of the U.S. Forest Service, sponsored biennial conferences on forest tree improvement and served as a coordinating body for regional research efforts in the areas of seed source (also known as provenance), forest genetics and breeding, and the production of superior seed.[116]

In one of its first and most important undertakings, the SFTIC established the Southwide Pine Seed Source Study under the direction of Philip Wakeley—the man known as "Mr. Southern Pine."[117] Up until this time, disregard for seed source in seedling production was the rule rather than the exception among seed collectors. Most state nurseries obtained loblolly pine seed (the most widely used planting stock) from the cheapest source available, without making any real effort to match seed source to the particular region where seedlings would be planted—an oversight that led to unnecessary planting failures.[118]

Wakeley's seed source study, which involved provenance research for the four major pine species of the South (loblolly, slash, longleaf, and shortleaf), marked a significant departure from traditional views in that it explicitly identified genetic factors within tree species as a source of variation.[119] It was a massive undertaking, requiring the cooperation of federal, state, industry, and university actors in spreading the costs and dividing the labor. Such a study, moreover, proved vital not only to efforts to plant "unimproved" seedlings in areas to which they were adapted but also to designing future tree improvement programs. Without attention to provenance, planting stock could all too easily end up on sites to which it was ill suited. Likewise, without knowing the geographic origin of particular seeds, forest geneticists would be unable to make accurate assessments of breeding successes and develop reliable pedigree analyses.[120] According to Wakeley, "genetic differences between individuals or local strains within a geographic race . . . seem likely to be overshadowed by the genetic unsuitability of the race as a whole when stock is transferred to a less favorable place. The inescapable conclusion is that selections and hybrids must be made separately region by region, within the framework of existing geographic races. To the extent that this is true, provenance studies designed to identify such races and define their territorial boundaries are fundamental to other phases of tree improvement."[121] Put crudely, geography mattered. Provenance studies would henceforth be the foundation of any successful tree improvement program. As a result of Wakeley's work, the importance of seed source was widely accepted and southern states adopted formal seed certification programs. By the end of the 1950s, virtually all state and industry efforts in artificial regeneration drew seed from local sources.[122]

As artificial regeneration efforts increased during the 1950s, early tree improvement supporters also began to argue for selecting seed stock from superior trees. Impressed by the productivity gains being recorded in agriculture, particularly with hybrid corn, some suggested that an intensive breeding program might also bring large benefits to industrial forestry. If you were going to plant, they argued, why not plant the best. Intensive breeding with forest trees, however, represented a substantially greater challenge than that involving agricultural crops, and it was far more complicated than provenance studies. Procuring seed from phenotypically superior trees and using this seed as the basis for the commercial production of seedlings represented a major hurdle facing those who wanted to operationalize tree improvement on an industrial scale. Such a program would have to be carefully planned and, due to the biological time lags involved, would necessarily take more than a decade to achieve any tangible results. Writing in 1954, Clemens Kaufman, the director of the School of Forestry at the University of Florida, reminded his fellow foresters that they were still operating in "what might be termed the fire protection-planting stage of forestry practice . . . We must remember that at this time we have neither super trees nor seed from trees of proven quality, or even stands of average quality or better, phenotypically, which have been selected to supply the quantity of seed required annually." Kaufman thus suggested that the principles of genetics could be employed immediately in an *extensive* manner by selecting and leaving phenotypically superior trees as seed trees for natural regeneration.[123] The *intensive* application of genetics to tree improvement, however, would have to wait until reliable volumes of improved seed could be produced for artificial regeneration—a challenge that, more than any other single factor, led to the creation of the industrial tree improvement cooperatives during the 1950s.

The first university-industry tree improvement cooperative in the United States was established at Texas A&M University in 1951. Initially, the cooperative included eight companies, which together provided most of the financing, while the university provided facilities and staff. Bruce Zobel, who had just received his PhD in forest genetics from the University of California, was hired as the first director. Influenced by Scandinavian and Australian research in forest genetics, most of the early work focused on developing an applied breeding program oriented to the members' areas of operation in Texas, Louisiana, and Oklahoma. Because of the relative lack of basic knowledge on tree biology and forest genetics, however, participating firms saw the whole effort as something of a gamble.[124] These companies were investing in a long-term effort that no one could say for sure would actually result in tangible gains.

Shortly after the establishment of the Texas cooperative, industry leaders, in concert with members of the forestry community, initiated two more tree improvement cooperatives—one at the University of Florida (1955) and one at North Carolina State University (1956).[125] In 1956, Zobel moved to Raleigh and became the director of the new NC State cooperative as well as a professor of forestry at the university. The cooperative at NC State, which focused primarily on loblolly pine, quickly emerged as the largest and best known of the three. Its principal objective was to develop strains of trees with desired characteristics—yield, quality, and adaptability—and to produce seeds of these strains on a commercial scale. Reducing turnover time was the critical challenge. As Zobel put it, "The objective of our tree improvement program is to get as much improvement as possible as quickly as possible . . . We are interested in gain per unit of time."[126]

From an organizational standpoint, the program was unique. There were no formal contracts or written agreements. Members (about thirty-five in the early 1970s: thirty-two from industry and three from state agencies) paid yearly dues and participated on an equal basis. Any member could withdraw at any time, and any member could be asked to withdraw if its contribution was considered inadequate. Members who withdrew, however, would not be allowed to return. The cooperative's university-based managers designed, analyzed, and interpreted field tests while professional foresters employed by the member firms performed fieldwork under the guidance of cooperative staff. More fundamental research, which was not a primary focus of the cooperative, enlisted associated graduate students, some of whom would later become employees of member firms. The entire effort was multidisciplinary, drawing on several fields outside of forestry, including pulp and paper science, genetics, botany, statistics, soil science, plant breeding, and plant pathology.[127]

As noted, the major operational objective of the cooperative was to provide commercial quantities of improved seed to meet the annual planting needs of members (approximately 400,000 acres or 300 million seedlings in the early 1970s). At the outset, the tree improvement cooperatives promised members a 5 percent gain in volume from improved seed. Member firms would select seeds from superior trees from their own lands and exchange them with the cooperative. These "superior trees" (approximately 3,000 for each cooperative) provided the genetic base from which gains could be realized in both the long run (through advanced generation breeding) and the short run (through intensive selection). Based on selected trees, the cooperatives established breeding or research orchards to maintain diversity and support long-run breeding efforts as well as first-generation commercial seed orchards to begin producing improved seed.[128] In the process, researchers performed progeny testing to establish pedigreed lines of improved trees, which in turn became the genetic foundation for industrial forestry in the South.

Given the turnover times associated with trees, however, applying the principles of quantitative genetics to select for and produce commercial quantities of improved seed was a long and laborious process. Indeed, even though the southern tree improvement cooperatives began operating in the 1950s, it was not until the early 1970s that commercial quantities of improved seed became available. By 1973, the NC State cooperative was only producing about half the planting stock needed by its members. Not until the end of the decade did the cooperative meet all its members' needs. For the region as a whole, the proportion of improved seedlings produced in southern nurseries (public and private) increased from about one in four in 1976 to more than 90 percent in 1986 (out of a total 1.6 billion seedlings).[129] By this time, NC State cooperative members had established some 4,000 acres of seed orchards that yielded some 630 million seedlings per year (almost 40% of the regional total)—enough planting stock to regenerate 900,000 acres annually.[130]

As for the actual gains from the program, the first generation of improved trees developed through the NC State cooperative showed an average increase of 7 percent in height growth, 12 percent in stem volume, and 32 percent in harvest value. Real after-tax returns from the total investment on tree improvement were estimated to be between 17 percent and 19 percent, quite healthy by any standard. By the mid-1990s, the NC State cooperative was moving into its third generation of selection and breeding. In roughly twenty-five years of seed production, the member organizations harvested sufficient seed to plant more than 13 billion genetically improved loblolly pines—enough to cover 19 million acres.[131] Biological intensification had become the driving force of industrial forestry in the region.[132]

The system of tree improvement cooperatives, however, was not the only game in town. The federal government was also an active player. The Southern and Southeastern Forest Experiment Stations of the Forest Service, for example, undertook research on forest disease and insect problems.[133] The Southern Institute of Forest Genetics, established by the Forest Service in 1954 in Gulfport, Mississippi, also developed a long-term project to investigate "the fundamental principles of forest tree inheritance."[134] In addition, several university departments of forestry carried out research on various issues associated with tree improvement.[135] And, of course, major firms developed their own tree improvement research programs. International Paper ran a research program and experimental forest at Bainbridge, Georgia. Union Camp operated forestry research programs in Bellville, Georgia, and in Princeton, New Jersey. Weyerhaeuser initiated a program in eastern North Carolina, and Westvaco ran a research facility in Summerville, South Carolina.

In the 1980s and 1990s, some firms began experimenting with techniques derived from molecular biology. As these "new" biotechnologies were employed in forest tree breeding, proprietary issues began to impact the nature of cooperative research in the industry. Indeed, although wide-scale use of recombinant DNA and other techniques derived from molecular biology has been slower to take hold in industrial forestry than in agriculture, it is clear that as particular processes or products (i.e., new genetically engineered seeds) become strategic resources and subject to the laws governing intellectual property, the cooperative arrangement that worked so well in the past will likely undergo substantial change. As one industry executive put it, "Now when you begin to look at the role of biotechnology and gene insertion and cloning your best materials you step farther away from the cooperatives. You're getting into intellectual property at that point."[136]

Notwithstanding the embrace of genetic engineering at the end of the 1990s, however, the cooperative system continued to provide much of the foundation for tree improvement efforts at the beginning of the twenty-first century. Looking back over the last half century, one is thus struck not only by the remarkable success of these efforts in transforming large acreages into highly productive industrial timber plantations but also by the hybrid public-private character of the enterprise and, perhaps most surprising of all, by the extensive cooperation between firms that competed against one another. Part of this may stem from what some have identified as the "cooperative culture" of the forestry profession in the South during the middle decades of the twentieth century. A more important factor has surely been the complexity and long-term nature of the enterprise, which militated against any single firm embarking unilaterally on tree improvement.[137]

The Forest Is the Future

> To meet permanently the increasing requirements for pulpwood will necessitate placing our forests upon a systematic basis of management for the production of wood as a crop. Only by so doing can manufacturers be assured of supplies within reasonable distances of their mills. Under such conditions the future growth of the industry must necessarily take place *in those regions in which pulpwood can be grown in short rotation with a high rate of growth.*
>
> —*John D. Rue, "The Development of Pulp and Paper-Making in the South," 1924.*[138]

> Today the South has come to count the meaning of the pulp and paper industry as something more than jobs in its towns and money in its woods. The greatest significance of this industry, which cannot move its multimillion-dollar plants like "peckerwood" sawmills, but which provides a market for trees a man can grow and sell in his own years, is already evident. It has made creative forest management in the South not only possible but imperative.
>
> —*Jonathan Daniels,* The Forest Is the Future, *1957.*[139]

In the thirty odd years separating John Rue's assessment of the challenges facing the development of a pulp and paper industry in the South from North Carolina newspaperman Jonathan Daniels' retrospective, the southern forest underwent a profound transformation. By the late 1950s, when Daniels penned his celebratory account of the forestry revolution transpiring in the South, the region had emerged as the center of the U.S. pulp and paper industry. Given its favorable climate, the vast acreages available for growing trees, and the relatively short turnover times associated with southern pine, it seems quite plausible (even logical) in hindsight that the South was destined to become a major national and international force in the production of pulp and paper. When Rue issued his assessment in the 1920s, however, such an outcome was hardly assured. Throughout the 1910s and 1920s and well into the 1930s, members of the forestry community voiced grave concerns over the future of the southern forest resource and its capacity to support a large forest products industry.

In essence, for a large-scale timber products industry to develop, a new regime of forest management would have to be inscribed on the southern landscape—a big part of which involved subordinating timberlands to the dictates of continuous-flow, industrial production. At the most basic level, this involved shifting from the earlier logic of extraction to a logic of intensive cultivation that treated timber as an industrial crop. Such a transition, driven in part by economic imperatives, required new institutional arrangements to deal with the enormous challenges of making timberlands amenable to rational economic calculation, regenerating cut-over and marginal agricultural lands, and developing and producing improved genetic stock. In all of these areas, extensive cooperation would be required—between government agencies, private landowners, and private industry; between universities and firms; and between competing companies within the industry. Through a process of regional collective learning, all of these actors played vital roles in constructing a highly competitive industrial system.[140]

As for the southern timberlands, they underwent a profound ecological transformation—part of a larger agrarian transition initiated by the New Deal that radically altered southern land use patterns. In effect, the southern forest was made over into a highly managed ecosystem more closely akin to a crop monoculture than a biologically diverse forest. Through a process of ecological simplification, timber plantations came to provide an increasing part of the biological foundation for a highly successful industry. Between 1940 and 1990, more than 38 million acres of land were artificially regenerated in the South. During that time, southern nurseries produced over 23 *billion* tree seedlings, the majority of which, since the 1980s, were genetically improved. As land values rose across the region, the premium placed on intensive management increased significantly. Through biotechnology (old and new) timber plantations were subjected to a sort of biological time-space compression.[141] Nature was (re)made to work harder, faster, better.

Despite such remarkable success, however, new challenges emerged during the final decade of the twentieth century. The spread of intensively managed pine monocultures throughout the South carried with it the potential for increased vulnerability to a variety of forest pathogens. The loss of species diversity and habitat associated with the ongoing transformation of mixed forests to plantation monocultures raised concerns among environmental groups, transforming southern timberlands into a political battleground. Finally, the spread of short-rotation pulpwood plantations in the southern hemisphere combined with persistent overcapacity in the global pulp and paper industry undermined the profitability of southern producers and raised questions about the long-term viability of the industry in the region.

Still, the American South continued to dominate both domestic and global pulp and paper production through the end of the twentieth century. Based on a distinctive set of institutional arrangements that facilitated the intensive cultivation of timber as a crop, the South emerged as a global leader in industrial forestry and the production of pulp, paper, and solid wood products. Whether the region will be able to meet the challenges of the twenty-first century and remain competitive will depend fundamentally on how effective these institutional arrangements are in adapting the existing industrial forest to the demands of the future.

CHAPTER TWO

Logging the Mills

> The relations men establish among themselves to make land productive go far to determine the quality, reach, and tempo of their lives. Their relation to the land poses basic issues of social organization because it involves the physical basis of life. For more subtle reasons, also, the relation is one of those which fix the framework of society. The terms of access to the land affect the practical power of decision which some men enjoy over the lives of others. —*James Willard Hurst, 1964*

For pulp and paper firms establishing operations in the South during the interwar and postwar years, subordinating the southern forest to the dictates of continuous-flow industrial production constituted a long and laborious process. Given that farmers and other small landowners controlled the vast majority of southern timberlands, the transition from the mining of forests to the farming of trees required new institutions and incentive structures that reached well beyond the boundaries of corporate organization. Imposing a new forest management regime on company lands, in other words, did little to facilitate timber farming on the vast acreages owned by nonindustrial actors. Individual landowners also had to be persuaded of the benefits of treating timber as a crop. Government intervention, whether in fire control, reforestation, agricultural policy, or tree improvement, played a key role in the whole process. As incentives changed and as the demand for wood grew, many landowners realized that timber could and did provide a viable alternative to row crop agriculture. By the early 1990s, timber covered almost three out of every five acres in the South and was the region's number one cash crop.[1]

Accomplishing this transformation, however, only solved part of the overall supply problem facing the newly established pulp and paper mills. Firms also had to establish cost-effective and reliable ways of procuring and delivering sufficient quantities of wood to the mills. Given the sheer volume of wood consumed by a modern pulp and paper mill, this posed a difficult problem anywhere (figure 2.1). But the problem proved especially challenging in the South due to fragmented landownership, the reluctance of local landowners to do business with outsiders, and the complex interpenetration of race and class in regional labor markets.[2] Simply put, to procure the pulpwood they needed, mills had to engage the institutions governing land and labor in the rural South. Strategies would have to be developed to gain access to the timber controlled by countless farmers and other small landowners. Local social and political networks would have to be tapped without eliciting widespread discontent and resistance. At the same time, tens of thousands of itinerant workers, many of whom were black and most of whom came from sharecrop agriculture or some other low-wage industry, would have to be recruited, organized, and managed, often in crews of five men or less, to harvest and transport an ever-increasing volume of wood. In the process, older work rhythms and labor regimes would have to be assimilated and adapted to the highly uneven and often dangerous realities of "woods work."

As if all of this were not challenging enough, certain biophysical characteristics of southern pine placed further constraints on wood procurement. Because southern pine deteriorated rapidly in storage, especially during the hot, humid summers, mills could only maintain a few weeks inventory at most—levels that northern mill managers considered dangerously low.[3] Moreover, although wood could be harvested in the South year-round (in contrast to the Northeast and the

Figure 2.1. Southern pulp and paper mills consumed massive quantities of wood. This photograph shows the woodyard at International Paper's Springhill, Louisiana, mill, circa 1948. Photograph by Gabriel Benzur, courtesy of Gabriel Benzur Photography LLC.

Lake States), wet weather effectively halted logging operations and thus mandated that mills procure their wood from a geographically diverse territory in order to avoid disruptions.[4]

Wood flow also had to accommodate certain requirements of pulp and paper manufacturing. Because the Kraft or sulfate pulping process required that the wood chips used in a particular "cook" be of the same basic age (juvenile and mature wood could not be mixed together), wood deliveries to the mills had to be coordinated in order to ensure that enough wood of the "proper age" would be available for pulping. And because papermaking machines could produce several different grades of paper, their fiber requirements often varied considerably.[5] Depending on the needs of particular customers or markets, the same machine might be used to make bag paper one week and linerboard (cardboard) the next. Since thicker paper obviously required more fiber, overall wood demand fluctuated in tandem with the thickness or grade of paper being made and the length of the production run. Wood requirements at Union Bag's Savannah mill, for example, varied by as much as 20 percent from week to week due to grade switching and differences in run times.[6] When added to the biophysical constraints on inventories, such variations in wood demand meant that the pure logistics of procuring the right amount of wood at the right time required considerable flexibility. In many respects, the underlying principles of southern wood procurement (even as early as the 1930s) more closely resembled a flexible, "just-in-time" inventory management system than the "just-in-case" approach typical of other mass-production industries. Once embedded in the institutional context of the postbellum rural South, wood procurement proved very challenging indeed.

This chapter focuses explicitly on this question of how pulp and paper firms constructed a viable wood procurement system in the face of these various challenges. The intention is to unpack the ways corporate managers coordinated with other economic agents, often in the context of vast differentials of wealth and power, to create new institutions for accessing land and labor. The resulting system, which varied over space and time, built on multiple layers of informal contracting and highly localized markets. Depending on one's theoretical leanings, one might argue that the particular institutional forms that emerged reflected the imposition of class interests on the one hand or efficient responses to transactions costs on the other. But although class interests and efficiency both played important roles in the story, neither carries sufficient explanatory weight. Indeed, on closer inspection, it is clear that the southern wood procurement system emerged as much from strategic political and legal concerns as from purely economic ones. Despite their privileged access to vastly superior financial and organizational resources, pulp and paper firms moving into the South could not simply impose their will on a prostrate region. Nor could they assume that "properly" functioning markets for land, labor, and timber would spontaneously emerge. Instead, they had to develop strategies to construct such markets that would truck with the complex regional politics of land and labor. Flexibility, in this case, derived not from a choice among exogenously posed competing technological paradigms but rather from a historically specific process of social and political construction.

To say then that there was no one right way of organizing wood procurement in the South might be a bit redundant, for procurement varied between and within firms and changed over time in response to new circumstances. Its inherent plasticity reflected variations in locale, corporate culture, technology, and the relations between different actors. Still, despite such variation, the basic institutional contours of the system proved remarkably durable. After moving into the region in the 1930s, pulp and paper companies relied almost exclusively on "independent" contractors for their wood supply. In practice, such an arrangement provided the mills with access to small landowners while maintaining flexibility in inventory control and freeing up the capital associated with buying and carrying wood. At the same time, it absolved them of the various costs and liabilities associated with recruiting, organizing, and managing the labor necessary for pulpwood "production."

Thus, beginning in the 1930s, virtually all mills used independent contractors known as "wood dealers" to handle their procurement needs. Recruited by the mills, these local power brokers possessed the social, political, and economic resources necessary to procure timber from local landowners and organize the labor necessary for harvesting and transporting pulpwood. They might be businessmen, merchants, county officials, or prominent landowners—anyone who was good at "hustling wood"—and they were always white. In the early days, each dealer had an exclusive territory and received a guaranteed commission on every cord of wood delivered from that territory. Dealers typically contracted out harvesting and transport operations to small independent loggers. These "pulpwood producers" usually operated in crews of five men or fewer and often received financing and equipment from their dealers. Paid on a piece-rate basis, most of these men, many of whom were black, occupied a social position more closely akin to sharecroppers than to wage laborers.[7] Viewed as a whole, the system appeared highly informal and deeply embedded in the political economy of race and class in the rural South. Written contracts did not exist. Producers had limited mobility and little recourse to formal legal protections. They depended on their dealers, who in turn depended on the mills. With wood orders changing from week to week, dealers and producers had to maintain sufficient flexibility in order to adjust or risk going out of business. For those at the bottom of the hierarchy, economic insecurity was a way of life.

Over time the system changed. As regional pulping capacity expanded, increased competition for pulpwood began to erode the exclusive territory arrangement. Meanwhile, new opportunities associated with postwar industrialization drew increasing numbers of workers out of the woods. Markets for stumpage and labor became more competitive. After two major wood shortages during the 1950s, firms began to push for mechanization. Scientific management and forest engineering emerged as watchwords in an effort to boost productivity and rationalize the procurement system. Seeking to gain more control over procurement, some companies abandoned the dealer system and went with an approach known as "direct purchase," using company employees to purchase timber directly from landowners, while outsourcing logging to independent contractors. A few firms even experimented with company harvesting operations, primarily as a means of facilitating mechanization.[8] By the 1980s, timber harvesting had become a highly capitalized endeavor. As productivity rose, excess logging capacity became endemic. Prices stayed low, and wood flowed to the mills. Seduced by easy credit, producers suddenly found themselves burdened with heavy debt loads and more dependent than ever on wood dealers and mill officials to "keep them in wood." Though some tried to organize and others sought recourse in the courts, such efforts rarely succeeded. Their insecurities and dependencies provided much of the social basis of flexibility in the system.

Thus, regardless of whether mills procured their timber through dealers, direct purchase, or some combination of both, they continued to contract out logging operations, even on company lands. Given the highly uneven nature of woods work and the many dangers associated with logging, contracting proved far more cost-effective than full integration. By maintaining an "arm's-length" relationship between the company and the loggers, the contract-based system insulated the mills from many of the risks and liabilities associated with logging. As long as loggers were independent contractors rather than employees in the eyes of the law, they were responsible for complying with wage and hour laws, tax codes, insurance requirements, and the growing number of safety and environmental regulations.[9] At the same time, however, because they worked as "independent contractors," loggers faced obstacles in their efforts to organize for collective bargaining purposes. They were effectively caught in a web of social relationships that, while deriving much sustenance from the informal networks and social hierarchies of southern society, were also deeply imbued with formal law. Law, in this context, did not simply condition the evolution of these relationships; rather, it played a fundamental role in *constituting* them.[10]

This chapter charts the historical development of the southern wood procurement system. The first section shows how the mills used a spatially demarcated dealer system in combination with their own land acquisition strategies to gain access to small landowners' timber, construct workable markets for pulpwood, and maintain low wood prices. This part of the story stretches from the 1930s to the 1950s, when the first major wood shortages occurred and the exclusive territory arrangement began to break down. The following section turns to the organization of logging, backtracking a bit to trace the ways rural workers were recruited, organized, and integrated into the specific labor regimes that marked the procurement system. The aim here is to situate pulpwood producers within the context of the rural South, describe the nature of their work, and explain how their relationships to the dealers and the mills changed in response to some of the larger political, legal, and economic transformations that transpired during the interwar and postwar years. The third section picks up the story in the late 1950s and early 1960s, when wood shortages and labor problems pushed firms to explore ways of mechanizing harvesting operations and reorganizing pulpwood markets. This essentially brings the narrative up to the 1990s. The concluding section analyzes how the procurement system and the use of contract logging in particular have functioned as a way of maintaining economic flexibility while displacing many of the risks and liabilities associated with logging onto pulpwood producers.

Constructing Markets

> [T]he paper and pulp mills have a monopoly in dealing with small timberland owners. Recently when I refused an offer made by a pulpwood company's agent on some timber in south Georgia, the agent told me I would come back to them because that was the best I could do. And it was true because there was no one to compete with them. —*Annie Mae Strickland, 1940*[11]

Annie Mae Strickland was angry. She had heard the boosterism surrounding the coming of the pulp and paper industry, and she looked forward to a new source of revenue from her south Georgia timberlands. Although she did not particularly care for the price she received from the mills, her only other option—to forego selling altogether—proved even less attractive. She needed the money, and it was a take-it-or-leave-it situation. Unaware of just what kind of market she had entered into, Mrs. Strickland soon realized that she was, in fact, a price taker and had very little bargaining power. In crying monopoly, she joined a growing chorus of landowners who were unhappy in their dealings with the pulp and paper mills. Her concerns, along with those of others, would soon become the subject of congressional hearings.[12]

By virtue of their existence, the mills created a huge demand for pulpwood within their immediate procurement areas. Much of the wood that they purchased had little previous economic value.[13] Expectations for the new "cash crop" were hardly in short supply. Given the prevailing pattern of land tenure in the rural South and the inherent spatial and temporal constraints involved in timber production and marketing, however, properly functioning pulpwood markets did not spring up overnight. Because of their need for continuous, year-round supplies of wood subject to particular inventory constraints, the mills could not simply rely on the open market. If they did, they would have no way of controlling wood flow and would likely be forced to reduce capacity or shut down altogether during the busy farming season.[14] More importantly, even if they could maintain sufficient wood flow year round, going with a spot market system also ran the risk of opening up stumpage to competitive bidding, leading to price increases. In the early years, when mills were few and far between, they might have been able to sit back and let the wood come to them. By the 1930s, however, such an approach proved too risky. Indeed, early mills, such as International Paper's plant in Panama City, Florida, which started out using the open-market system, soon abandoned it in favor of the contract-based system.[15]

Contracting was not the only alternative to an open-market arrangement. Company representatives and company loggers could also be used to procure and harvest wood from the thousands of small landowners in a mill's procurement area. While such an arrangement might provide greater control over wood flow, it came at a cost. Having a paid employee constantly available in all of the small towns (some no bigger than a railroad depot and a post office) from which the mill purchased timber would be rather expensive. As for harvesting, the recruitment and monitoring costs associated with the salaried men and day labor necessary to keep sufficient wood moving to the mills were prohibitive. Finally, using company employees to procure timber directly from small landowners also ran the risk of eliciting resistance to the mills, many of which were controlled by northern corporations. As strangers in a strange land, these mills could not afford to alienate local landowners.

Full-scale integration represented the other possible approach to procurement, and there were precedents. Lumber companies operating during the late nineteenth and early twentieth century often bought vast tracts of land to supply their mills, using large-scale company harvesting operations to maintain a steady wood flow. Pulp and paper mills in northeastern states such as Maine also procured most of their timber from company-owned lands and used company-employed loggers for harvesting. With prices for timberland at rock bottom, firms moving into the South during the 1920s and 1930s could certainly have acquired enough land to establish fully integrated operations. Such a strategy, however, carried its own risks. Fire and other destructive agents threatened timberland investments, while taxes and other carrying costs often made landownership a relatively poor use of capital. Simply put, pulp and paper mills did not need land. They needed wood. And until wood became a limiting factor, the capital costs of owning huge tracts of land had to be weighed carefully against other more productive investments.

Still, almost all southern mills did acquire significant land bases during the interwar and postwar years. Their goal was never self-sufficiency in fiber supply. Rather, they used "wood security" to insure price stability. A timberland base close to the mill, for example, could compensate for disruptions in wood flow. More importantly, such a reserve also provided a means for counteracting higher pulpwood prices. If prices rose too high, mills could simply cut more timber from their own lands, thereby reducing requirements for purchased timber. Owning timberland, in short, provided a hedge against higher prices.[16]

This hedging capacity represented a critical component of a firm's procurement strategy. Yet no one could say with any certainty how much land was needed to provide an effective cushion.[17] One government forester who surveyed the southern wood procurement situation in the 1930s suggested that "mills locating in the South should acquire a sufficient reserve of forest lands to ultimately insure two-thirds to three-fourths of the mills' pulpwood requirements. Such a forest reserve is very effective in controlling the market price of pulpwood. Without a reserve, the mill is more or less at the mercy of the wood producers. With a reserve, the company can, whenever higher wood prices threaten, start logging operations on its own lands." Obviously, the actual amount varied from region to region and mill to mill. In areas dominated by many small farmers and landowners such as the Piedmont, it would be very difficult for timber owners to influence stumpage prices. Thus, a large forest reserve might not be as important there as in areas such as the coastal plain, where individuals who owned or controlled larger unbroken tracts might be able to "control the cut" and thus influence the price of pulpwood. There were also significant first-mover advantages in land acquisition. During the late 1920s and 1930s, prices paid for southern pine timberlands averaged around five dollars per acre.[18]

For those who moved in early, large forest reserves could be amassed quickly and cheaply. By 1937, for example, Union Bag had essentially completed the land acquisition program for its Savannah mill, owning or controlling "considerably more" than the 400,000 acres of timberland it had estimated would be needed to supply the mill. Such a reserve allowed the company to maintain "a proper balance between supply and demand . . . a safeguard against possible collusion by land owners and contractors which might raise the price of stumpage beyond reasonable and competitive levels."[19]

Outright purchase, of course, was not always feasible or desirable. Landowners might be reluctant to sell, or the carrying costs of full-blown ownership might prove too high to bear. To get around these sorts of problems, firms such as Union Bag developed leasing strategies that gave them long-term control over stumpage. Known in the industry as the "southern lease," these arrangements typically lasted for ninety-nine years.[20] Most often, such a lease committed the property owner to sell the timber on his or her land for the duration of the agreement. For its part, the company acquired the surface rights associated with the land, agreed to manage the land for "maximum timber growth," and was bound to return it to the owner with a stand of timber in "good condition." Should the owner decide to sell before the end of the lease, the company usually retained a purchase option.[21] During the 1930s and 1940s, many mills employed such instruments to gain access to strategically located properties as a means of shoring up their land bases.

Controlling stumpage, through ownership or lease, thus represented a critical step in the construction of pulpwood markets. More than simply a hedge against supply disruption and price volatility, this strategy of backward integration also provided mills with considerable leverage in their efforts to establish a viable procurement system in which they would be the dominant players. In effect, mills could use their pulpwood reserves not only to mitigate against supply problems, price increases, or possible collusion by landowners and contractors but also to create more favorable terms of access. By altering the distribution of assets, they could use their supply-side position to influence demand and thus leverage their position as buyers.

The actual process of negotiating and institutionalizing such access, however, represented a significant challenge in and of itself. Clearly, people wanted to sell their wood, especially during the 1920s and 1930s, when times were tough.

Yet there was no guarantee that the mills would actually get access to the timber controlled by landowners on terms they found favorable. Aside from the many transactions costs associated with procuring timber from thousands of small landowners scattered over a vast area, mills had to be careful not to elicit undue local resistance. Too many Annie Mae Stricklands and the government might step in to try to regulate pulpwood markets.[22]

At the same time, competition needed to be controlled. With more mills moving into the region, procurement managers had to be careful to avoid (where possible) situations in which they would be bidding against other mills for the same timber. Too many buyers might drive up the price of stumpage. Mill managers needed a strategy for demarcating local pulpwood markets in a way that would facilitate access to local landowners while minimizing competition. By contracting out wood procurement operations to local businessmen and granting them exclusive territories, they solved both of these problems while maintaining flexibility in wood flow and removing themselves from the task of organizing the labor necessary for harvesting operations. This was the so-called wood dealer system, and by the 1930s it was in widespread use throughout the South.

Although the precise origins of this particular institution are somewhat obscure, it is clear that contracting had been used (albeit sparingly) for procurement in the southern lumber industry and by pulp and paper firms in other regions.[23] As the industry expanded in the South during the 1930s, managers and executives turned to a contract-based system as a means of avoiding the various costs and risks associated with either the open-market approach or full integration. Firms granted dealers exclusive territories or franchises and paid a flat commission for all wood shipped out of their territories. This allowed mills to regulate wood flow by shifting wood orders up or down among the various dealers and to keep prices low by minimizing competition. Landowners who wanted to sell their wood to the mills had no choice but to go through the dealers. During this time, pulpwood markets were highly localized, narrow, and exclusive.

Union Bag, for example, established such a system during the 1930s to service its Savannah mill. The overall objective was to confine pulpwood operations within a radius of 130 miles and to maintain pulpwood prices at a set level. Wood located farther from the mill would receive a lower price to compensate for higher haul rates. In order to maintain stable prices, the company recruited thirty or so contractors (also known as dealers) and allotted them exclusive territories. In the words of the company, this was done "not only to keep the contractor's operating costs at a minimum by localizing the extent of his supervisory activities, but also to eliminate the competitive bidding for stumpage between . . . contractors." The size and location of each contractor's territory would be determined by experience, dependability, managerial capacity, financial resources, and "local community prestige and political influence that will not only open up stumpage otherwise not available, but also create local good will and political influence against the enactment of unfavorable legislation."[24] Fearing the "possibilities of collusion," as well as the potential that an individual dealer might become too powerful, Union Bag took care to allot territories to "a reasonable number of contractors rather than a select few"—even though "it would be more economical to deal with a few well-financed contractors of long experience and with adequate financial resources and managerial capacity."[25]

In establishing its system of dealer franchising, the company also sought to spread its procurement operations over an extensive area. Not only would a geographically diverse procurement territory provide some insurance against the vagaries of weather (i.e. local rainy seasons would not materially affect the overall flow of wood), but it would also serve to keep prices low and stable. Rather than exploiting cheaper wood in adjacent areas for a few years and then moving into more distant areas subject to higher haul rates, a geographically extensive approach to procurement would, assuming natural rates of reproduction, ensure a consistent price level over a longer period of time. Pulpwood markets thus had an important spatiality, which was used to overcome the inherent time lags involved in tree growth and avoid granting too much leverage to any one supplier—a distinct geography that was carefully constructed to keep prices low and maintain wood flow.

Like other companies, Union Bag took an active role in building its organization of contractors, recruiting men who appeared to possess the political and financial resources necessary to succeed as wood dealers. Because most of these men had no prior experience in pulpwood procurement, they often required financial support as well as training in management and wood production techniques. If a particular contractor did not perform adequately, he was quickly replaced. For those considered capable and dependable, however, the company often provided financing and support until they could get up and running on a profitable basis. In some cases, large landowners who refused to sell their wood through a dealer would themselves be given "dealerships" as a way of gaining access to their timber.[26]

Based on his skills and experience, a contractor or dealer might obtain wood from as few as five to as many as fifty different locations in a territory ranging anywhere from 200 to 3,000 square miles. Once his territory was set, a contractor then recruited subcontractors for cutting and hauling operations or, if he was one of the smaller contractors, arranged to do it himself. Subcontractors, also known as pulpwood producers, usually received credit from the contractor for equipment, fuel, stumpage purchases, and even groceries. Each contractor received a weekly wood order from the mill and was responsible for wood purchases, harvesting, and transport. Some contractors, who operated in nearby "truck-wood" areas, had their producers deliver wood directly to the mill. Others, such as those who logged more distant territories located along railroad lines, had their producers deliver wood to a rail outpost where it was loaded and shipped to the mill. Each contractor received a set payment per cord of wood delivered from his territory. Subcontractors likewise received a certain price for every cord.[27] Wood was generally purchased from local landowners on a "pay-as-you-cut" basis. Prices paid for stumpage varied by location and cost of transport, with the difference deriving from the location of the timber and its accessibility. By adjusting prices to equalize for differences in transportation costs, the mill effectively guaranteed a stable overall cost per cord of wood delivered—despite variability in wood orders and in the conditions facing contractors and subcontractors.[28]

For contractors and their producers, of course, meeting the changing wood needs of the mills on a timely basis required considerable flexibility. For any given contractor, weekly wood orders might vary by 20 to 50 percent. In Union Bag's view this required that a contractor "maintain an organization of sufficient flexibility so that extra profits on the 'good' weeks will offset the losses on the 'bad' weeks."[29] In short, a contractor had to be able to extend "credit favors" to his drivers, helpers, and cutters, finance stumpage and equipment purchases for subcontractors, "carry" his organization through periods of low production, pay cash immediately on wood delivery, and "do many of the things that require *extreme flexibility of business policy* that could not possibly be done by a large corporation whose expenditures must be on a less flexible basis and whose extension of credit must be on a secured or sound credit policy basis." Union Bag was thus by no means unaware of what it called the "hazards of the wood contracting business."[30] Indeed, this was precisely why the company preferred contracting to full integration.

When functioning properly, the advantages of the contract system were substantial. Good contractors provided dependable supplies of wood at a set cost. As local businessmen, they could often obtain stumpage that might otherwise not be available to mill representatives. In the words of one former executive:

"The dealer was local, out in the country. He knew people. He wasn't here today gone tomorrow. That's where he lived, that's where his business was, and he was able, particularly during the early days, to buy timber from people simply because he knew them, he was trusted, he had been in the community for a long time and what not . . . The dealer was able to negotiate to buy wood that would be very, very difficult, if not impossible, for the paper companies to get."[31] Simply put, wood dealers could tap into local social networks in order to gain access to wood. They had a political and social cachet among rural landowners that mill employees would never be able to match. As a former procurement manager put it, "Crackers or country people . . . they'll sell wood to somebody they know a lot quicker than they'll sell it to me."[32]

Because they were self-employed and paid on a piece-rate basis, moreover, contractors and subcontractors had incentives to work harder and longer than company employees. Given their "flexible business policy," they could also operate at significantly lower costs than a corporation "with its necessary bookkeeping, financial, and credit procedures."[33] Finally, contractors with a good reputation could be a source of social and political capital for the mills. In addition to their ability to "procure stumpage without undue local resistance," well-respected contractors served as company allies in pursuing local political objectives and in the more general process of "building . . . good will."[34]

During the late 1930s, however, goodwill seemed to be in short supply. As the industry expanded in the South, unfavorable publicity and the very real possibility of federal regulation threatened to put a damper on the boosterism of years past. Many lumber and naval stores operators, for example, voiced concern that the pulp and paper mills would cut young second-growth timber before it had a chance to mature. Conservationists and government officials protested that firms were further destroying the forest resource.[35] And local landowners complained about the prices they received for their pulpwood. Part of the problem, according to Union Bag, was that these landowners had "obtained exaggerated ideas as to what their pulpwood would be worth . . . When the expected financial returns were not realized, there was a feeling of resentment against the paper mills."[36] Many blamed the contractor system, assuming that the middleman was taking most of the profits. One government report from 1939 referred to the contractor system as the "sore spot" of southern forestry.[37]

Responding to such concerns, Rep. Hampton Fulmer of South Carolina introduced a resolution in Congress in January 1941 calling for an investigation of "the apparent monopolistic purchasing of pulpwood by pulp and paper mills under a contract purchase system from farmers and other owners."[38]

That February, Fulmer convened hearings under the auspices of the House Committee on Agriculture and received testimony on the pulpwood procurement system in the South.[39] Landowners such as Annie Mae Strickland testified about the challenges they faced in getting a "fair" price for their pulpwood. Government investigators also offered evidence on the workings of southern wood procurement. One such investigator who had conducted extensive field research throughout the region noted how "the territory for any given contractor . . . assigned to him by the company was strictly limited so that no two contractors for any one company competed in any one territory. That meant that, not only did a farmer have only one mill to sell to, but he had only one agent of that mill through whom he could market his wood."[40]

Others spoke of the "ruthless destruction" being visited on the southern pine forests, and charged the mills with price fixing. No one, however, presented any hard evidence to substantiate the price-fixing charge.[41] The state of South Carolina conducted its own investigation of pulpwood pricing during the early 1940s, concluding that there was no explicit price-fixing arrangement among mills.[42] Despite the lack of hard evidence, however, there did seem to be a de facto "one-price" system in operation throughout the South during the 1930s and 1940s. Given its size and the distribution of its mills across the region, International Paper was widely regarded as the "price leader" when it came to setting pulpwood prices.[43] As one former procurement manager put it: "For years and years and years, there was a cost of wood in this part of the world. Everybody paid the same price . . . And IP [International Paper] pretty much set the price . . . This was just the way it was done."[44] Whether or not there was an explicit agreement to follow International Paper's lead, the geographic isolation of local pulpwood markets and the contractor system together meant that farmers and other landowners could either take the price they were offered or leave it.

With growing U.S. involvement in the World War II, however, the system began to show signs of strain. Running at full capacity in order to meet wartime needs and facing war-induced labor shortages in the woods, mills scrambled to maintain wood flows. In some areas, German and Italian prisoners of war were used to supplement the logging force.[45] Through their contractors, mills began paying higher prices for pulpwood. Between 1940 and 1945, southern stumpage prices rose by more than 20 percent in real terms.[46] Criticism of the pulpwood marketing situation soon abated.

After the war, as the industry resumed its expansion in the South, pressures on the pulpwood marketing system grew. With old mills expanding and new mills being built, procurement areas began to overlap and competition for wood increased. The advent of the mechanical pulpwood loader in 1950 exacerbated these trends. Developed by International Paper, the device allowed for the bulk handling of wood between trucks and rail cars, greatly reducing the labor needed for such tasks. Mechanical woodyards (also known as concentration yards) began springing up along rail lines.[47] By reducing costs associated with loading and transporting wood, these concentration yards allowed mills to expand their procurement areas. As new areas were opened up, this relieved some of the pressure on existing procurement regions. At the same time, however, overlap increased and competition between mills intensified. By 1954, very few pulpwood areas remained that were not subject to competition between at least two large mills.[48] This effectively marked the end of the exclusive dealer franchise. What had been a series of highly localized monopsonistic markets gave way to more regionalized oligopsonistic markets.

Against this background, the entry of six mills into southern pulpwood markets during 1954 created a surge in demand. The new mills alone added some 1.5 million cords (the equivalent of 10 percent) to regional pulpwood requirements. Meanwhile, capacity expansions at existing mills added another 1.6 million cords. Within the span of a year or so, annual wood demand had increased by some 3 million cords, or 20 percent. This unprecedented rise led to severe inventory problems during 1955, precipitating the first real pulpwood "crisis" in the industry. At most mills, pulpwood inventories shrank to zero. Some mills even reported being unable to buy enough wood to keep their machines running at full capacity. The conventional view held that the crisis resulted largely from a woods labor shortage caused by the migration of loggers to other jobs.[49]

The root of the problem, however, lay in the structure and operation of the pulpwood marketing system and its lethargic response to large increases in demand. Because of the rigidities of the "one-price" system combined with the highly flexible arrangements between mills, dealers, and producers, there were inevitable lags in the adjustment of supplies to large increases in demand. In other words, even though dealers and producers proved adept at adjusting wood flow to meet the changing needs of the mills, they could do nothing to meet demand that exceeded their own capacities. Until the mills raised the price, moreover, new producers were unlikely to enter a profession marked by low margins and unpredictability. Initially, mills were reluctant to raise pulpwood prices, perhaps because they assumed that the increases would be captured by landowners who could take advantage of the increased competition between mills, dealers, and producers. Instead, some mills attempted to lure dealers and producers away from other mills. But this had no effect on aggregate capacity and thus did nothing to solve the larger structural problem of undercapacity. In effect, the de facto "one price" system stifled the feedback necessary to induce sufficient capacity expansion. In late 1955, Champion International broke ranks and raised its unit price of wood. International Paper and other firms quickly followed. New producers rapidly entered the pulpwood business. The output of wood rose so fast that inventories were back to normal within a few weeks and wood flows actually had to be curtailed in some cases.[50]

The 1955 pulpwood "crisis," therefore, illustrated some of the weaknesses embedded in the procurement system. As exclusive dealer territories eroded and as new mills locked up more and more acreage in their forest reserves, competition over stumpage increased. Between 1940 and 1955, stumpage values rose by more than 110 percent in real terms while prices paid by mills for pulpwood barely rose at all. Thus, as long as mills held firm to their one-price system, dealers and producers would be increasingly squeezed between rising costs for stumpage purchases and stagnant prices for pulpwood delivered. Raising the price, as the events of late 1955 illustrated, was one obvious way of bringing more producers into the market and alleviating the shortages. But such action represented a dangerous precedent for mill managers who feared rising raw material costs almost as much as they did running out of wood. Though companies could and did use their own forest reserves to meet some of their needs, they still drew roughly 80 percent of their pulpwood from nonindustrial private landowners during the mid-1950s.[51] Moreover, regardless of where the wood actually came from, if producer capacity could not adequately meet demand, wood shortages, however periodic and spotty, would almost certainly occur.

The underlying lesson of 1955, then, was that increased harvesting productivity would be necessary to maintain adequate wood flows in a period of rising demand while keeping pulpwood prices under control. The conviction among industry leaders that they faced imminent labor shortages, as loggers left the woods in search of better opportunities, thus contained a kernel of truth. For even if the threat of outmigration turned out to be exaggerated, the longer-term issue of labor productivity in the woods would have to be addressed. By 1959, when another wood shortage briefly shook the industry, manpower and harvesting productivity had emerged as major issues of concern for procurement managers and mill officials. Woods work would soon undergo a dramatic transformation.

Woods Work

> A loaded double barreled shotgun is facing all employers in the forest products industries in the form of Senator John F. Kennedy's bill . . . to change the Fair Labor Standards Act. The one barrel is loaded with the proposed increase in the minimum wage . . . The other barrel . . . consists of the proposed elimination of the small forestry or so-called 12-man exemption which now exempts all employers from the [minimum wage and overtime] provisions of the Wage and Hour Act . . . Unless you men of the Forest Products Industries of the South, and in Georgia particularly, quickly make all your Senators and Congressmen in Washington D.C. aware of this threat to one of your largest cash crop industries . . . , this loaded gun will go off and the Southern Forest Industries will be dealt a blow that will be felt seriously for many years. —*Willard S. Bromley, 1959*[52]

> We feel that many workers in the logging industry are suffering from the following problems: they are being denied a decent wage and therefore a decent standard of living; they do not have many of the rights and privileges to which employees are entitled in modern times; and they are denied their right to try to improve their economic status by collective bargaining . . . The primary problem is that these logging workers are doing the work for the big paper companies . . . , yet they are not considered to be the employees of the companies. This means that the big companies gain the benefit of the work that these workers perform, but these same companies do not have to live up to the responsibilities which would be required of them if these workers were recognized as the employees of these companies. This may provide an extra profitable situation for the companies, but—in plain words—the workers are being exploited. —*A. F. Hartung, 1961*[53]

Speaking to the Georgia Forestry Association in May 1959, Willard Bromley, executive secretary of the American Pulpwood Association, evoked the specter of government regulation. As he delivered his remarks, the U.S. Congress was debating the labor situation in the forest products sector. Beginning in the mid-1950s, hearings were held over the minimum wage and the so-called twelve-man exemption to the 1938 Fair Labor Standards Act (FLSA). This provision, which was enacted in 1949 at the urging of southern timbermen, exempted virtually all of the logging operations in the South from the minimum wage and overtime provisions of the act. As a result, pulpwood flowed to mills well below what it would have cost had the full weight of the FLSA wage and hour provisions been enforced. Controversial from its inception, the exemption came under vigorous assault from progressive legislators and organized labor during the 1950s. In their view, the large forest products firms reaped the benefits of the exemption through cheaper wood while small loggers engaged in a self-exploiting race to the bottom. Industry leaders, by contrast, claimed that small loggers did not possess the financial and managerial skills necessary to meet the provisions of the law. Echoing the paternalism of postbellum industrial boosters, they argued that the exemption allowed these loggers to stay in business. Working through trade groups such as the American Pulpwood Association, which received the majority of its funding from corporate sources, these men adamantly opposed any government intervention in the pulpwood logging sector. The crux of the issue, as A. F. Hartung of the International Woodworkers of America suggested, lay in understanding and defining just what kind of workers these loggers actually were and the nature of their relationship with the mills.

Seen in a larger context, these debates reflected the ongoing attempt to reconcile New Deal labor protections with the reality of southern woods work. While much has been written about the New Deal assault on the South's low-wage economy and the resistance mounted by southern industrialists, very little attention has been directed to the implications of New Deal labor legislation for southern woods workers. With the signing of the FLSA in 1938, President Roosevelt and his allies hoped to put an end to the isolated, low-wage labor markets that marked so much of southern industry. Textiles became a focal point for reform, and, after a significant boost from World War II, southern wage rates did begin to converge with those in the rest of the country.[54] Yet loggers and other woods workers remained somewhat immune to the direct effects of New Deal labor legislation and the larger integration of the "isolated" southern labor market into national labor markets.

Indeed, despite their overall numbers and their importance to the regional economy, loggers were largely invisible to those accustomed to viewing a worker as someone who labored in a large factory. They tended to operate in small crews scattered over large areas, and they participated in labor markets that were highly localized and often separate from those servicing the mills and factories of the region. According to industry leaders, these men resembled agricultural laborers more than factory workers and thus they were to be treated as such (i.e., exempted from "factory-type" labor protections). The nature of woods work, combined with the multiple layers of contracting and dense social networks characterizing the procurement system, made it very difficult to define the precise status of southern woods workers.

In this respect, the forest products industry had its own special version of the "labor question" that occupied so much attention during the New Deal period.[55] Straddling the divide between industry and agriculture on the one hand and between employee and independent contractor on the other, woods workers all too often fell victim to the ambiguities inherent in labor law and thus were unable to claim many of the basic protections accorded to other American workers. That so many of these workers were black, moreover, made the situation more challenging. Racial barriers often inhibited them from seeking other forms of employment and denied them many of the rights exercised by their white coworkers. Their political and economic vulnerabilities made it all too easy for them to end up in ties of dependency to their white employers. As the production base of one of the New South's most promising industries, pulpwood loggers occupied a socioeconomic status closer to sharecroppers than to factory employees. Their insecurities and vulnerabilities provided much of the social basis of "flexibility" that marked the wood procurement system.

Viewed in historical perspective, their situation had deep roots in the postbellum southern political economy. Many of those who became pulpwood loggers, for example, had previously worked in the woods for lumber companies or naval stores operations. They participated in the same kinds of labor markets—highly localized and tightly linked to agriculture—and the divisions of race and class operated in much the same way regardless of whether one was cutting sawtimber or pulpwood. On the whole, the labor markets that grew up around the various wood products industries tended to be flexible, secondary labor markets marked by part-time labor and occupational shifting.

Because of this, those who employed woods workers, like their contemporaries in farming and other industries, were deeply concerned with labor scarcity. Indeed, part of the paradoxical nature of the labor question in the postbellum South rested in this juxtaposition of low-wage, flexible labor markets on the one hand with the constant threat of labor scarcity on the other. The explicit and implicit institutions of labor control that marked the regional economy during the postbellum period derived in part from this juxtaposition, as well as from the competition between agriculturists and industrialists for labor. Like other industrialists, employers in the forest products industries sometimes supported coercive institutions of labor control. Some employers in the early lumber industry, for example, took advantage of the convict-lease system.[56] Others used debt peonage, which, though certainly not the norm in these industries, was by no means unfamiliar, particularly in the naval stores region.[57] Vagrancy laws and

Jim Crow legislation also aided employers seeking extra-economic means of retaining scarce labor, while the paternalistic nature of many employment relations in the woods further limited the mobility of timber workers. Finally, of course, the infamous hostility of employers and employer associations to all efforts to organize timber workers made collective bargaining a distant dream for all but the very few.[58]

During the 1920s and 1930s, as pulp and paper firms moved into the economic space vacated by a declining lumber industry, they recruited their wood dealers from among local elites with prior experience in the timber industries. In the naval stores belt of the coastal plain, for example, some of the dealers had previously operated turpentine camps and owned or controlled large acreages of pine stumpage. Similarly, local lumber barons, who also controlled large acreages and possessed the requisite social and political capital, often became dealers. These men were obvious choices to be wood dealers because of their access to timber and their experience in recruiting and organizing local labor.[59] Indeed, they acted as a sort of carrier class, adapting the labor practices of the earlier timber industries to the new requirements of pulpwood production.

The contract-based approach to wood procurement used by pulp and paper mills also had precedents in the lumber industry. During the 1910s and 1920s, with the shift to second-growth timber and the concomitant decline of large-scale railroad logging operations, lumber mills began using contract logging operations.[60] Taking advantage of improved highways and roads, truck logging became increasingly attractive to smaller sawmills subsisting on thinner stands of second-growth timber. Because of their greater mobility and adaptability to different terrains, trucks were often profitable where railroad logging operations were not. The net effect was to decentralize and deindustrialize logging operations. By reducing the costs of recruiting and monitoring labor, this move to smaller, more labor-intensive, logging operations made contracting more attractive than full integration.[61]

The same sort of dynamic operated in pulpwood production, where logging crews were typically quite small and spread over large areas. The major difference, however, was that the majority of pulpwood came from noncompany lands and that the work itself was typically more labor intensive than cutting and hauling larger sawtimber. As a result, contracting through dealers, rather than directly with loggers, proved to be a more attractive way of organizing pulpwood procurement. Because of the fragmented nature of southern landownership, the small volume of timber involved in individual sales, and the dispersed nature of rural labor, organizing the workforce necessary to log the many small tracts of timber spread throughout a mill's procurement region greatly exceeded the institutional capacities of the early pulp and paper mills. Moreover, the ability of the industry to utilize smaller trees also meant that much of the labor could be performed by hand rather than with the large steam-driven machines used in the early lumber industry. The logging operations servicing a particular mill typically employed thousands of workers operating in small crews spread over a wide area. Most of these workers, particularly in the coastal plain, were black, and few had any formal education. Mill managers moving into the rural South during the middle decades of the twentieth century had very little knowledge of the region's institutional realties and cultural mores. Many had probably never had any significant interaction with African Americans, and few, if any, were familiar with the highly localized labor markets. More likely than not, they were only too happy to let someone else deal with the labor problem. As one industry executive put it: "The whole issue of dealing with a relatively uneducated manual labor workforce wasn't something that the industry was attuned to. The industry culturally was attuned to running productive plants with well-trained union labor, and the world out beyond the mill gates was something totally different."[62]

Because dealers were already involved in procuring timber from thousands of small landowners, their role in mobilizing and organizing the labor necessary to harvest and transport this timber represented a logical extension of their responsibilities. A dealer was "faster on his feet" than a large company. He was good at "hustling wood," and he could tap into long-established social networks to organize the labor necessary to deliver wood in the quantities needed. His dual role as timber broker and labor contractor made him the critical link in the procurement system and served to further insulate the mills from the risks and liabilities involved in logging. According to one former procurement manager, "We tried to establish this arm's length relationship with the producers cutting the wood. We didn't want them to be considered as employees of the company for various reasons. So we operated through the dealer system of wood procurement, which was standard for just about everybody."[63] Another procurement manager noted, "The dealer system put the dealer between [the company] and the producer. We weren't even supposed to talk to the producer."[64]

Given the many dangers associated with logging and the various employer responsibilities incorporated in the emerging regime of New Deal labor regulation, such an arrangement proved quite desirable for the mills. As one former executive recalled:

> Paper companies were quite properly scared to death to get involved directly with this kind of a labor situation where these people might wind up being construed as their employees, so they very much wanted a buffer . . . The dealer, he could deal with the low-paid, un-mechanized, bottom-rung-on-the-economic-ladder wood producer and sort of finance him a little bit, advance him a little bit, pay him sort of rock-bottom rates, but just barely keep him afloat and keep him going in a way that a big company just couldn't quite do. Back in the early days, there were probably an awful lot of producers not being in compliance with all of the wage and hour laws, and [they] weren't paying the social security and withholding taxes and workmen's comp and all of that. The paper companies could not get anywhere close to that.[65]

Contracting thus insulated the mills from the risks, liabilities, carrying charges, and monitoring costs associated with employees.

Indeed, the nature of the economic relationship between the mills, wood dealers, and pulpwood producers and the implications of this for taxes, workers compensation, and tort liability were the subject of considerable litigation during the post–World War II period. Many of the early cases focused on which party was responsible for unemployment and workers compensation. In virtually all cases, the courts had little trouble finding that pulpwood producers operated as independent contractors, thus absolving the mills and the wood dealers from paying the various taxes and compensation claims associated with normal employees.[66] But there were limits to how far courts would go, and not all cases resulted in decisions against pulpwood producers. In a particularly stark case from the late 1960s, the Mississippi Supreme Court reversed an order of the state Workmen's Compensation Commission in favor of a wood dealer, L. A. Penn & Son, and against a pulpwood producer, Napoleon Brown. The court found that Brown, a black, illiterate pulpwood cutter who was permanently disabled due to a work injury, was not an independent contractor but an employee under the control of the dealer and thus entitled to workmen's compensation.[67] One can only wonder how many other cases like this never made it to court.

In addition to insulating them from certain forms of liability and taxes, contracting also allowed the mills to forego the investments in fixed capital associated with logging and trucking, while maintaining low and stable wood costs despite fluctuations in wood demand, constraints on inventories, and the uneven labor demands that resulted. In an era of surplus rural labor, company harvesting operations would never be able to compete effectively with small contract-based crews. As Union Bag noted in the late 1930s, "The contractor system is the only practical means by which wood can be obtained at the time wanted and in the amounts and age required."[68]

By tying producers to dealers, the contract system also relieved producers of the challenges of finding working capital and assured them of a market for their wood. At the same time, however, it eliminated alternative outlets for their product, severely restricted their ability to bargain over price, and undermined any incentives that they might have to experiment with alternative methods of harvesting. Operating on the basis of short-term oral contracts and bound through ties of debt, these producers were as effectively constrained in their ability to seek other opportunities as many of their predecessors who worked in the logging camps and company towns of the lumber industry.

As for the workers themselves, many, if not most, were black, and almost all probably had some sort of connection to agriculture. Indeed, throughout the southern pine region, the various forest products industries together employed more black workers than any other industry outside of agriculture, and the specific labor regimes and work rhythms in these industries bore the strong imprint of plantation agriculture.[69] When Union Bag established its Savannah mill, for example, the company reported a "plentiful supply of inexperienced Negro labor . . . of a rather low grade, both physically and mentally." According to the company, such workers "require[d] constant supervision to keep them working, even tho [*sic*] they are usually paid on a piece price basis."[70]

Unlike the more industrialized logging operations of the early southern lumber industry, pulpwood logging entailed a very labor intensive work regimen. In general, pulpwood harvesting was "a hand labor process with cheap and simple tools," and was organized on the basis of a task system that echoed the work regimes of plantation slavery (figure 2.2). Most crews consisted of five men or fewer, and logging camps were seldom, if ever, used. These men typically worked out of their own households, often traveling many miles every day to and from their logging sites. Many probably came from tenant or sharecropper households. Like so many in the rural South, they often had one foot in agriculture, moving back and forth between field and forest depending on seasonal and economic conditions. When there was work, hours were long and wages low. In order to maintain work incentives and minimize monitoring costs, logging crews were most often paid on a piece-wage basis. As a profession, pulpwood production paid some of the lowest wages in the South. During the summer of 1934, for example, researchers from the southern forest experiment station calculated that the earnings of pulpwood cutters averaged about thirteen cents an hour.[71]

Figure 2.2. Northern Florida logging crew, circa 1948. Early pulpwood logging crews typically consisted of only a few men, most of whom were black, working with limited tools. Courtesy of the Forest History Society, Durham, NC.

For these men, the work must have been incredibly difficult. Added to the physical demands of logging went the vagaries of weather, bugs, isolation, and the very real possibility of serious accidents. Indeed, ever since reliable worker injury statistics have been recorded in the United States, logging has consistently ranked as one of the most dangerous occupations.[72] One investigation conducted by the U.S. Department of Labor during the early 1940s, for example, found that worker injury rates for pulpwood and sawtimber logging led all other sectors and exceeded the all-manufacturing average by a factor of four. In 1944, approximately one in seven pulpwood loggers experienced a disabling work injury, compared to one in twenty-four for manufacturing as a whole. Overall, an injured pulpwood logger was six times more likely to suffer a severe injury, and nearly twice as likely to suffer a fatality. Temporary disabilities required an average recovery time of more than three weeks. The most common injuries involved slips or falls, falling limbs or trees, rolling or sliding logs, the malfunction or mishandling of power-driven machines, and overexertion. Given the strenuous nature of the job, the variability of weather and work conditions, and the fact that most loggers carried axes, saws, or other sharp tools, even a small mishap could end in severe injury or death. Because they were usually paid according to how much timber they cut, moreover, loggers sometimes gave inadequate attention to safety and maintenance concerns. Finally, since most loggers worked in unsupervised crews in remote locations, access to safety guidance and professional medical assistance was limited, to say the least.[73]

With few employment alternatives and little recourse to the protections of occupational health and safety laws or workers' compensation insurance, loggers operated in a landscape of risk and vulnerability that is difficult to comprehend. Even as late as 1957, according to one study in Florida, fewer than 15 percent of the state's pulpwood loggers were covered by worker's compensation insurance.[74] In 1970, with the passage of the Occupational Safety and Health Act (OSHA),[75] the Department of Labor listed pulpwood logging as a target industry for special scrutiny because of its high rate of worker injury.[76] Part of the problem turned on the status of loggers as independent contractors. Reluctant to undermine their "arm's length relationship" with these producers, mills refused to provide the worker protections or benefits that might be appropriate for regular employees. Contracting, it seemed, trapped many of these men in a very dangerous and low-paying occupation while denying them some of the basic protections accorded to other workers. For much of the post–World War II period, the government, despite good intentions, did little to improve their situation.

Beginning in 1940, the Department of Labor's Wage and Hour Division initiated proceedings against the American Pulpwood Association and several pulp and paper firms operating in the South for conspiracy to violate the wage and hour provisions of the FLSA.[77] The crux of the issue was whether these companies had the responsibility for ensuring that the pulpwood consumed at their mills had been produced in accordance with the law. Department of Labor representatives claimed that because the mills sold their products in interstate commerce they were responsible for ensuring that all materials used in the manufacturing process had been produced in accordance with the law and could thus be enjoined from selling their products until their various suppliers were in compliance. Industry representatives, of course, claimed that the independent status of dealers and producers meant that the companies did not have primary responsibility for ensuring compliance, and, in reality, that there was no effective way for them to ensure that every piece of wood they used was "legally" produced. During the 1941 pulpwood hearings, several Labor Department officials reported on their investigations and offered testimony in support of the contention that pulp and paper mills used the contractor system to get around wage and hour laws.[78] These investigators argued that the workers' converted hourly wage was well below the minimum required by the FLSA. Paper companies responded by pointing out that they had no employees in the woods—that they bought their wood through contractors and thus had no legal responsibility to ensure that woods workers were being paid in accordance with the statute. Many companies apparently required that their producers sign statements that the wood was produced in conformity with the FLSA, contending that this ended their responsibilities, and that if the law was violated it was the producer's problem.[79] Labor department investigators, however, questioned the nature of the relationship between pulpwood loggers and the mills. They argued that mill managers exercised "very close control over the production of wood, [and] over the people that the company contends and says 'are not our employees'; whereas, as to the ones that they acknowledge to be their employees, which are the people they employ in their own mills, they have been careful . . . to comply with the statute."[80] Here was a classic manifestation of a segmented labor market.[81]

While the Labor Department did succeed in putting the industry on notice, resolution of the issue would have to wait for legislative action. In the meantime, mills continued to use the wood-dealer system, avoiding many of the responsibilities and costs associated with normal employees.[82] The key advantage, of course, was cheap wood. As one government investigator noted in 1941:

> It has been explained to me by company officials and producers that the benefits to the companies of the contracting system are numerous, and various companies may have adopted the system for different reasons. Among the reasons that have been given are the following: The companies are no longer liable for employee suits for accident compensation. They do not have to pay social-security taxes. They have relatively low supervisory costs. They can keep an excessive number of producers on their lists and thus get them to bid against each other for work much as sweatshop contractors used to bid against each other in the garment industry. The producers form a cushion against community resentment over the low stumpage prices and low wages. The larger producers may even speak for the companies on legislation and other matters. Workers are less likely to organize into unions and strike against small producers than against the company. But the basic reason given is to keep the prices of pulpwood from rising.[83]

As long as there was surplus labor, the contract system worked well for the mills. With few employment alternatives, many pulpwood loggers had little choice but to continue working in the woods at whatever price they could get. Those who tried to find other jobs usually ran into barriers of debt and racial discrimination. Along with agricultural labor, woods work was one of the few occupations in the rural South open to blacks. Most loggers, moreover, quickly found themselves indebted to their dealers for equipment and supplies. Since they depended on these same men for work, they had very little leverage in negotiating the terms and conditions of their employment. Few, if any, dealers would be willing to take on someone else's producer without prior agreement. Even World War II, which did so much to brighten the prospects of southern workers, had little impact on the conditions facing pulpwood loggers. Though labor markets tightened to the point where German and Italian prisoners had to be used as woods workers in some places, the general prospects for small loggers improved very little.[84] By the end of the 1940s, their status as workers, particularly in the context of federal labor law, was increasingly in dispute and, as it turned out, on shaky ground.

As suggested, much of the controversy revolved around the wage and hour provisions of the FLSA. The last major piece of New Deal legislation, the FLSA was a direct attack on the low-wage economy of the South. Intended in part to stimulate the economy, the law ostensibly sought to boost purchasing power by raising wages and by increasing employment through the elimination of child labor and other forms of sweated work. Some have characterized the act as a confrontation between North and South—between southern business leaders, such as those assembled under the Southern Pine Industry Association and the Southern States Industrial Council, seeking to maintain the South's "comparative advantage" of cheap labor, and northern business and labor interests distressed by the flight of northern capital (most prominently in textiles) to the South.[85] Although southern business leaders failed in their efforts to include a regional wage differential in the law, they did succeed, in alliance with others, in exempting millions of workers from many of the provisions of the act.[86] The most famous example of this was the agricultural labor exemption, which exempted "any employee employed in agriculture" from the wage and hour requirements. As defined by the FLSA, agriculture encompassed "any practices (including any forestry or lumbering operations) performed by a farmer or on a farm as an incident to or in conjunction with such farming operations."[87] This particular catchall phrase provoked considerable controversy regarding the scope of the exemption and its applicability to logging operations. Because the courts generally held that statutory exemptions to the act were to be strictly and narrowly applied, only those loggers working as employees of a farmer or on a farm in a manner that was both incidental and subordinate to established farming activities would be exempt.[88] The ambiguities were palpable.

Advocates on both sides of the issue wanted resolution. Starting in 1945, extensive congressional hearings were held to debate amendments to the FLSA and reconsider the scope of its coverage. On one side, organized labor and its allies pushed for an increase in the minimum wage and extension of the provisions to cover more workers. On the other side, a coalition of Republicans and southern Democrats, supported by various business interests, urged that the act's coverage be narrowed, both in general terms and through specific exemptions. Representatives of the forest products industry argued against any increase in the minimum wage and for a more specific logging exemption. Speaking for the Southern Pine Industry Association, R. M. Eagle noted, "The southern pine industry believes that the Fair Labor Standards Act can interfere with individual freedom and can lead to a controlled economy . . . The harvesting of timber is closely akin to agriculture and, in fact, is performed to a wide extent in the South by farmers in their off seasons. The inclusion of logging in the coverage of the law and the overtime provision does not appear to be benefiting anyone . . . [L]ogging workers in the South should be exempted from the act."[89] Southern pulpwood dealers, industry leaders, and trade association representatives offered similar arguments in favor of an exemption.[90]

In October 1949, after much political maneuvering and some 11,500 pages of testimony, Congress finally passed the Fair Labor Standards Amendments of 1949. Quickly signed into law by President Truman, the amendments became effective January 25, 1950.[91] Timber interests got their wish in the form of section 13(a)(15), which exempted from the minimum wage and overtime provisions of the law "any employee employed in planting or tending trees, cruising, surveying, or felling timber, or in preparing or transporting logs or other forestry products to the mill, processing plant, railroad, or other transportation terminal, if the number of employees does not exceed twelve."[92] Known as the "twelve-man rule," this exemption applied to virtually all loggers in the South, where small crews predominated, as well as to many loggers in other parts of the country.[93] With the stroke of a pen, tens of thousands of American workers were effectively excluded from enjoying such basic rights as a mandatory minimum wage and overtime pay. Already on the economic margins, these men and women now faced a situation of severely diminished legal protections and political power.

Although timber industry representatives hailed the exemption as providing relief to small employers in the logging industry (a victory for small business), the underlying intent and the overall effect was to maintain cheap wood for the mills. Needless to say, the amendments further consolidated the advantages of contracting. Companies now had less incentive than ever to use company harvesting operations that would be subject to the full weight of the federal wage and hour law. As long as loggers were independent contractors working in crews of twelve men or fewer, the added costs associated with the minimum wage and overtime provisions of the law would not translate into higher wood prices. If for any reason loggers were declared to be employees of the mills, however, the exemption would no longer hold.

Not surprisingly, controversy surrounding the twelve-man rule and other exemptions to the FLSA continued to fester during the early 1950s. By the middle of the decade, legislation had been introduced to repeal many of the exemptions and extend the coverage of the act to millions of unprotected workers. Hearings were convened, and partisans on both sides of the issue pressed their case. In 1956, Senator John F. Kennedy, opened hearings on the FLSA by pointing to the "shocking fact that nearly two-thirds of the 67 million people working in this country have no Federal protection against substandard wages . . . Continued failure to broaden the act's coverage not only denies decent living standards to the unprotected workers, but tends, by encouraging sweatshop competition and depressing purchasing power, to undermine the economic status of the entire nation. Unfortunately, those workers who currently lack the protection of the act are those who need it the most. Their wages are low and their bargaining power is weak."[94]

Kennedy sounded an alarm among timber industry representatives, putting them on notice that the small forestry exemption was under assault. Throughout 1956, 1957, and 1958, bills were offered up in both houses of Congress, several of which proposed to eliminate the twelve-man rule. In support of those who sought to repeal the small forestry exemption, A. F. Hartung, president of the International Woodworkers Union of the AFL-CIO, argued that the provision subsidized large, highly profitable corporations at the expense "of a group of helpless workers." Using some rather powerful language, Hartung appealed to basic principles: "I submit to you that the exemption of employees of logging operations of 12 men or less is morally wrong, economically unsound, and unjustifiable from a legal standpoint. The exemption arose out of political logrolling of the worst kind. Its effect is to condemn thousands of American workers to a life of misery and poverty, in order to further inflate the profits of an already prosperous industry. Traditional American justice and the welfare of our free enterprise system call for the abolition of this injustice." On the other side, men such as R. E. Canfield and Willard S. Bromley of the American Pulpwood Association argued that the FLSA was intended to apply to factory workers and that pulpwood logging operations hardly conformed to "factory conditions."[95] R. D. Wilcox, a pulpwood dealer from Laurel, Mississippi, argued, "Without the twelve-man exemption, . . . the small producer would gradually be forced out of business." Letters from forty-nine other pulpwood dealers and pulpwood producers from southern states were added to the record of those opposing elimination of the twelve-man rule.[96]

For a while, at least, it seemed as if timber industry representatives had succeeded in stifling efforts to repeal the twelve-man rule. Then, in April 1959, Senator Kennedy and others introduced yet another bill to repeal the twelve-man rule and other exemptions.[97] Timber industry representatives mounted a vigorous response. Led by Willard Bromley, the American Pulpwood Association's FLSA task force organized a highly effective campaign to combat Kennedy's bill.[98] Eliminating the exemption, they argued (again), would crush small operators with excessive administrative and monitoring costs, forcing them out of business. Such a move, in their view, ran counter to the principles of free enterprise. In his May 1959 speech to the Georgia Forestry Association, during which he referred to Senator Kennedy's bill as "a double barreled shotgun . . . facing all employers in the forest products industries," Bromley took the argument a bit further, playing on cold war fears: "If the labor unions are over-successful in reaching their objectives on small forest operations, we will in this country be just one step away from Socialism. At that point, all forest land will become the property of—or under control of the State. The company will have no operations at all and the State or labor union cooperatives will control all cutting and logging and deliver wood in accordance with government edicts or quotas. The final step from socialism would be communism—where all the companies, all dealers and producers—and all workers—even labor unions—would have one boss, the government."[99]

This was strong stuff. Labor representatives, however, employed rhetoric that was equally strong. Harry Scott of the United Papermakers and Paperworkers Union argued that the exemption had "given rise to a feudal system in the industry which effectually [*sic*] denies the protection of the Fair Labor Standards Act to many thousands of exploited workers." Echoing the concerns of others, Scott continued: "The basic issue at stake in this case is the payment of living wages and social costs of unemployment, accident and old age insurance to people who work in the production of pulpwood and other timber products. The combination of market power over prices of pulpwood exercised by the paper industry which keeps them abnormally low is aggravated by an exemption in logging to make for exploitation of both people and forests."[100] Feudalism or communism;

exploitation or government control—the choices were stark. By the summer of 1960, however, the timber lobby had prevailed, successfully removing language that would repeal the twelve-man exemption from FLSA legislation.[101] Not until 1966 were any changes made in the provision, when the twelve-man limit was reduced to eight. And it was another eight years after that before the minimum wage exemption was finally repealed.[102] After twenty-five years (1949–1974) of being denied federal protection against substandard wages, the employees of small logging operations could now enjoy the basic right of a mandatory minimum wage that millions of other American workers had come to take for granted.

As to the question of what actually was going on in the woods during this time, no one really knew, except the loggers themselves. Despite all of the debate bearing on their legal and economic status, not a single logging employee testified during the thousands of hours of FLSA hearings. Though many claimed to speak on their behalf, none could say for sure what the conditions were facing those who made their living working in the woods. Moreover, because the small logging exemption inhibited systematic record keeping, data regarding wages and working conditions for logging employees were virtually nonexistent. In the eyes of the administrative state, these workers were largely invisible.

Some things were clear, however. First, the vast majority of loggers worked in small crews eligible for the exemption. Of the 137,000 workers employed in U.S. logging operations during 1963, 87,000 or two-thirds worked in crews of twelve men or less, and most of those loggers working in larger crews were in the Pacific Northwest, where large company-run logging camps predominated. In the South, where the average number of employees per logging establishment was five, virtually all loggers fell under the exemption.[103] Second, by all accounts, southern loggers received very low wages. An internal Union Bag document from 1949 indicated that the average wage rate for Georgia pulpwood producers was around sixty-seven cents an hour for the coastal plain region and around seventy-eight cents an hour for the central part of the state.[104] If one assumes that these figures were roughly accurate, they bracketed the minimum wage of seventy-five cents an hour as enacted by the 1949 FLSA amendments.[105] Fourteen years later, in 1963, a Department of Labor report found that the hourly earnings for nonsupervisory workers in small logging establishments averaged $1.26 an hour for a composite sample of loggers from four southern states (Alabama, Arkansas, Mississippi, and North Carolina). This compared to an average of $1.79 per hour in the north-central states and was only slightly above the minimum wage of $1.25 an hour enacted in 1963. Moreover, according to the report, more than half of southern logging employees sampled received less than the $1.25 minimum.[106] Third, southern woods work rarely provided a full-year's worth of employment—a fact that further diluted the compensation of logging employees. Because of weather and changing demand, most loggers worked fewer than 200 days per year.[107] When they did work, loggers typically were on the job far more than eight hours a day. Yet they could not collect a premium for overtime. Fourth, logging was a very dangerous profession, with a worker injury rate that exceeded all other manufacturing sectors. Fifth, the majority of pulpwood loggers in the South were black, making them even more vulnerable than their white counterparts, and few had much formal education.[108] And finally, unlike their counterparts in the Pacific Northwest, southern woods workers enjoyed little if any recourse to the benefits and protection of union representation.

In sum, logging was a dangerous, low-paying job marked by extreme physical demands, highly variable work rhythms, and massive economic insecurity. Southern pulpwood loggers labored at the bottom of the region's social and economic hierarchy, despite providing the production base for one of its most important industries. They moved in flexible, secondary labor markets riven with ties of dependency and vulnerability. Employment alternatives were few and far between. Because of the layers of contractual relationships embedded in the procurement system, loggers could not access the privileges and protections enjoyed by those working inside the mill gates. Federal labor law offered little assistance, denying them some of the most basic rights of the New Deal regime of labor protection. These men were very much on the outside looking in.

As for the mills, through their use of the contractor system, they effectively relieved themselves of the direct burdens of the FLSA and other "carrying costs" associated with normal employees. Through the twelve-man exemption, they ensured that wood costs would not be further inflated by increases in the minimum wage and overtime pay. Given the incentives built into the procurement system, mills could rely on the dealers and producers to procure and deliver wood on demand. Costs remained low and relatively stable, while wood flowed to the mills in accordance with changing demand. Viewed in the aggregate, the system resembled a flexible, just-in-time method of inventory control.

Because such flexibility was rooted in the economic vulnerabilities of those working in the woods, however, the system sometimes had trouble adjusting to large and sudden surges in demand. During the wood shortages of 1955 and 1959, rapid increases in pulpwood demand outstripped the harvesting capacity of the existing logging force. In effect, although the procurement system provided flexibility up to a point, the substantial variability of wood demand combined with low prices inhibited the capacity expansion and productivity increases needed to meet significant demand growth. Embedded within the system's advantages were potential risks.

By the end of the 1950s, these problems had come to occupy a prominent place in the thinking of industry officials. Wide-ranging discussion regarding the so-called labor problem focused less on the structural vulnerabilities of loggers and more on the factors underlying constraints on wood flow and rising procurement costs. In this context, debate over the FLSA exemption concerned industry leaders only insofar as it affected wood costs and corporate liabilities. Meeting the increasing demand for pulpwood without excessive price increases dominated the agenda. By the early 1960s, scientific management had become a watchword among procurement managers as they sought to rationalize wood flow and increase harvesting productivity. "Forest engineering" emerged as a new discipline seeking to integrate forest management with pulpwood harvesting systems. Loggers would eventually become part of a larger, more capital-intensive system focused on the mutual adaptation of forest and machine. Achieving this, while holding on to the advantages of contracting, would prove challenging indeed.

Engineering Wood Flow

> What is an ideal system of wood procurement? Essentially, it is nothing more than a technique of purchasing or producing wood in a steady flow of sufficient quantities to assure full mill requirements at all times, at the lowest possible cost. Such a system, of course, assumes a cadre of efficient managers capable of assembling the inputs of money, men, machines, and timber to produce the required volumes of raw material, again at the lowest possible cost, from any specific location . . . Our wood harvesting systems can no longer afford the luxury of mechanization only in stands and on terrain which must be tailored to the machine. The Forest Engineer must team up with the equipment manufacturer to design and build machines which can economically operate in today's stands on today's terrain.
>
> —*Thomas A. Walbridge, 1966*[109]

Up until the end of World War II, southern pulpwood logging changed very little. Given the particular characteristics of the timber resource and the presence of surplus rural labor, pulpwood production persisted as a predominantly hand-labor operation with little incentive to mechanize. When compared to the industrialized logging operations of the earlier lumber industry and those in regions such as the Pacific Northwest, such practices appeared primitive. By the end of the war, however, some logging crews had begun to experiment with power saws of various kinds. Gasoline powered chainsaws along with wheel-type "bicycle" saws, which were particularly well adapted to the flatwoods of the coastal plain, represented the first wave of mechanization. By the early 1950s, power saws had become standard equipment for most logging contractors, doubling the productivity of pulpwood cutting.[110] Yet no matter how fast a crew could cut timber, there was a limit to how much wood it could load and haul in a day. Because loading was still performed by hand and hauling done with small single-axle "bobtail" trucks, the overall efficiency of the procurement system did not improve by much. In terms of cost per cord of wood delivered, a typical five-man logging crew in 1950 was about as cost-effective as one from 1940.[111] With demand for pulpwood increasing rapidly, this posed a problem.

The wood shortages during 1955 and 1959 frightened procurement managers. If a mill ever had to curtail production capacity for lack of wood, the company could lose millions of dollars and the procurement manager might well lose his job.[112] An effective procurement system had to deliver wood cheaply and on demand, but it also had to be capable of adjusting to circumstances. During the 1950s, as new mills established operations and old mills expanded capacity, the procurement system strained to meet increased demand. With the advent of concentration yards, exclusive dealer territories began to break down. Isolated, monopsonistic pulpwood markets, where landowners had only one outlet for their wood, gave way to larger, more competitive markets where several buyers might be competing for the same wood.

Labor markets were also changing. During the 1950s, the impact of government farm programs (such as the Soil Bank Program) and mechanization (particularly of cotton harvesting) forced many southerners off the farm. Some certainly found work in the woods, but many others left to find opportunities elsewhere. As the New South industrialized, opportunities for those at the bottom of the social and economic hierarchy improved, and pulpwood logging faced new competition for workers. By the 1960s, the civil rights movement and various Great Society programs also began to affect rural labor markets. Despite all the obstacles standing in their way, many loggers had more incentives than ever to leave the woods in search of something better.

Even in the absence of a substantial exodus from pulpwood logging, however, the major challenge facing procurement managers was less one of maintaining a stable workforce in the face of tightened labor markets than one of substantially increasing logging capacity to meet the rapid growth in demand. Raising prices, as Champion and other mills had done in the mid-1950s, was one way of inducing people to enter the profession, thereby increasing aggregate capacity, but it did not represent a viable long-term solution in the eyes of most procurement managers. In their view, such price increases could all too easily end up in dealers' pockets or in higher stumpage prices, never making it to the producer level. Instead, procurement managers began to push for a program that would increase logging productivity, primarily through mechanization, and rationalize the overall procurement system to make it more responsive to demand increases.

Such efforts faced several obstacles. Small, widely dispersed tracts with low timber density made it difficult to capitalize on the economies of scale necessary to justify large investments in mechanized logging. Credit was often in short supply for those who needed it. And the contractor system offered few incentives to invest in mechanization and provided limited information on broader conditions to those operating in the woods. Given the variability of the work, the lack of "business" experience, and the technical difficulties involved in timber harvesting, dealers and producers had neither the means nor the incentive to experiment with mechanization. Because they operated on a tract-to-tract basis with no long-term contracts, moreover, these men often had no idea if demand increases were merely temporary or more permanent. The system's inherent flexibility mitigated against the investments in fixed capital necessary to make mechanization a reality. As one procurement manager noted in 1957, "In the South . . . the presence of scattered stands and light cuts per acre, with relatively no assurance of stumpage ahead, puts mechanization other than a power saw and a bobtail truck in the realm of near impossibility."[113]

In response to these challenges, the American Pulpwood Association, acting on behalf of nineteen companies operating in the South, commissioned the Battelle Memorial Institute to conduct a three-year study of the southern wood procurement system. Composed of ten separate reports compiled between 1960 and 1963, the Battelle report provided the first systematic appraisal of the procurement system and, in the process, catalyzed a debate in the industry over the best way to rationalize wood procurement, improve harvesting productivity, and integrate these activities with forest management.[114] For the first time, the principles of scientific management and industrial engineering were applied to wood procurement. Very much in vogue among corporate management during the post–World War II decades, Battelle's "systems approach" to operations research viewed wood procurement like any other industrial or military problem. The challenges facing mills, in other words, could best be understood and overcome by breaking the system into its constituent parts, analyzing them, and recombining them into an "optimal" system. Wood flow had become an engineering problem.

Battelle's mandate was to determine the underlying causes of the wood shortages of 1955 and 1959 and develop ways to avoid such shortages in the future. During the three-year study, the researchers focused primarily on the dealer-producer system and its capacity to respond to demand increases, the harvesting system and its component costs, and the potential for improving harvesting technology and management. Their conclusions, while not exactly earth-shattering, did challenge the conventional wisdom of the day. Periodic wood shortages, they argued, did not derive from a shortage of labor in the woods, as so many in the industry believed, but rather from "the inability of the dealer and producer to distinguish between temporary changes in orders and those of a permanent nature."[115] The sluggish response of the pulpwood procurement system in responding to permanent demand increases, in other words, stemmed from the incentives embedded in the system itself.[116] Given the normal fluctuations in wood demand from an individual mill, dealers and producers were understandably hesitant to expand their production capacity. As the report noted, "Both the dealer and producer often regard increases in wood orders as temporary in nature and, therefore, are extremely cautious about making long-term financial commitments dependent on higher orders."[117]

In analyzing the various components of the harvesting system, the Battelle researchers identified loading and hauling as major bottlenecks.[118] Until these barriers could be overcome, harvesting productivity would be stuck at around two cords per man-day.[119] Given their engineering bias, the authors preferred mechanization as a solution. Yet it was by no means a straightforward one. Indeed, the authors of the Battelle report identified several factors contributing to the "slow rate of innovation in the pulpwood harvesting industry." Cheap labor, the scattered nature of the resource, and a dearth of research were all cited as reasons. More importantly, they argued, the industry did not have a proper basis for analyzing and evaluating the economic and technical merits of alternative systems—something they hoped to correct.[120] From their perspective, the proper application of Taylorist principles could transform woods work from a "backward," agrarian-like system of task labor to one more closely resembling an assembly line.

With these ends in mind, the Battelle analysts, armed with an arsenal of flow charts, time-and-cost studies, dynamic models, and statistical regressions, compared alternative harvesting systems. Longwood harvesting systems, in-woods chipping operations, even balloon transport were a few of the harvesting concepts analyzed in the report. When compared to the standard practices of the day, however, the researchers concluded that "none of the [proposed] systems appears to offer large cost savings on a per cord basis, given present labor rates." Until labor costs rose by a sufficient amount, in other words, substantial investment in mechanization was unlikely. In the meantime, the report suggested, companies should begin experimenting with alternative methods of harvesting and develop a framework for assessing their economic viability.[121]

Finally, in keeping with a systems approach, the Battelle report also urged companies to reorient their internal organization in order to develop the tools and expertise required to integrate harvesting, procurement, and wood-growing activities. With the use of linear programming, a minimum-cost approach to wood flow could be developed that would optimize various investments in land acquisition, forest management, manpower recruitment, and mechanization.[122] Under a regime of scientific management and operations research, land, labor, and capital could be brought together into a smoothly functioning industrial system.

More than anything else, therefore, the Battelle report stimulated a new way of thinking about wood procurement. Not only did it challenge the conventional wisdom regarding the nature and underlying causes of wood shortages, it also pushed procurement managers to view wood flow as an engineering problem—a conception that has persisted ever since.[123] Approached in this manner, the task of procuring and harvesting enough pulpwood to meet the needs of the mills was simply one big optimization problem in which the various components of the harvesting operation could be rationalized and reintegrated into a more efficient system. Once the rural, secondary labor markets that served the pulpwood logging industry converged with larger regional and national labor markets, mechanization would be sure to follow.

It sounded simple enough—perhaps too simple. Indeed, for all of their emphasis on a systems approach, there were some serious gaps in the researchers' analysis. Despite all of their research on logging productivity and alternative harvesting systems, for example, they missed what was perhaps the single biggest obstacle to mechanization in the rural South: the lack of credit for small producers.[124] As shoestring operators facing a highly variable demand structure with little economic security, pulpwood producers hardly qualified as exemplars of financial stability. Given that these producers had little to offer in the way of collateral, financial institutions were not racing to lend them money. Finally, because many of these producers were black, prevailing patterns of racial discrimination further exacerbated the challenges of accessing credit. In short, the highly localized credit systems supporting wood procurement did not have the capacity to facilitate full-scale mechanization. Dealers and producers faced an uncertain situation regarding wood demand and thus were understandably hesitant to leverage themselves any more than necessary. Their access to credit and working capital was itself quite limited. It was one thing for a dealer to loan his producers money for a truck, a couple of chainsaws, and other supplies. It was another thing altogether to find the credit necessary to purchase and adapt the machines to mechanize the entire harvesting system. For this to happen, the companies would have to get involved.

During the late 1950s and early 1960s, wide-ranging discussion within the industry emphasized the challenges facing mechanization and the need for company involvement. At the 1957 annual meeting of the American Pulpwood Association, for example, the Hiwassee Land Company's Tom Walbridge, a leading proponent of mechanization, noted that "if further mechanization is to take place, it will be largely through the efforts of our [company] participation in a program of financing, building, and testing experimental machines. We cannot expect the machinery manufacturers or the producer to enter this field actively enough to solve the problem of producing more wood with less men. This is our problem to accept and solve." From Walbridge's perspective the issue was clear: mechanization was desirable for all concerned. For the woods worker, it "offer[ed] the chance to step up in life; to become an operator and gain prestige among his fellow workers." For the producer and the dealer, mechanized operations would "attract and hold a better class of labor," increase productivity, and generate higher rates of return. And for the mills, it promised to meet increasing wood demands in the face of a decreasing and potentially unstable labor supply.[125]

None of this would happen, however, unless the companies stepped in with a serious commitment of institutional and financial support. Beginning in the 1950s and accelerating in the wake of the Battelle report, several companies began moving in this direction through a dual program of experimenting with mechanized logging operations and developing a system of financing capable of stimulating mechanization among producers. International Paper quickly emerged as a leader in the field of "forest engineering." Under the direction of Tom Busch, the same man who developed the mechanized woodyard in 1950,

Figure 2.3. A "feller-buncher," at work, circa 1974. Mechanized pulpwood logging operations resulted in significant increases in productivity. Courtesy of the Forest History Society, Durham, NC.

International Paper's harvesting operations developed and adapted several new harvesting machines.[126] During the early 1960s, for example, the company introduced the so-called Busch combine. Named for its developer and based on twelve years of research and development, this machine combined three harvesting steps—cutting, felling, and bunching—and was capable of operating in wet conditions and on uneven terrain (figure 2.3). By 1966, the machine allowed for productivity rates of up to sixteen cords per man-day and provided the foundation for company harvesting crews that could compete on a cost basis with the standard dealer-producer system then in use.[127]

A similar system introduced during the same period, the Beloit harvester system, employed three different machines—a mechanized harvester similar to the Busch combine, which combined cutting, felling, and bunching; a rubber-tired grapple skidder or logging tractor for hauling logs to loading points; and a hydraulic knuckle-boom loader for loading logs onto trucks. Each machine was operated independently by a single man. Functioning as a whole, the Beloit system proved to be highly productive and, along with the Busch combine, provided much of the basis for subsequent mechanization in southern pulpwood harvesting.[128] The key to these systems lay in their capacity for operating in various conditions and terrains, their relatively low manpower requirements, and their ability to harvest tree-length stems or longwood rather than the five-foot shortwood that was standard in traditional pulpwood harvesting.[129]

Following the lead of International Paper, which had about a hundred company logging crews operating throughout the South in the late 1960s, other major firms began establishing company harvesting programs. Union Camp, for example, initiated a pilot program at its Savannah mill in 1963, which it subsequently expanded to its other mills. By 1975, the company had formed a harvesting department, and by 1980, the peak year of the program, some twenty-five different company crews worked in the Savannah area alone. For the Savannah mill, which accounted for the bulk of company harvesting operations, these loggers provided about one-seventh of the mill's fiber supply during the late 1970s and early 1980s. Most crews consisted of three men operating machines in the woods, supported by company trucks. In order to maximize productivity, the company restricted these logging operations to large tracts of company-managed timber. Overall, the productivity of these crews averaged around one cord per man-hour—far above the two cords per man-day rate of average logging contractors. Like other company programs, Union Camp used its harvesting operations to push mechanization, develop a basis for calculating manpower needs and costs, and provide a limited amount of fiber security in times of tight wood markets.

As important as company harvesting was in pushing mechanization, however, it never provided more than a small fraction of the total fiber delivered to the mills. By the end of the 1970s, most companies had phased out their company crews.[130] In the process, many shifted their support to the harvesting research cooperative that was established in 1973 at Virginia Tech's school of forestry. Like the earlier tree improvement cooperatives (see chapter 1), the harvesting co-op, known formally as the Industrial Forestry Operations Research Cooperative, provided an institutional mechanism for individual firms to spread the costs and risks associated with the development and implementation of new timber harvesting systems. From the beginning, the cooperative, which included both forest products firms and equipment manufacturers, focused on training future industry employees and developing machines with the aim of encouraging mechanization among the existing logging force.[131]

Once pulp and paper firms realized that mechanized harvesting was possible and that it could greatly increase logging productivity, they turned their attention to the far more challenging task of facilitating mechanization among small contract loggers. At the end of the day, none of these companies wanted to take on the costs associated with full-scale integration into logging. Contracting simply had too many advantages. The problem lay in the highly localized circuits of credit that marked the existing dealer-producer system, which could not provide the level of investment required for producers to make the transition from chainsaws and trucks to fully mechanized operations. Whereas traditional operations might require an investment of $10,000 at most during the 1960s, a mechanized operation could easily require $100,000 or more.[132] As Tom Walbridge and others suggested, companies would have to get more actively involved in financing. In establishing new financial arrangements with dealers and producers, however, these firms had to be careful not to undermine the arm's-length relationship they had labored so hard to establish.

Companies dealt with this challenge in different ways. Firms that preferred the extra protection and flexibility afforded by the dealer system, such as Union Camp and International Paper, set up financial arrangements with dealers and local banks, who then provided financing to producers. In general, dealers, with backing from the companies, would either lend money directly to producers, as they had done in the past, or co-sign notes with their producers at local banks. As long as the big companies approved of the arrangement, the producer would get his loan. Other firms, such as Bowater, Brunswick Pulp, and Westvaco, developed a system of dealing directly with their contract loggers, eliminating the "middleman" operations of the wood dealer altogether. Under this system, the company took the place of the dealer in arranging for producer financing, most often through local banks. While providing somewhat less insulation than the traditional dealer system, this approach allowed for closer management of contract loggers and facilitated more rapid technology transfer.[133] In the long run, however, the end result was the same. Mechanization was on its way. For those loggers willing and able to mechanize, logging would soon become a very capital-intensive operation. Whether they were backed by dealers or directly by companies, these loggers would eventually have to take full title to all equipment. They would ultimately be responsible for making the business work.

With this new capital intensity came changes in the incentive structure of the procurement system. Mechanized producers, who were now highly leveraged, faced increased pressure to "move wood" as often and as quickly as possible. This further undermined their ability to bargain with mills or dealers over prices, especially during slack periods. Indeed, even though they now had substantially more at stake, mechanized loggers rarely received any sort of guaranteed contract. Because their operations were still "tract-to-tract," many found themselves more dependent than ever on the mills and the dealers to "keep them in wood." At the same time, however, those dealers and mills that provided significant financial backing to newly mechanized loggers also faced a new set of incentives. Afraid that they might have to cover some of the losses if a logger went broke, these mills and dealers had a certain vested interest in keeping less efficient producers in business by pumping more money into their operations and giving them priority status in the allocation of wood orders.[134] The bonds of debt and financial dependency worked in both directions. Yet the loggers clearly had the most to lose. With debt burdens in the hundreds of thousands of dollars, no real long-term contractual arrangement, and little if any alternative use for their machines, these men found themselves at the mercy of those who controlled their access to wood.[135]

Given this situation of "bonded performance," the newly mechanized loggers pushed harder than ever to increase the productivity of their crews. By the late-1960s, mechanized logging contractors could achieve productivity rates of almost six cords per man-day, more than triple that of unmechanized crews. Overall, the average productivity of southern pulpwood harvesting at the end of the 1960s was 3.3 cords per man-day—65 percent greater than the two-cord-per-man-day average of 1960.[136] By 1979, average productivity had increased another 47 percent to almost five cords per man-day, and it had doubled again to almost ten cords per man-day by 1994.[137]

These figures reflected the slow but steady adoption of mechanized equipment by producers. Even as late as 1979, though, some three-fourths of the southern pulpwood logging force continued to operate with chainsaws and trucks. For whatever reason, these men were unable or unwilling to take on the capital risks associated with mechanization and continued to operate much as they had before. With the advent of mechanization, however, their share of pulpwood production had fallen to about a third of total volume. At the other extreme, the highly mechanized producers, accounting for slightly more than

6 percent of the total logging force, delivered about a third of total volume as well. Pulpwood logging had thus become a two-tiered system, with small "shortwood" producers accounting for the majority of the workforce and highly mechanized, longwood logging crews accounting for the majority of the productivity increase.[138] Over time, increasing numbers of these smaller producers mechanized or dropped out as logging came to be dominated by the more efficient, capital-intensive operations. By 1987, shortwood producers accounted for only about a third of the total workforce; by 1994, the figure had fallen to around 20 percent.[139]

One impact of increased harvesting productivity can be illustrated by the decline in pulpwood prices since the mid-1970s. Between 1974, when prices peaked, and 1992, real prices declined by more than 30 percent.[140] During the same time period, stumpage prices, which are a component of pulpwood prices, declined by 19 percent in real terms.[141] One way of interpreting this is that roughly one-third of the overall decline in pulpwood prices can be attributed to the increased productivity stemming from mechanization. Because fiber accounted for the largest overall share of mill production costs, this represented considerable savings for the mills.

As for the loggers, the benefits from mechanization were more elusive. Clearly, a mechanized logging contractor could produce more wood per unit of labor, thereby saving considerably on labor costs. At the same time, however, the capital costs associated with the new machines greatly exceeded anything they had seen before. In the late 1960s and early 1970s, for example, a fully mechanized logging operation could easily cost more than ten times as much as a more traditional shortwood operation. Given the uneven nature of the work and loggers' inexperience with the new machines, recovering these costs proved difficult. In fact, in the early years of mechanization it was not at all clear that mechanization actually reduced harvesting costs. As the American Pulpwood Association's 1968 survey of southern loggers put it: "On the basis of the limited data obtained by this survey we can again state that mechanization has not resulted in appreciably lower harvesting costs. The assumption that it should is a trap that many forest engineers have fallen into. It has been used to justify heavy expenditures for capital equipment when in fact mechanization most often increases harvesting costs. Loggers are usually forced into mechanization because of a shortage of labor or a lack of stumpage that can be produced with a low degree of mechanization. They mechanize to stay in business—not to lower production costs."[142] In theory, as loggers learned how to run their machines and manage a more capital-intensive operation, they would begin to capture some of the savings generated by the increase in productivity. Yet this assumed that mechanized loggers would be able to operate at close to full capacity—something that was very difficult to do given the spatially extensive nature of the work, variability in site conditions and weather, and fluctuations in the mills' demand for wood. Even though the machines were sometimes capable of operating in conditions that traditional, shortwood operators could not handle, most mechanized loggers rarely worked more than forty-five weeks per year.[143]

Part of the problem stemmed from the inherent flexibility of the procurement system. Given the variability in their demand for wood combined with the unpredictability of weather and other conditions, the mills preferred a certain amount of excess logging capacity to meet peak demands. As one former procurement manager explained: "There always will be [excess logging capacity] as long as the paper mills have anything to do it with. Because that's the way they can operate where they've always got an out. That's their security blanket. It's not going to go away because it's about the only security the pulp and paper industry has to give them the slack they need when the going gets tough. Now, it's rough on the guy that produces."[144] Rough, indeed. For loggers, excess capacity translated into a situation of underutilization of capital and "destructive" competition with other loggers for available wood. By going with mechanization, loggers effectively tied their fortunes to "dedicated" machines in a profession that required "extreme flexibility of business policy."[145] Facing a situation of financial insecurity, many pulpwood producers often had little choice but to take whatever work they could get.

To make matters worse, logging contractors typically had little if any experience with standard accounting practices. With limited formal education, many of these men suddenly found themselves managing businesses that required careful evaluation of cash flow, discounting, and capital depreciation.[146] Given the large increases in cash flow that came with increased productivity, some probably thought they were getting rich when in fact they were living off the depreciation of their equipment.[147] As one procurement manager put it: "For a long while most loggers lived off the depreciation. Now they're beginning to realize that you can't do that, you've got to plan for the future, you've got to depreciate your stuff, you've got to buy new equipment."[148] Most machines had a working lifetime of five years or less. Alternative uses were nonexistent, and there was no real market for used equipment.[149] Contrary to the arguments of some forest engineers, loggers were not building equity. If they didn't take care to account for equipment depreciation, they would find it very difficult to get out of debt, which is precisely what seems to have happened.

As mechanization proceeded, equipment suppliers standardized financing, providing package deals backed by local banks. Mills and dealers stepped back and removed themselves from any sort of financial liability associated with the producers' debt. Access to credit no longer presented a problem. In fact, easy credit had actually become part of the problem by further fueling excess capacity. As one logger described it: "You've got so many companies out there financing equipment, anybody will take you. I don't care whether you've got good credit or bad credit or whatever, you can get anything you want today. Easy financing, nothing down, just step in."[150] Loggers were encouraged to invest in new machines without any sort of binding commitment from mills and dealers to "keep them in wood." Yet if a logger could not produce on a sustained basis, he stood to lose his entire operation. Debt service became an overriding imperative and made it difficult for loggers to seek out alternatives. In the words of one former logger: "It's a very, very, very hard thing to get out of. You get hooked into logging and you can't hardly get out because your obligations are so great and your cash flow is so important that if you stop, you lose everything you got."[151] In sum, mechanization created a technological treadmill for many loggers and, in the process, radically altered the landscape of financial risk and liability in which they operated.[152]

The advent of mechanized logging also changed the way companies approached wood procurement. During the initial push for mechanization, the role of companies in providing or guaranteeing financing meant that company managers had to pay closer attention to the producer and his organization. With the emergence of high-capacity loggers, moreover, some companies felt that the dealer system was no longer the most effective way to procure timber. Starting in the late 1960s, several companies began to experiment with a system of procurement known as direct purchase.[153] In essence, direct purchase eliminated the wood dealer and replaced him with company employees who assumed responsibility for making wood purchases. Loggers were assimilated from dealer organizations and provided with written contracts establishing the terms of the relationship—emphasizing that the logger was an independent contractor or vendor just like any other supplier. Company timber buyers purchased most of the stumpage and contracted with the logger for a set fee to cut and haul the timber.[154] Overall, "going direct" offered more control over wood flow, closer contact with loggers, and more efficient technology transfer. At the same time, however, by requiring a company to tie up capital in buying and carrying stumpage, direct purchase meant less flexibility in managing inventories. By eliminating the wood dealer, the direct approach also removed a layer of protection between the mills and the producers. Though most companies that used direct purchase went to great lengths to emphasize the "independent" status of their producers, their so-called arm's-length relationship was more intimate than it would have been under a dealer system.[155]

In the end, whether a mill procured its timber through direct purchase or through dealers depended on a host of factors. For those companies with substantial land bases, the dealer system often made sense. These companies preferred the increased flexibility and added protection afforded by dealers, and because they had their own stumpage to fall back on, they did not necessarily need the extra control afforded by direct purchase. For those without large acreages of company controlled stumpage, however, direct purchase often provided a more effective way of ensuring that the mill did not run out of wood. This was especially true in competitive woodbaskets such as south Georgia, where "going direct" allowed companies to stay in closer touch with wood markets and avoid becoming hostage to large dealers. Advocates for each approach argued that their way was the cheapest way to log the mills. Those who preferred direct purchase felt that they saved money by cutting out the middleman. Those who favored the dealer system pointed out that the dealer had a greater incentive to buy wood as cheaply as possible and was therefore much more efficient than company employees. Most companies approached the matter pragmatically. Some used a dealer system at one mill and direct purchase at another. Others used a combination of both at their mills. The choice often depended on the style or preference of a mill's procurement manager, the amount of company-controlled stumpage, and local circumstances.[156]

The move to direct purchase, then, like the push for mechanization, was part of a larger effort by pulp and paper firms to rationalize wood flow and develop a procurement system capable of logging the mills in the face of a growing regional demand for fiber. Yet despite the increase in harvesting capacity that came with mechanization and the better control over wood flow afforded by direct purchase, neither system could alleviate pressures on the existing supply of pulpwood in highly competitive woodbaskets. Simply put, direct purchase and mechanized logging did not make trees grow any faster, nor did they allow for significant expansions of a mill's procurement area without adding considerably to overall costs. Firms, of course, had been hard at work since the 1950s trying to increase the biological productivity of southern timberlands. Beginning in the mid-1980s, they also began to extend the boundaries of certain competitive procurement areas through the use of satellite chip mills.

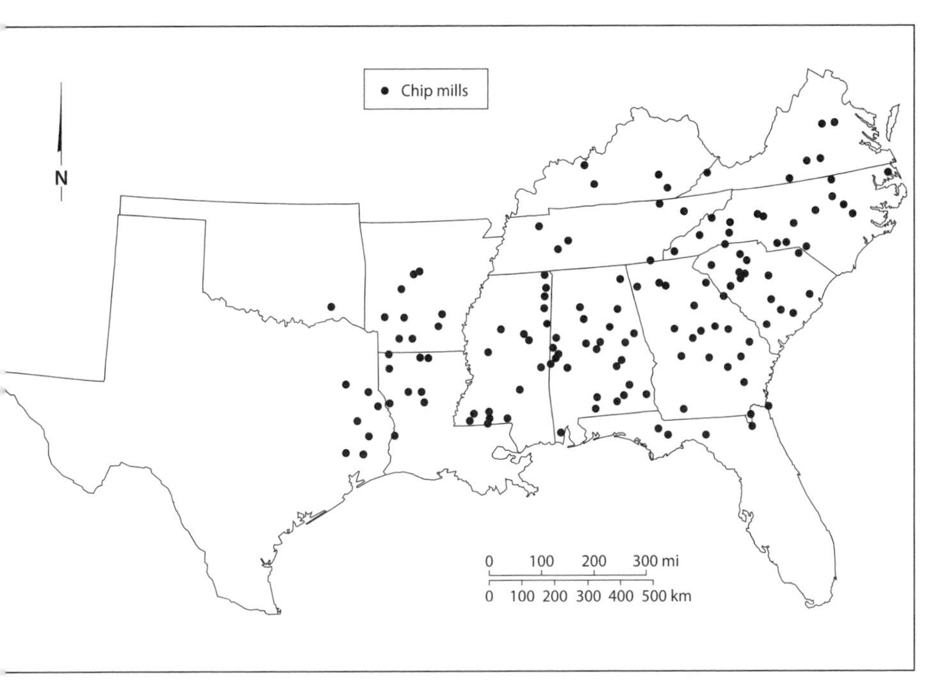

Figure 2.4. Satellite chip mills expanded dramatically in the region during the 1980s and 1990s. This map shows the location of chip mills in the South circa 2000. Adapted from data compiled by Jeffrey Prestemon et al. at the USDA Forest Service Southern Forest Research Station, Asheville, NC.

Although chip mills first appeared in the South in the late 1950s, they did not become permanent features of the procurement system until the 1980s, when the advent of new chipping technologies combined with increased competition for wood pushed firms to expand their procurement areas.[157] As part of the larger effort to rationalize wood flow, satellite chip mills allowed firms to extend procurement into new, less competitive areas and relieve pressure on mill woodyards.[158] With increased competition over softwood timber, moreover, chip mills have also provided access to previously inaccessible hardwood stands in the interior regions of the South. By the mid-1990s, there were some 140 chip mills operating in the South, fed by the timber cut from more than 1 million acres a year (figure 2.4).[159]

Some mills preferred to operate their own facilities. Others contracted out the operation to third parties. In general, loggers delivered roundwood logs to a satellite facility just as they would to a pulp and paper mill. Chips were then sent via rail, barge, or truck to the mills. By reducing the volume-to-weight ratio of the transported fiber, chip mills thus allowed firms to expand procurement areas well beyond what would be possible with standard log transportation systems. A typical chip mill in the 1990s, running on two ten-hour shifts, could produce about 2,000 tons of chips per day. This translated into roughly 100 truckloads of wood a day. Whole logs were devoured in a matter of seconds. As one chip mill operator put it: "Seventy-five year-old trees gone in two or three seconds—that's what gets me."[160]

Starting in the late 1980s, some chip mills also began exporting excess chips to East Asia, particularly Japan. This was driven on the demand side by the sourcing strategy of Japanese pulp and paper firms seeking to reduce their dependence on imported market pulp by increasing their imports of virgin fiber.[161] On the supply side, increased chip exports, consisting primarily of hardwood, were made possible by the completion of the Tennessee-Tombigbee waterway in the mid-1980s, which opened up the hardwood regions of Alabama, Mississippi, and Tennessee by providing cheap river transport to export facilities in Mobile, Alabama. As a result, many chip mills began to export their additional supply. At the same time, several chip mills were built, particularly along the Tennessee-Tombigbee, solely to serve the export market.[162] Between 1989 and 1995, hardwood chip exports from the Southeast grew fivefold, with the port at Mobile accounting for almost half of the total volume.[163] Scott Paper Company, which invested heavily in chip export facilities in the early 1990s, exported more than a million tons of hardwood chips, priced at roughly fifty dollars a ton, from Mobile to Japan and other countries in 1994. According to Scott manager Bobby Burke, "Our exports of chips each year . . . offset about 2,700 imported Toyotas, and that, I think, is healthy for our economy."[164]

Needless to say, not everyone agreed that chip mills and chip exports were economically or environmentally beneficial for the region. In fact, the explosion of satellite chip mills in the 1990s elicited criticism from local and national environmental groups, small sawmill operators, and others concerned about these mills' potentially detrimental impacts on local communities and local environments. Much of the controversy was driven by the fact that a substantial proportion of the chipping was being done in some of the last significant stands of mixed-mesophytic forests in the South and by the image of hardwood chips being exported to Japan without any further value-added processing.[165] Environmentalists raised concerns about loss of habitat and species diversity, particularly in light of the fact that much of the hardwood area being cut was replanted as pine plantation.[166] Small sawmill operators were concerned about losing the raw material supply that their livelihoods depended on, and labor representatives and local economic development groups felt that exporting unprocessed raw materials led to lost jobs at home.

For all the controversy surrounding chip mills, however, little effort was made to place them in their proper historical and economic context. Some environmentalists argued strenuously that chip mills represented a new and highly destructive force leading to extensive clear-cutting—the "butcher shops of forestry," as one reporter put it.[167] In reality, clear-cutting had been standard practice in southern forests for more than a century, and chip mills were little more than extensions of the operations already taking place inside mill woodyards. Furthermore, hardwood chip exports to Japan, which elicited considerable controversy in the 1990s, accounted for only a small portion of the total output from chip mills, the bulk of which continued to feed the appetite of the region's massive pulp and paper industry.[168] At the end of the day, satellite chip mills were simply part of an ongoing effort by firms to log their mills as cheaply as possible.[169]

In retrospect, the southern wood procurement system of the 1990s appeared considerably different from that of the 1950s and 1960s. Mechanization, the rise of direct purchase, and the use of satellite chip mills all changed the complexion of wood procurement. Yet the basic social and legal relations that constituted pulpwood logging remained largely unchanged. Loggers still operated as independent contractors and were, in many respects, as dependent as ever on those who controlled their access to wood. Impeded by their own structural vulnerabilities and economic insecurities, these producers continued to face substantial obstacles in their efforts to improve their situation through political and legal channels.

Contracting, Risk, and Liability

> The actual, effective control of the dealer over the producer is virtually absolute . . . There is no bartering or discussion concerning the price the producer will receive; he may either take the dealer's offer or leave it. There are no written stumpage contracts between dealers and producers, all agreements are oral. Since most producers cannot afford trucks, saws, and other necessary equipment, the dealer either leases this equipment or co-signs a mortgage or financial agreement with him . . . In this way, the dealer exercises virtually total economic control over the producers . . . [A]lthough nominally free in many cases to go elsewhere, . . . the economic reality of the situation requires the producer to haul exclusively for his dealer in order to obtain the necessary credit to keep up with equipment maintenance and in order to pay off the often sizable debt owed

> to the dealer . . . Unlike a truly independent businessman, the producer usually has nothing to offer but his labor and has no control of the amount of money he will receive for his labor . . . He is, in the words of § 2 of the Norris-LaGuardia Act, "commonly helpless to exercise actual liberty of contract and to protect his freedom of labor." Viewing the "totality of circumstances," the court finds that the control exerted by the dealer is of such a nature and amount as to render the producers the employees of the dealers for the purposes of this suit.
>
> —Scott Paper Co. v. Gulf Coast Pulpwood Assoc'n Inc., *1973*[170]

During the mid-1960s, a group of black and white pulpwood producers from Alabama and Mississippi began to organize in pursuit of better wages and working conditions. In 1971, these producers formed the Gulf Coast Pulpwood Association—the first such organization in the South. Two years later they presented formal demands to pulpwood dealers and pulp and paper mills in the area. "The Gulf Coast Pulpwood Association represents the interests of the majority of pulpwood producers in Alabama and Mississippi," the demands read. "We seek to do away with the injustices and wrongdoings that a pulpwood producer has to put up with every day of his life. In short, we want a better living and working condition for the people who cut and haul pulpwood for a living."[171] Their issues were basic: higher prices, standard measurements for pulpwood purchases, accident insurance, accountability, and face-to-face bargaining with representatives from the industry.[172] If the industry rejected these demands, the association would initiate a strike.

On Friday, September 7, 1973, with their demands still unmet, the producers began their strike, refusing to cut, load, or haul pulpwood for any dealer or company. The following Monday, they began picketing woodyards in Alabama and Mississippi, and on Thursday, September 12, they started picketing the entrances to the Mobile, Alabama, pulp and paper plants owned by Scott Paper and International Paper. These actions resulted in a substantial reduction of pulpwood deliveries to the mill, which threatened capacity reductions and shut-downs at the plants. Consequently, Scott and International immediately filed a joint complaint in federal district court requesting that the court enjoin the strike. Claiming that the producers were independent contractors rather than employees, the companies argued that such a strike was not permitted under the federal antitrust laws.[173] The court issued temporary restraining orders against the picketing and held a full evidentiary hearing. On September 21, Judge Virgil Pittman ruled that the producers were in fact employees of the dealers, thereby dissolving the temporary restraining order and denying plaintiffs' request for an injunction.[174] Six months later, Scott and International lost their appeal to the U.S. Court of Appeals for the Fifth Circuit.[175] Rather than pursuing their case further, they agreed to increase pulpwood prices by five dollars per cord and pay Gulf Coast's attorney fees.[176]

Pittman's ruling got the attention of the pulp and paper industry. At the 1974 Yale Industrial Forestry Seminar, which took labor as its theme, participants discussed the implications of the case. One forest economist suggested that the court's ruling offered "a strong precedent if similar facts are present in subsequent cases," thereby opening the door to collective bargaining for woods workers.[177] The key issue, of course, turned on the status of the pulpwood producer and the nature of his relationship to those he supplied with wood. Was he, in short, an independent contractor, as wood dealers and paper companies claimed. Or, rather, was he actually an employee of the dealer or the mills. In the eyes of Judge Pittman, the situation was unambiguous. The pulpwood producers who had organized under the Gulf Coast Pulpwood Association were clearly operating in an employee-like relationship with their wood dealers. As such, these producers were well within their rights to organize and strike in pursuit of better wages and working conditions. Under the provisions of the Norris-LaGuardia Act, according to Pittman, this was a labor dispute, and the federal courts had no authority to intervene.[178] In rejecting the companies' claim that the producers' decision to strike violated the antitrust laws and that they should thus be enjoined from striking, the judge sent a strong message to the forest industry: "The economic relationship between the wood producers and the wood dealers . . . is more like that between a piece-work employee and his employer than that of an independent contractor."[179]

More than anything else, though, this case illustrated the complexities of pulpwood contracting and the importance of situating the producer in his larger political-economic context. By attending to the "non-contractual elements" of contract,[180] Pittman exploded the myth of the "independent" pulpwood producer, pointing to the massive asymmetries of power that existed in the southern wood procurement system. According to the court's findings, "The wood producers, white and black, generally have less than a high school education, are economically deprived 'hip pocket' bookkeepers who are employed about forty weeks or less in the year, with take home pay generally less than one hundred dollars a week." Burdened with ties of debt and dependency, these producers had "little freedom in moving from one dealer to another or in serving two or more dealers at the same time because of the economic domination of the dealer over the producer." Given the producers' economic situation and the dealers' control of access to stumpage, the court found that "there is no real bargaining between the producer and the dealer for the producer's labor and services. The increases the dealer has paid the producers during the past ten years have been negligible. The situation really consists of a quotation by the dealer to the producer that he can take or leave. The producers, who live from hand to mouth, are operating in a condition where they work for the dealer's price or are unemployed. Their education is limited, their skills are few, and they have few, if any, real work alternatives."[181] Pittman's ruling represented an important victory for pulpwood producers, demonstrating the potential of an organized campaign in pursuit of better wages and working conditions.

The victory turned out to be short lived, however. Even though the paper companies lost this particular battle, they had already won the war. *Scott v. Gulf Coast* would be the only successful court case ever won by a pulpwood association. By the 1980s, with mechanization in full swing, the small pulpwood producer had become something of an anachronism, and efforts to leverage previous gains came to an abrupt halt. As logging capacity expanded, pulpwood prices declined and producers scrambled to adjust. Although Gulf Coast and other organizations such as the Southern Woodcutters Assistance Project and the United Woodcutters Association persisted into the 1980s, the obstacles they faced in trying to organize pulpwood producers proved too difficult to surmount.[182] Referring to the settlement reached with Scott and International in the mid-1970s, one Gulf Coast Pulpwood Association leader remarked in 1981, "It was a significant step forward at the time. Unfortunately . . . that was the last we heard from the companies. We got the increase. They could not deny us that. But the negotiations stopped there. That was the last raise we received."[183]

In reality, collective bargaining had always been a rather distant prospect for pulpwood producers.[184] Even if the courts agreed that these men operated as de facto employees and thus were within their rights to organize in pursuit of better wages and working conditions, any effort to organize woods workers faced immense challenges. Most producers worked in small crews spread over large areas. They typically worked for only part of the year, often moving back and forth between pulpwood logging and various other part-time jobs. In many areas of the rural South where these producers lived and worked, substantial pools of unemployed and underemployed labor were available. Most of the men who worked as pulpwood producers did so because they had few if any employment alternatives. Serving a single dealer or company, their capacity for organized resistance was limited—all of which made the accomplishments of Gulf Coast that much more remarkable and that much more difficult to replicate.

As mechanization proceeded in the 1970s and 1980s, the legal obstacles to forming effective bargaining associations grew. Mechanized loggers, although facing substantial debt burdens and still dependent on those who controlled access to stumpage, appeared to be more "independent" than traditional pulpwood producers. These loggers took full title to their own equipment and, as time passed, received all of their financing from local banks or equipment manufacturers. Though they could and did form logging associations, their status as independent contractors restricted them from bargaining collectively over price.[185]

Notwithstanding the *Scott v. Gulf Coast* decision, then, efforts to challenge the "independent contractor" status of producers under federal labor law ultimately proved unsuccessful. During the late 1970s, however, the independent contractor issue came under scrutiny in another context—tax law. In an effort to improve compliance rates and bolster tax revenues, the Internal Revenue Service launched an investigation of various sorts of independent contractors in 1978. The basic issues involved income and social security tax compliance by workers treated as independent contractors and whether such workers should be reclassified as employees under the "economic reality" doctrine in U.S. tax law.[186] Findings indicated widespread noncompliance—tantamount to tax evasion.[187] Such practices, argued IRS officials, not only deprived the Treasury and social security system of revenues, but they also deprived many workers of basic social security coverage. Compliance rates, moreover, correlated positively with the level of income. In other words, low paying industries, such as logging, had some of the highest rates of noncompliance. In response, the IRS filed lawsuits against logging contractors and wood dealers, propelling the issue into the ideological maelstrom of antigovernment sentiment that would soon sweep Ronald Reagan into the White House. In fact, none other than Reagan himself addressed the issue in a February 1978 radio broadcast, entitled "Independents vs. IRS," in which he castigated the IRS for "hounding" logging contractors throughout New England and the Deep South. Arguing that these contractors were being "harassed into insolvency," Reagan joined the chorus calling for congressional relief.[188]

Responding to such concerns, Congress passed interim legislation in 1978 that temporarily limited IRS enforcement actions until a long-term legislative solution could be hammered out. To that effect, House and Senate committees convened hearings over the next several years to debate the issue. Timber industry interests were well represented, with testimony coming primarily from trade associations, southern wood dealers, and large independent loggers and timber truckers from New England and the Pacific Northwest.[189] As in earlier hearings over the twelve-man exemption under the FLSA, the exceptionalism of logging was a common refrain. Witness the words of Ken Rolston, president of the American Pulpwood Association: "Efforts to try to create an employment relationship for a logging contractor are just going to be futile. He is not going to be an employee. He is that type of person. Logging does not lend itself to large-scale enterprise. It is way off in remote regions. You just can't get this whole corporate structure out there gathered around some trees, and have successful logging. They try it in the Soviet Union and they do it terribly."[190] Citing the rising costs of litigation facing logging contractors battling the IRS and the desire among these men "to stay at peace with their government," Rolston, along with others, pleaded with Congress to enact legislation that would protect independent contractors from IRS enforcement actions.[191] In August 1982, Congress extended its 1978 legislation indefinitely, effectively halting IRS actions to redefine so-called independent contractors as employees.[192] In the eyes of the federal government at least, the ambiguities inherent in the status of pulpwood producers had been resolved. They were independent contractors, just as industry representatives had claimed all along.

At the same time that their legal status was being resolved, however, pulpwood producers found themselves facing an onslaught of new responsibilities in the areas of worker health and safety, worker's compensation, and environmental protection—adding substantially to their existing liabilities and compliance costs. Beginning with the passage of the Occupational Safety and Health Act (OSHA) in 1970, which identified logging as one of five target industries in need of substantial safety record improvement, loggers faced a wide array of new safety regulations.[193] Given their "independent" status, moreover, there was no one with which to share the compliance costs associated with these liabilities. During this time, some southern states also began to revisit their existing worker's compensation laws, tightening what had been fairly loose requirements for logging contractors. Independent loggers now had to pay weekly for worker's compensation and liability insurance.[194] Finally, new environmental protection laws enacted in the 1970s also added to the liabilities of logging contractors. Logging practices had to be adjusted to meet regulations on water quality, land use, and endangered species—all of which added to the overall costs of harvesting timber.

For loggers, the upshot of these increased costs and liabilities was increased pressure to move wood—an imperative that often ran counter to such goals as improving safety and environmental performance. As one logger put it: "Your insurance, your workmen's comp—it'll eat you alive. My workmen's comp runs around six or seven hundred dollars a week. My liability and equipment insurance runs about four hundred dollars a week. So you see, you got to produce."[195] Given these constraints, it becomes easier to understand why some loggers have appeared so hostile to government regulations. If they can't move wood, they stand to lose their businesses.

Nowhere has this been more apparent than in the context of the Endangered Species Act (ESA).[196] Like the controversy in the Pacific Northwest over the spotted owl, the southern forests were the site of conflict between loggers and environmentalists over the fate of the red cockaded woodpecker.[197] The issue proved particularly complex in the South, moreover, because the vast majority of timberland was privately owned. During the 1990s, logging and timber industry advocates argued vigorously against federal efforts to regulate logging practices on private lands under section 9 of the ESA, which bans the "taking" of a listed species on public or private land. In *Babbitt v. Sweet Home Chapter of Communities for a Great Oregon*, a 1995 case involving logging and timber representatives from the Pacific Northwest and the South, the U.S. Supreme Court upheld the Department of the Interior's reading of section 9, thereby allowing the secretary of the interior to continue regulating activities on private land that "harm" an endangered species through "significant habitat modification or degradation."[198] The implications for loggers were clear. If an endangered species was found on a tract of timber, logging operations would have to stop until federal authorities gave the go-ahead. For a highly leveraged logger, any such downtime could spell disaster. Needless to say, loggers were angry and frustrated. As one put it: "The environment—we needed to have done something. But when they talking 'bout the woodpecker and things like that—hell with it—I mean bull crap on that. We had a tract of timber that one was spotted on . . . We shot it so they wouldn't find it. They talkin' about extinct, they fixin' to be extinct because the loggers are getting rid of 'em."[199]

In addition to the ESA and other environmental regulations, the industry's leading firms, through the American Forest & Paper Association (AF&PA), also began adding to loggers' environmental responsibilities in the 1990s through various programs such as the Sustainable Forestry Initiative (SFI).[200] This ostensibly voluntary initiative called for limiting the size of clear-cuts, improved protection of water quality and critical habitats, protection of biodiversity, enhanced logger education and training, and outreach to nonindustrial private landowners. Adherence to the principles of SFI became a requirement for AF&PA membership, and several companies dropped out of the trade association in protest.[201] For those companies that stayed in, SFI represented a departure from past environmental practices and a significant effort to bolster their public image. For loggers, however, the bottom line was more hassle and more cost. Although some companies and universities began working with loggers to establish logger certification programs, most loggers were effectively on their own. Given their economic constraints, few had the requisite financial and technical resources necessary to practice "sustainable forestry."

Even as liabilities and compliance costs increased for loggers, logging continued to be one of the most dangerous occupations in the United States. In 1974, four years after the passage of OSHA, an American Pulpwood Association report found that the disabling injury rate for pulpwood logging was the highest it had been since 1951, ranking second in the United States after the meat products industry. Moreover, a logger injured on the job in 1974 was twice as likely to suffer a fatality than the average U.S. manufacturing worker. As a whole, logging led all other industries in terms of frequency and severity of occupational accidents.[202] Twenty years later, there had been little if any improvement in logging's relative position. Between 1992 and 1997, for example, logging had the highest rate of fatal occupational injuries in the country (over 128 deaths per 100,000 workers), and loggers faced a risk of fatal work injury that was approximately twenty-seven times greater than the average for all occupations. As a region, the South accounted for almost 54 percent of logging fatalities, despite containing less than 44 percent of loggers. North Carolina led all states, followed by Mississippi, Kentucky, Virginia, and Washington.[203] After more than two decades of mechanization and in spite of new safety regulations, southern woods workers continued to operate in a landscape of extreme occupational hazard.

Looking back, then, the contract-based system of wood procurement that emerged in the southern pulp and paper industry played a critical role in distributing the many risks and liabilities associated with logging. If the system worked primarily to the advantage of pulp and paper firms, this was because of the enormous imbalances in wealth and power that existed between them and those who logged their mills. Though loggers were and are ostensibly "free" to contract with the mills or their dealers, they have had little control over the terms of the relationship, and they have faced structural vulnerabilities that severely limit their ability to seek alternatives.

Nonetheless, it bears repeating that these firms did not simply impose their will on those they encountered when they moved into the region. They still had to coordinate with others to construct arrangements that fit with the distinctive institutions and social relations that marked the rural South. In many ways, the resulting system represented a remarkably effective way of tapping the social hierarchies and inequalities of the region to deal with the economic and logistical challenges of logging the mills. As suggested above, moreover, this was as much a process of active political and legal construction as it was of economic organization. That is, industry representatives actively shaped the institutional contours of the pulpwood procurement system through political and legal channels at the same time that industry managers were constructing these institutions on the ground.

Contracting, in sum, proved to be highly effective for the mills precisely because it was built on the structural vulnerabilities of loggers. Flexibility derived not from a choice among competing industrial paradigms but rather from the political-economic marginality of pulpwood producers. Instead of carrying large inventories, mills and dealers preferred to maintain excess logging capacity, which allowed them to meet their changing wood demands while maintaining price stability. Loggers, for their part, assumed responsibility for absorbing the fixed capital costs associated with this excess capacity and, because of their "independent" status, took on the many risks and liabilities associated with woods work. Their dependence on those who controlled access to wood greatly constrained their bargaining ability. As mechanization proceeded, their debt burdens increased, reinforcing existing economic dependencies.

Procurement managers played a key role in maintaining this relationship, and some were clearly aware of the situation facing loggers and their own power to affect a logger's livelihood. As one former procurement manager confessed,

> I personally put a man out of business because I wouldn't buy his wood. I couldn't buy his wood. I had encouraged him to get mechanized because I had been told to do that, and then we got into a period where we weren't running good, couldn't sell the product, and the poor guy went broke because he had done what I told him to do . . . Producers were expendable. I mean they were the buffer, and if you didn't need the buffer, he would go out of business, and when things picked up again you would just go out and get somebody else and go again. It just beat people

> to death. It was pretty bad. The early years—I wasn't very proud of what we were doing, but that was the way it happened.[204]

Viewed in this context, contract logging provided the mills with an institutional vehicle for accessing cheap, casual labor without taking on any of the burdens of equity or the costs and responsibilities associated with wage labor.

As to why these men continued to work as pulpwood producers in the face of such inequities and economic hardship, there is no single answer.[205] Given their indebtedness and the relation-specific nature of their assets (logging machines have very few alternative uses), most loggers found themselves in a situation with few viable exit strategies. In a classic case of what A. V. Chayanov, in his study of peasant households, referred to as self-exploitation, debt and financial insecurity worked to reproduce the contractual relationship.[206] Most loggers found themselves in a situation of "doing what it takes" to hold onto their operations. At a basic level, they were "bonded" to the dealers and the mills through their large investments in logging equipment, a situation that effectively transformed a short-term tract-to-tract contractual relationship into a long-term arrangement. Still, there were obviously limits to how far these loggers could be pushed. Indeed, even industry executives expressed concern about the situation facing loggers and whether there will be enough loggers in the future. As one executive put it, "It's a little bit scary to us frankly about where the next generation of logger is going to come from."[207]

Stepping back, then, it is clear that the contractual relations between southern pulpwood producers and those who have historically controlled their access to wood cannot be understood simply as discrete transactions between autonomous economic actors. Instead, they must be seen as institutional relationships deeply embedded within regional and local social networks and shaped by prevailing distributions of wealth and power. This is hardly a novel observation. As Emile Durkheim pointed out long ago, a contractual relationship implies much more than a voluntary arrangement between free individuals; rather, it depends on a whole structure of "non-contractual" norms and regulations, some of which could lead to or reinforce coercive and "unjust" contracts.[208] Likewise, Max Weber, in his sociology of law, emphasized the ways "freedom of contract" provided opportunities for some to exercise power over others as a result of preexisting endowments of property and resources.[209] Building on these insights, legal scholars have remarked on the pervasiveness of coercion and dependence in ostensibly "free" marketplace relationships and on the inability of "law" to provide ready-made remedies to ensure "actual liberty of contract." In an early article on contracts of adhesion, for example, Friedrich Kessler noted that "the law, by protecting the unequal distribution of property, does nothing to prevent freedom of contract from becoming a one-sided privilege."[210] From this perspective, the capacity for bargaining in a contractual relationship derives not from the "market value" of the goods or services for which the parties are transacting but rather from the various resources (legal, political, cultural, and economic) that each party brings to the relationship. As legal scholar Robert Lee Hale argued: "The market value of a property or a service is merely a measure of the strength of the bargaining power of the person who owns the one or renders the other, under the particular legal rights with which the law endows him, and the legal restrictions which it places on others. To hold unequal bargaining power economically justified, merely because each party obtains the market value of what he sells, no more and no less, is to beg the question."[211] By emphasizing the broader social conditions and norms that provide the basis for different parties' bargaining capacities, this relational perspective on contracting provides a far more realistic account of the southern pulpwood procurement system than a purely "transactional" approach.[212]

Southern pulpwood contracting, in this view, did not simply emerge to meet the procurement needs of pulp and paper mills. Indeed, the distinctive social, economic, and legal relationships that constituted the procurement system were actively constructed, drawing sustenance from the hierarchies of race and class that marked so much of the rural South and from the institutions that governed access to and control over land. As James Willard Hurst observed in his legal history of the Wisconsin lumber industry, "The terms of access to the land affect the practical power of decision which some men enjoy over the lives of others."[213] And so it has been with southern pulpwood producers. Although legally "free" to enter and exit the relationship, these men found themselves in a situation of extreme dependence on those who controlled their access to wood. At the same time, however, they were denied some of the most basic protections afforded to American workers. Their economic marginality was matched by their political invisibility. Not surprisingly, some loggers expressed a bitter pessimism about their prospects of ever being truly independent. As one logger concluded, "We're slaves. I'm controlled by that mill out there . . . just as sure as I'm sitting here. It's just like they're my God."[214]

Making Paper

> The fact is that the paper industry is run, more or less logically, to conform to such a completely cockeyed set of economics that most people outside the industry believe that all paper men are crazy. As individuals, they are far from crazy. But they are living with an industry that has often been very sick even when industry in general has been very well; that seems almost to have a mental illness of its own—a sort of manic-depressive cycle through which it inevitably runs, and by which prosperity immediately creates the first symptoms of disaster, and the depths of disaster provide the first means to a recovered prosperity. If that looks like a business cycle, it is no ordinary one. —Fortune, *1937*

In 1937, *Fortune* magazine published a series of articles on the American paper industry.[1] Intended primarily as a survey for the average investor, it offered a revealing portrait of the economic challenges then facing the industry. Although the depression had taken its toll on pulp and paper firms just as it had on so many others, the paper industry seemed to be suffering from a particularly intractable set of problems that transcended the downturn of the 1930s. Since the turn of the century, in fact, overcapacity, crowded markets, "ruinous" competition, and bad management had converged repeatedly to create situations that *Fortune* referred to as "industrial insanity." Papermaking was apparently plagued by a "cockeyed set of economics" that all too often translated into "dismal" returns for investors.[2] Why, the authors asked, was such a large and vital industry unable to operate on a sound and rational basis? What were the obstacles that prevented the industry from settling into a more stable organizational form? Why was it so hard to generate consistent, long-term profits in the paper business?

Part of the problem stemmed from the ways that pulp and paper firms managed their fixed capital requirements (both individually and collectively) and the implications of this for industry structure. While the high fixed costs associated with papermaking exerted a dominant influence on the make-up of the sector, several features of the business rendered the question of industrial structure somewhat more complicated. The challenges associated with managing fixed capital, in other words, did not reduce simply to questions of developing (or failing to develop) an optimal organizational form to capture the inherent economies of scale, scope, and speed in paper-making. As in all industries, the search for order and stability in pulp and paper had a distinctive set of meanings that shaped and was shaped by the larger political-economic context.[3] Like the challenges associated with constructing a new regime of forest management or developing a viable procurement system, the actual process of making paper involved significant institutional challenges that were as much about politics as economics. This chapter explores some of these issues, focusing specifically on how the search for order and stability affected the organization of production within firms, the structure of the industry, and the complex relationship between race and class in the recruitment and management of mill labor in the industry's southern production complexes. The point of departure is the problem of fixed capital.

Among the many problems that *Fortune* magazine's 1937 series highlighted, none proved more central than that of fixed capital. The massive investments embodied in pulp and paper facilities created a set of economic imperatives that drove endemic overproduction and the cyclical nightmares that ensued. On the surface, the problem seemed relatively straightforward. Because "the combination of tremendous fixed charges and a comparatively low-cost product demand[ed] volume and more volume,"[4] firms pushed to accelerate throughput in order to recover fixed costs more rapidly. In doing so, they inevitably overshot demand and exacerbated sectoral overproduction problems. Once prices began to fall, firms responded by slowing production, devaluing assets, and waiting for demand to catch up with supply—at which point they started expanding capacity again and the whole cycle began anew.

Although this appeared to be somewhat typical for a commodity industry, there were some peculiar features associated with the "economics of paper" that rendered the industry's problems more difficult. In addition to the very high rate of fixed capital investment relative to the value of the product, the industry also suffered from a "deadly slow" rate of capital turnover and from the fact that paper machines lasted "virtually forever." Old paper machines, in other words, rarely died but instead tended to be devalued, sold off, refurbished, and cranked back up as soon as prices rose high enough. At the same time, new machines, particularly those added during the 1920s, could achieve much faster speeds. Between 1919 and 1931, for example, the maximum speed of a paper machine increased from 735 feet per minute to 1,200 feet per minute—an annual rate of increase of more than 4 percent.[5] And as new machines made paper faster and faster, old machines continued to operate on the margins.

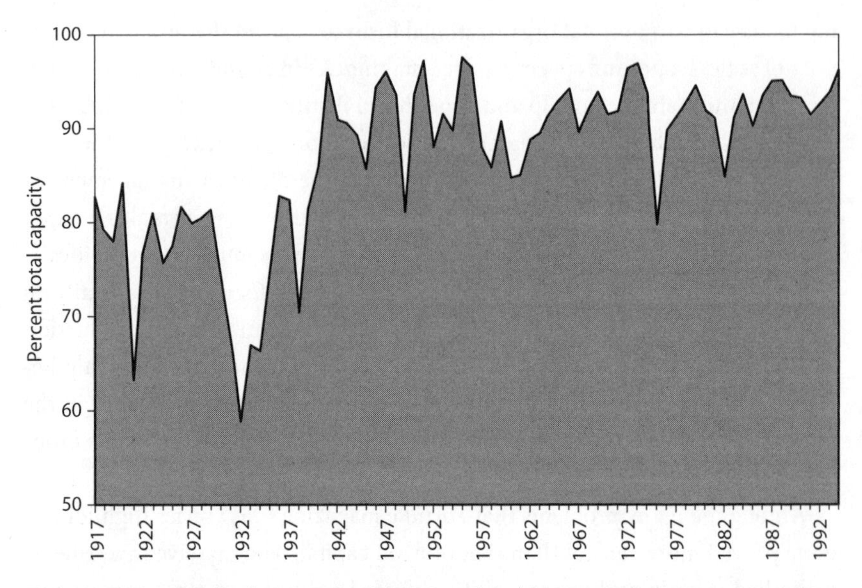

Figure 3.1. Capacity utilization for the U.S. pulp and paper industry, 1917–1994. Bureau of Census data; 1917–1959 data compiled by the American Paper and Pulp Association, *Statistics of Paper 1964* (New York, 1964); 1960–1994 data compiled in *Pulp & Paper: North American Fact Book 1996* (San Francisco: Miller Freeman, 1997).

As a result, the industry had difficulty operating at full capacity, despite rising consumption, throughout the early twentieth century. Between 1917 and 1939, production capacity never exceeded 84 percent and frequently fell well below this amount (figure 3.1). During this time, the industry's operating capacity averaged a paltry 75 percent. Even during the "bright boom years following 1926," according to *Fortune*, "paper swam against the current of prosperity, held back by price cutting and too much production . . . already groggy and insecure, an industrial invalid long before that memorable day in October 1929 when the economic world came to an end."[6]

When the depression did hit, capacity utilization plunged, sinking to an all-time low of 58 percent in 1932. By 1936, the situation had begun to improve, yet not until World War II did the industry achieve a capacity utilization rate of over 90 percent. Since that time, capacity utilization has been higher on average than it was during the prewar decades, though it has continued to fluctuate, sometimes dramatically (see figure 3.1).

Vertical integration, in and of itself, did little to stem these problems of overcapacity and "ruinous competition." Since the end of the nineteenth century, new chemical pulping processes along with the increasing speed of paper machines had generated strong incentives for firms to combine various aspects of pulp and paper production under a central administrative structure. By the 1930s, some 80 percent of pulp production took place in a single organized structure.[7] Moreover, many of the leading firms had integrated backward into timberland ownership and forward into distribution and marketing.[8] If anything, the large investments in dedicated production systems and relation-specific assets embodied in such a strategy merely increased the incentives to "run full" and maintain high-volume throughput. In the absence of significant first-mover advantages and structural barriers to entry, vertical integration could not serve to control competition and manage sectoral production problems.

Efforts to create a more stable industrial structure through horizontal combination also failed to solve these problems. Despite significant consolidation during the "great merger movement" at the turn of the century, large combinations such as International Paper and Union Bag were unable to hold on to their respective market shares and control the "destructive competition" that drove the industry's business cycle.[9] In contrast to their counterparts in steel and oil, paper men failed to maintain a level of concentration needed to insure price stability. By the 1930s, for example, Union Bag's share of the paper bag market had fallen to around 15 percent, down from almost 80 percent in 1899. Likewise, International Paper, which controlled some 70 percent of domestic newsprint capacity in 1900, saw its market share drop to less than 15 percent by 1920.[10] Though large firms such as International Paper sometimes acted as price leaders in particular grades, they were unable to check price competition effectively for any extended period of time. Contrary to received wisdom, the industry never really exhibited a stable pattern of price leadership and oligopolistic competition over the long term.[11] In spite of having one of the highest capital-to-output ratios in U.S. manufacturing, it persisted as one of the least consolidated sectors of heavy industry. The obvious question is why.

Part of the reason derived from the long-lived nature of paper machines and their ability to shift from one grade to another. This capacity for grade shifting—yet another example of flexibility within the paper industry—allowed mills and firms to move in and out of different markets depending on prices, thus undermining attempts to control competition and exert market power. As *Fortune* magazine noted, the "flexibility of paper machinery" acted as "the great price leveler" in the industry.[12] Because paper was not a true simple commodity but rather a spectrum of commodity variants, the flexibility of paper machines made it virtually impossible to defend market niches. Consequently, even the largest firms had trouble exerting market power in any particular grade.

Given the difficulties in establishing a stable oligopolistic structure with dominant firm pricing, pulp and paper firms sometimes succumbed to the temptation of price-fixing as an alternative way of controlling competition. During the 1900s and 1910s, for example, a series of investigations brought by the Department of Justice and the Federal Trade Commission (FTC) ended in consent decrees, fines, and pleas of nolo contendere by manufacturers in several branches of the industry. During the late 1930s, both newsprint and Kraft paper manufacturers were indicted on price-fixing charges.[13] Some of these episodes became the subject of congressional investigations. Between 1904 and 1951, Congress investigated newsprint manufacturers on price-fixing charges on twenty-two separate occasions.[14] By the 1970s, renewed investigations of price fixing in the industry led to several convictions and jail terms for executives, more than $500 million in fines, and the dubious distinction of being labeled the "nation's biggest price fixer" in the popular press.[15]

The challenges associated with fixed capital in the industry and the concomitant search for order and stability also had implications for industrial relations and mill labor. Compared to workers in other U.S. manufacturing sectors, pulp and paper workers unionized early, and most firms followed leaders such as International Paper and did not offer any real resistance to collective bargaining. Given the industry's capital intensity and the need to maintain high-volume throughput, executives and managers were more than willing to trade high wages and union benefits for a stable, amicable labor force. Commitment to a Fordist model of labor relations ran deep in the industry. By the late 1930s, in the wake of the National Labor Relations Act, virtually all of the pulp and paper mills in the country (even in the South) had been unionized.

Thus, as the industry expanded in the South, it brought with it elements of a national industrial culture that were somewhat foreign to the region. Unlike many other southern industries, most notably textiles, the vast majority of pulp and paper mills operated as union shops from the beginning. Those lucky enough to get a job at a new mill received high wages, substantial benefits, and union representation—a striking contrast to the labor situation in the woods where wages tended to be very low, benefits scarce, and unions nonexistent. Still, the mills' occupational structure also reflected the racial cleavages that marked southern labor markets and society during the interwar and postwar decades. Many of the most lucrative mill jobs were reserved exclusively for whites, particularly when the industry was rapidly expanding.[16] Throughout the 1940s and 1950s, for example, white-dominated union locals, often with cooperation from management, erected systematic barriers to the hiring of blacks for the best jobs, highlighting how a particular model of industrial relations developed outside the South came to reflect the realities of racial discrimination within the region. Not surprisingly, jobs held by blacks tended to be dirtier, lower paying, and more physically demanding. With increased mechanization of certain mill operations during the 1950s and 1960s, many of these "black" jobs were eliminated. As a result, the proportion of blacks in the industry declined sharply from more than 20 percent in 1950 to less than 14 percent in 1960.[17] In fact, it was not until the late 1960s, after increased activism among black workers and their advocates led to a series of high-profile employment discrimination cases against several mills and union locals, that many of the previously "white-only" mill jobs became available to blacks and other minorities.[18]

By this time, however, the expansionary phase of job creation in the southern pulp and paper industry had passed. Beginning in the 1970s, employment growth in the mills slowed considerably, despite ongoing increases in production. Much of this stemmed from the use of automated systems and increased mechanization.[19] The resulting increases in labor productivity (output per man hour increased by more than 90 percent between 1970 and 1990) led to substantial reductions in employment at many mills.[20]

During the 1970s, moreover, challenges emerged that threatened the profitability of the industry. Environmental concerns created a whole host of pollution-control regulations for the industry that diverted capital from other uses (see chapter 4). At the same time, the general decline in economic growth and productivity that began in many industrialized countries during this period created a harsh macroeconomic environment for U.S. manufacturers. As the expanding markets and rising productivity of the so-called golden years of the post–World War II boom seemed to be coming to an end, pulp and paper companies found themselves in a very mature and crowded market. The old problems of excess capacity and poor returns reasserted themselves with a vengeance. Meanwhile, overseas markets, which had become an increasingly important source of expansion for the industry as the domestic market matured, also came under attack from low-cost producers in the southern hemisphere. All in all, American papermakers found themselves in the 1970s and 1980s facing a situation of heightened uncertainty.

Efforts to reimpose order and stability proceeded in several directions. As noted, during the 1970s, some firms in the industry engaged in price-fixing schemes. In the ensuing decades, moreover, leading firms also sought to deal with the industry's problems through consolidation. Growth through acquisition rather than through capacity expansion seemed to offer a possible solution to the problems of the past. But the great rationalization that so many hoped for did not materialize. Consequently, the industry badly underperformed relative to similar industries through the end of the twentieth century.[21]

As in other mass-production industries, pulp and paper firms had to confront the sometimes painful implications of their huge investments in dedicated production systems. Their search for order and stability, individually and collectively, constrained their ability to deal with certain kinds of problems. Making paper, in short, involved a host of political and economic challenges that had implications for the organization and evolution of the industrial system that stretched well beyond the boundaries of any particular firm and the sector as a whole.

Making Paper

> Paper machines have an unusually long life. The essential design of the two types in common use has not been changed since their invention nearly a century and a half ago, although their speed, size, and productivity have been greatly increased. Since investment in a paper machine is large, the machine is not quickly scrapped. During depressions many are sold at bankrupt prices and thereafter reappear as strong competitors to expensive new equipment. This situation is further exacerbated by the fact that machines used to produce one type of paper may with small additional investment and relatively minor adjustments be diverted to the production of other types of paper . . . The result has been excess capacity in every branch of the industry. —*John Guthrie, 1946*[22]

Although the art of making paper has been practiced for at least 2,000 years,[23] modern paper manufacturing dates from the nineteenth century and was made possible by two crucial innovations: the development of the Fourdrinier paper machine and the successful application of chemical-based pulping. Developed in the early 1800s, the Fourdrinier machine revolutionized papermaking from a labor-intensive "craft" process to one that could produce long sheets of paper having a highly uniform character with much less labor.[24] Similarly, chemical pulping, first introduced with the soda process in the 1850s, allowed wood fiber to be used as a feedstock on a significant scale and across a range of different paper grades. Starting in the 1870s, the sulfite and then the "Kraft" or sulfate pulping processes were developed and refined, both of which were well suited to long-fibered wood such as pine. The Kraft process in particular produced a very strong paper (*Kraft* is German for "strong") and allowed for greater speed and versatility in paper manufacturing. In 1909, the Roanoke Rapids Manufacturing

Company of North Carolina became the first U.S. mill to produce Kraft paper when it successfully reduced southern pine. With the advent of multistage bleaching in the 1930s, the Kraft process soon dominated both bleached and unbleached pulp and paper production in North America.[25]

When combined with the Fourdrinier paper machine, chemical pulping processes provided the technological foundation for the modern pulp and paper industry. Although there have been countless refinements through the years, mills have continued to employ the same basic processes developed during the nineteenth century. Wood chips are cooked in "digesters" in a chemical solution that separates the cellulose from the lignins or natural glues that give wood its strength. The pulp is then washed (and sometimes bleached) before being spread out over the thin wire net of the Fourdrinier machine and run through a series of rollers and dryers in order to remove the remaining water and press the fibers together into a roll of paper. In terms of basic technology, the industry has essentially followed a path of incremental or evolutionary change since the nineteenth century.[26]

What did change, however, was the scale and speed of these technologies. During the first three decades of the twentieth century, several key process innovations dramatically increased the speed of paper machines. Most significant was the application of electric power and, in particular, the use of sectional electric drives with automatic speed regulators that could be used to adjust the speed of separate sections of the Fourdrinier. Complementary innovations such as friction-reducing ball bearings, dynamic balancing, lighter table rolls, and improved suction allowed for further increases in speed and width. The overall result was substantial growth in paper machine capacity. Between 1899 and 1921, for example, average machine capacity increased by about 150 percent.[27] The capacity of pulping digesters grew accordingly, especially in the Kraft segment of the industry, where large-scale pulping operations maximized the economies associated with chemical recovery efforts.[28]

As pulping capacities grew and paper machines increased in speed, firms built large integrated complexes that allowed pulp digesters to feed directly into paper machines. When compared to the small paper mills of the late nineteenth century, these integrated pulp and paper complexes were gigantic, representing investments in fixed capital and appetites for raw materials that would have been inconceivable at the beginning of the century. By the 1930s, a new complex could easily cost $30 million (not including timberlands).

In the eyes of paper executives, profitability thus depended in large part on maintaining a high rate of throughput, what Alfred Chandler referred to as "economies of speed."[29] High fixed costs and the "relation specific" nature of pulp and paper complexes made increased velocity of throughput a key part of efforts to reduce unit costs and increase profitability, which created strong incentives for vertical integration. Joining pulp and paper mills in a single complex was merely the first step. Beginning around the turn of the century, leading firms also began integrating backward to secure access to raw materials, most conspicuously through their purchases of timberland. At the same time, firms also began to integrate forward into paper-converting operations and distribution channels in an attempt to secure and maintain markets for their products. By the early twentieth century, vertical integration had become the industry's dominant organizational strategy.

To say that there have been substantial technological and economic incentives for vertical integration in the pulp and paper industry, however, is not to imply that these have been sufficient to determine the structure of its leading firms. Nor does it suffice to argue that these vertically integrated firms represent efficient responses to transactions costs—that they somehow embody optimal solutions to the governance challenges associated with particular forms of economic activity.[30] Such a view, in fact, leads to the rather banal conclusion that the mere existence of this organizational form implies that it is efficient. In the process, many of the interesting questions about production, industrial organization, and the historical evolution of industrial systems go unexplored.

Clearly, though, these firms had to meet certain "tests" in the marketplace. They had to develop structures and organizational capabilities that would maintain an orderly flow of materials through the production process as quickly as possible. Part of this involved building an organization capable of exploiting the efficiencies embodied in particular technologies. Within the pulp and paper industry, the coordination problems and transactions costs of relying on markets to supply raw materials and deliver finished products grew as the scale and speed of operations increased. Vertical integration served as an effective strategy aimed at locking in first-mover advantages by increasing efficiencies and lowering costs—at least for a while.

The actual process of integration, however, is more complex than the simple logic of transactions costs would suggest.[31] Efforts to build a competitive organization did not merely follow a well-worn path to optimality dictated by the distinctive economies embedded in particular technologies. Instead, they involved strategic choices in the face of all sorts of uncertainties. Shifting political environments, changing markets, new sources of raw materials, and new technologies all shaped and were shaped by the investment decisions of these firms.

Vertical integration provided one way of capturing certain economies given a particular set of constraints and opportunities. But it was by no means the only, or even the most efficient, strategy available. Indeed, the cost advantages of vertical integration often proved ephemeral in the face of technological change, new markets, and changing raw material supplies. As Philip Scranton and others have pointed out, there was no one best way of organizing a particular industry.[32]

Once the die was cast, though, there was a degree of institutional lock-in among major firms in the industry. While smaller, less integrated, "independent" firms operated on the margins, almost all of the leading firms of the early twentieth century pursued a strategy of vertical integration. By building paper machines and pulp mills as part of integrated production complexes and integrating backward into timberland ownership and control and forward into converting operations, they hoped to capture first-mover advantages and erect barriers to entry for new firms. In the process, they reinforced their imperative to run "full" in order to capture economies of speed.[33] Thus, while a vertically integrated production complex may have allowed for better coordination of material flows, it also deepened the overall investment in dedicated production systems and relation-specific assets, thereby adding to the pressures to make paper as quickly as possible. Although this appeared to be a perfectly rational move for individual firms, it sometimes led to severe problems of excess capacity for the sector as a whole.

Part of the problem stemmed from the long-lived nature of paper machines. As *Fortune* magazine explained:

> There seems to be nothing you can do about paper capacity. The big, heavy machines in which pulp miraculously turns to paper last virtually forever: once installed they are nearly as permanent as any machinery can be. That makes for a situation that impedes the reduction of used capacity in response to falling demand. For when paper companies founder on all sides, as they did during the depression, the machines remain—somebody buys them at bankrupt prices, and presently they start up again, strong competitors of new and more efficient mills because of their low fixed charges. Thus, in spite of constant technical improvements (higher speeds, greater widths, both raising the productivity of the individual machine), there is no apparent obsolescence in a paper mill.[34]

Grade shifting also extended the economic life of paper machines. Instead of shutting down slower, less efficient machines, firms could simply reconfigure them to make different grades of paper where the cost advantages of the most efficient machines were not as important. The general bias in the industry toward incremental process improvements rather than radical innovations also meant that older machines were often rebuilt rather than scrapped in order to take advantage of particular improvements. The overall result was a rather dated capital stock. To take one extreme example, a machine originally built in 1857 was still making paper in 1935. For the industry as whole, some 22 percent of paper capacity in 1935 had been installed before 1900.[35]

Several other salient features of pulp and paper economics compounded this extremely slow rate of obsolescence. The largely inelastic demand for most products exacerbated market adjustment problems and intensified competition. For most grades of paper, moreover, there was very little product differentiation.[36] Paper was a commodity like oil or steel, and there was little difference between the products made by individual firms. As a result, there were few alternatives to competing on price—a fact of life in a mass-production industry based on a low-cost, nondurable, undifferentiated product. When combined with high fixed costs and dedicated machines, these conditions yielded an underlying propensity for excess capacity and intense competition.

During the 1920s and 1930s, these problems of excess capacity became acute and the industry sank into a "sick" state. Beginning in the 1920s, the substantial increases in the speed of paper machines started to outstrip existing demand and forced firms to slash prices. With the onset of depression, things only got worse. Facing massive reductions in demand, firms had little choice but to cut production substantially. In 1932, capacity utilization fell to an all-time low of 58 percent (see figure 3.1). Not until 1936 did production recover to predepression levels.[37]

Meanwhile, pulpwood supplies in the Northeast were dwindling, and many of the larger firms found themselves increasingly dependent on Canada and Scandinavia for fiber. Union Bag imported pulp from Sweden in the 1920s to service its bag plants in New York and other northern states. Facing a situation of excess capacity and declining profits on the one hand and increased dependence on raw material imports on the other, leading firms in the industry eagerly sought to develop a new production base.

It was against this background that firms accelerated their efforts to move South. Given the timing of this locational shift, the new southern mills were able to take advantage of recent process innovations that allowed for substantial increases in the scale and speed of operations. At the same time, these mills also benefited as new markets for Kraft paper products began to open up. In 1914, the Interstate Commerce Commission prohibited railroad discrimination against the use of cardboard boxes for transport, thereby opening up a vast potential market for Kraft linerboard. During World War I, Kraft wrapping paper and linerboard found a variety of new uses that persisted into the postwar period.[38] Finally, the rise of retail supermarkets during the interwar years created substantial markets for brown paper bags and boxes. For companies such as Union Bag, the timing could not have been better. By the end of the 1940s, with the supermarket revolution well under way, cardboard boxes and brown paper bags had become two of the more ubiquitous accoutrements of American consumer culture.

Constructing these southern production complexes, however, did not bring an end to the challenges of managing fixed capital and maintaining velocity of throughput. Quite the contrary. Given their size, the southern mills faced an imperative to keep their fixed capital in motion that exceeded that facing northern mills. For these new mills, speed was everything. International Paper, for example, rebuilt the paper machine at its Panama City, Florida, plant in 1931, after only two months of operation, in order to take advantage of process innovations and capture more economies of speed. For company president Richard J. Cullen, accelerating throughput represented the key to competitiveness in the industry. "His obsession is speed and rightly so," wrote *Fortune* magazine. "For Mr. Cullen, a paper technician if ever there was one, knows that every extra ton he can force through his machines without adding to his fixed charges for plant facilities comes to him virtually without cost."[39]

Like their counterparts at International Paper, Union Bag managers and technicians worked constantly to reduce the time it took to transform timber into paper bags. On the basis of process refinements and a vertically integrated production structure, the company soon achieved one of the fastest rates of throughput in the industry. According to *Fortune*, "Union's superior speed means that at top tempo a log leaving the storage yard is converted into finished paper within twelve hours. The process goes on for twenty-four hours of the day. The paper is then taken to the bag factories. Every second of the working day, Union makes 480 bags—28 million bags per day."[40] And this was just the beginning.

The capacity of southern mills increased steadily throughout the post–World War II period. Total regional pulping capacity grew from slightly less than 30,000 tons per day in 1953 to roughly 140,000 tons per day in 1996, an increase of more than 360 percent (figure 3.2). During this same period, average mill capacity increased by some 180 percent, from less than 500 tons per day to almost 1,400.

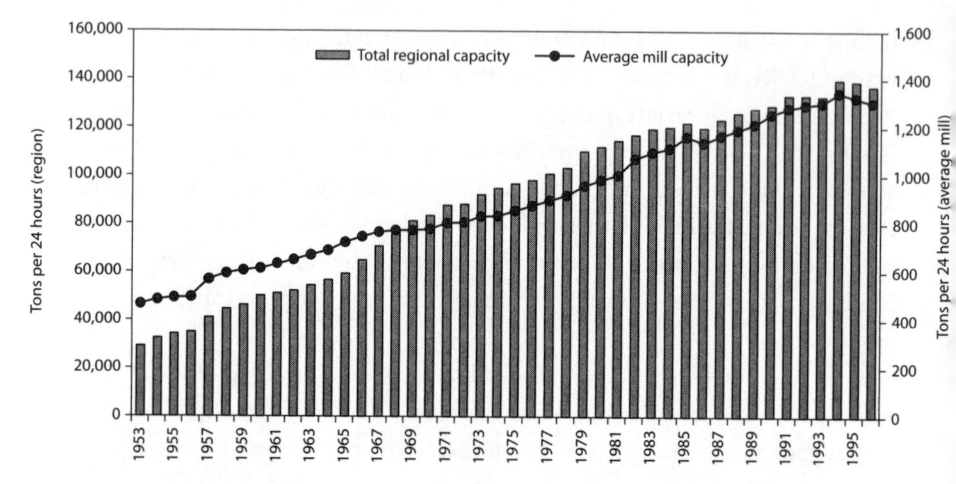

Figure 3.2. Total and average southern mill pulping capacity, 1953–1996. Tony G. Johnson, *Trends in Southern Pulpwood Production, 1953–1993*, Resource Bulletin no. SRS-3 (Asheville, NC: Southern Research Station, 1996), 6 table 4; Tony G. Johnson and Michael Howell, *Southern Pulpwood Production, 1994*, Resource Bulletin no. SRS-1 (Asheville, NC: Southern Research Station, 1996), 8; Tony G. Johnson and Carolyn D. Stephenson, *Southern Pulpwood Production, 1995*, Resource Bulletin no. SRS-8 (Asheville, NC: Southern Research Station, 1996), 7; Tony G. Johnson and Carolyn D. Stephenson, *Southern Pulpwood Production, 1996*, Resource Bulletin no. SRS-21 (Asheville, NC: Southern Research Station, 1997), 8.

Some of these southern mills attained truly gargantuan proportions. Union Camp's Savannah complex, which began operations at a capacity of 150 tons per day, was operating at an astonishing 3,500 tons per day by the 1990s. From 1938 through the end of the century, this mill was the largest pulp and paper complex in the world, consuming a quantity of raw materials that is difficult to comprehend. About 170 miles up the coast, the International Paper mill at Georgetown, South Carolina was the world's largest producer of Kraft linerboard for cardboard boxes. At midcentury, the famous Number 3 linerboard machine was producing a strip of Kraft linerboard 196 inches wide and one mile long every four minutes (figure 3.3). One day's production would cover a highway 16 feet wide from Mobile, Alabama, to Atlanta, Georgia.[41]

If anything, therefore, the throughput imperatives facing pulp and paper firms only increased with the shift in production to the South. As capital intensity rose with the push to build ever larger mills, managers found themselves under more and more pressure to increase the volume of production per unit of time. These managers were keenly aware that the constant use of existing capi-

Figure 3.3. The Fourdrinier section of the famous Number 3 linerboard machine at International Paper's Georgetown, South Carolina, mill, circa 1948. Photograph by Gabriel Benzur, courtesy of Gabriel Benzur Photography LLC.

tal was the way to drive down unit costs. Solving the industry's capacity problems thus required a level of coordination and collective action that far transcended the capabilities of any single firm.

Ruinous Competition

> Conspicuous among a long list of industries in which combination for market control has been repeatedly attempted and has frequently proven a failure is the paper industry. Hardly a branch of this industry has escaped a pooling agreement or a consolidation, yet in not a single instance has a monopolistic position been successfully maintained. The vitality and efficiency of independent producers,

> many of whom entered the field after combination had been effected, have prevented huge unwieldy consolidations or loose secret understandings from dominating the markets for any long time to the advantage of insiders and the corresponding disadvantage of the public. —*Myron R. Watkins, 1927*[42]

In their ongoing search for order and stability, firms in the paper industry, like those in other mass-production industries, attempted various forms of horizontal combination and coordination during the late nineteenth and early twentieth centuries. As Myron Watkins observed in the late 1920s, these efforts rarely worked for any extended period of time. In fact, notwithstanding the emergence of major consolidations such as International Paper and Union Bag during the great merger movement at end of the nineteenth century, paper companies repeatedly failed to create a stable oligopoly with dominant firm pricing. In some grades, such as newsprint, such a structure persisted for short periods of time. On the whole, though, the long-lived, "flexible" nature of paper machines allowed producers to move in and out of particular markets as prices dictated, undermining any attempt to establish a stable pattern of oligopolistic pricing.[43]

These problems of industry structure first appeared in the late nineteenth century when American papermakers began adopting modern pulp and papermaking methods. As paper machine speeds increased and pulping capacities grew, mills combining pulp and papermaking sprung up throughout the Northeast, where seemingly abundant timber resources and waterpower provided ample opportunities for expansion. By the early 1890s, several vertically integrated single-plant firms competed with one another in various branches of the industry. None of these firms, however, could secure significant competitive advantage. In terms of technology, they were evenly matched. In terms of raw materials and waterpower, there were too many good sites for any one firm to control. Because they produced a relatively homogenous set of products, resorting to product differentiation as a competitive weapon was also limited. Finally, given their large (and growing) investments in fixed capital, these firms had an incentive to "run full" in order to recover their costs. High-volume, continuous-flow production was considered the key to profitability.[44] Even as prices fell, most firms continued to run their machines in an attempt to cover variable costs and hold on to market share—a move that exacerbated overproduction and forced prices even lower. The more normal collusive arrangements of the day, such as pooling associations or gentlemen's agreements, proved "utterly incapable" of controlling competition. By the end of the decade, consolidation appeared to offer the only possibility of relief.[45]

Beginning in the late 1890s, a wave of consolidations swept through the various branches of the paper industry. Chief among these was the formation of International Paper in 1898, combining thirty-four of the largest newsprint mills in five northeastern states. Backed by massive landholdings acquired during the previous decades (the new company reportedly owned 1 million acres in the United States and 1.6 million in Canada), International seemed poised to corner the newsprint market. In his first annual report, company president Hugh J. Chisholm remarked that "competition was not of a serious nature." At the time of its consolidation, the company controlled some 90 percent of newsprint production in the East and 70 percent of the total output of newsprint in the United States. According to one historian, the formation of International Paper represented "a successful marriage of the Fourdrinier and the counting house; the future of the industry seemed to lie in this direction . . . The era of the big firm had arrived."[46]

The following year, the Union Bag Company, which had been started by a group of bagmakers as a tool to consolidate the early bag machine patents and license them to manufacturers, combined with several other bag manufacturers into a company that controlled eighteen paper mills, nine bag factories, several groundwood mills, waterpower facilities, and some 400 patents. In 1900, the new Union Bag and Paper Company sold some 4 billion bags, accounting for 80 percent of the U.S. market. Several months after the Union Bag consolidation, the American Writing Paper Company was incorporated. The new concern controlled thirty-two mills in five states. Then came the United States Envelope Company, incorporated in July 1899; the Union Waxed and Parchment Paper Company, incorporated in March; and, finally, the United Box Board and Paper Company, which combined some twenty-five companies in July 1902.[47]

By the early 1900s, every major branch of the paper industry had experienced significant consolidation. As in other industries, the underlying motive was to control severe price competition. The "basic" economic facts of high-volume, mass-production industries seemed to demand market power as a way of controlling competition. Consolidation apparently held the answer.[48]

But these vast consolidations also brought challenges. Chief among these was the issue of pricing. As in any oligopolistic market dominated by a single large producer, the so-called dominant firm had to determine what the "optimal" tradeoff was between price and market share. In the absence of significant cost advantages or barriers to entry, such a firm could only maintain a relatively high price level if it were willing to give up market share. Pushing prices up, in other words, would induce new firms to enter, thus sacrificing some of leader's market share. Keeping prices too low might mean forgoing some of the profits achievable to it. Under the strategy of dominant firm pricing, therefore, the leading firm would (in theory at least) trade market share for higher prices in a manner that maximized profit. In practice, of course, this entailed setting a price, allowing the smaller "independent" firms to sell as much as they wanted on the margins, and adjusting output accordingly.[49]

Maintaining profitability under such a structure proved to be no easy task, and in every branch of the industry the large consolidations formed during the great merger movement lost market share throughout the predepression decades. Some of the new combinations actually went under, and few were ever able to sustain a model of dominant firm pricing. In response, some manufacturers resorted to "pooling" associations, such as the Fibre and Manila Association or the American Paperboard Association, in an effort to manage competition. Most of these associations, however, were either dissolved or sharply constrained by the Federal Trade Commission (FTC).[50]

By 1920, then, the paper industry as a whole was decidedly less concentrated than it had been twenty years before. Newsprint production had shifted to Canada, forcing International Paper and other newsprint manufacturers to diversify into other grades. In the process, International became a major competitor to other large consolidations. Meanwhile, low-cost competitors taking advantage of faster paper machines entered all of the major branches of the industry, capturing market share from the larger firms. Price regulation, to the extent that it was even possible, took place less through dominant firm leadership than through the somewhat looser coordinating role of trade associations.[51]

In the end, the efforts of leading paper firms during the early twentieth century to develop an industry structure capable of controlling competition failed. Although the political and legal environment in the United States mitigated against combination and collusion in restraint of trade, the problem was more deep-seated. Because these consolidations proved to be no more efficient than their rivals, they quickly lost their positions of dominance and thus any ability to set prices. Barriers were not high enough to stop new firms from entering the market, particularly in the "downstream" converting sector. In addition, the grade-shifting capacity of paper machines made it difficult for one consolidation to capture any sort of monopoly rents by dominating a particular market. As one observer put it in 1940, "the paper industry is almost self-regulatory in regard to monopolistic attempts because of 'grade-shifting' . . . When a monopoly is attempted by a group of mills and the price is unduly advanced, it encourages 'grade shifting' and soon the would be monopolists find themselves in an oversupplied market."[52]

Consequently, problems of excess capacity continued to plague papermakers during the late 1920s and early 1930s. Several leading companies, including International Paper and Union Bag, entered bankruptcy. As in other industries, the massive deflation of the period called forth stronger forms of coordination and cooperation. The so-called Hooverian associationalism of the 1920s, a product of President Hoover's efforts to foster trade associations as instruments of industrial self-government, had clearly failed. Government intervention, it was argued, was necessary to put the ailing business system on more rational foundations. Through the National Industrial Recovery Act (NRA), the Roosevelt administration hoped to provide the institutional basis for a system of industrial governance capable of regulating prices and production.[53]

For the pulp and paper industry, the NRA experiment, despite its short life, provided a welcome departure from the instabilities of previous decades. Most firms had had their fill of "unregulated competition." Trade associations, though they had greatly improved the flow of information among competing firms and had sometimes even served as the basis for pooling efforts, generally proved unable to maintain order and stability for an extended period of time. In effect, the NRA represented what many paper executives had always wanted—stabilization of prices, rationalization of business conduct, the realization of a new industrial order. Cooperation would be the foundation for the so-called new competition. Witness the words of S. L. Willson, president of the American Paper and Pulp Association in 1933: "Most people today look upon the [National Industrial Recovery Act] as an activity that has brought many benefits which have developed from the cooperation of those engaged in the same competitive effort. We are on the road to much better business and under more favorable cooperative conditions than before."[54]

As for the actual NRA industry codes, three applied directly to the pulp and paper industry: a general code for the sector as a whole, a newsprint code, and a paperboard code. Written and administered largely by the American Paper and Pulp Association, each code contained general provisions regarding wages and hours, collective bargaining, standard cost methods, and guidelines for reporting of industry statistics. The central provision of the industry-wide code was an open-price filing plan whereby prices filed could not be below the "cost" of the filing company or the lowest price for the particular product or grade. Prices were to be set on the basis of geographic zones. In addition, the general code also contained provisions empowering the Paper Industry Authority to investigate filed prices; establish terms and conditions concerning sales to dealers; develop plans to bring about a "reasonable balance" between production and consumption of the industry's products; promote efforts "to conserve forest resources and bring about the sustained production thereof"; and place restrictions on the construction of new manufacturing capacity or the shifting of capacity to different grades. This last provision, known as the "birth control" provision, proved instrumental in stifling for a short time the southern expansion then under way. Firms interested in building mills in the South faced serious resistance from the American Paper and Pulp Association, which repeatedly invoked the provision to halt such "undesirable" expansion.[55]

In contrast to the general industry code, the newsprint code did not contain any serious price-regulating clauses but instead focused specifically on curtailing production. In particular, it prohibited newsprint paper machines from operating on Sundays. The code also empowered the Newsprint Authority to confer with the manufacturers and their customers "in respect of stabilization of the industry and the elimination of unfair practices and destructive competitive prices."[56]

On the whole, the general paper industry code succeeded in limiting production in most grades to the orders received. As a result of the "birth control" provision, southern expansion was effectively put on hold. Paper prices began to rise. Between April 1933 and May 1935, for example, the price index for paper products increased by some 14 percent.[57] While part of this increase clearly derived from the general economic upturn of 1934 and 1935, the NRA paper code also played a role by "rationalizing" the capacity situation in the industry. In the case of newsprint, however, efforts by manufacturers to raise prices above the depression low of forty dollars per ton failed. The primary reasons were lack of cooperation by Canadian producers and the political power of the newspaper publishers.[58]

In the wake of the Supreme Court's 1935 *Schecter* decision invalidating the NRA,[59] paper manufacturers attempted to maintain the relative stability and price increases achieved under the codes through voluntary compliance with price zoning and open-price filing. Although the open-price filing system soon fell out of practice, most manufacturers continued to adhere to the price zones. established under the NRA. Moreover, some firms also persisted in following the pattern of closer cooperation permitted (and even encouraged) under the NRA, particularly in the context of discussing common problems and developing alternative ways to control competition.[60] By fostering a culture of cooperation within the industry that had been building since the 1920s, the NRA reinforced habits and preferences favoring cooperation over the "cutthroat competition" of the past.

Still, the political landscape was changing. By the second half of the 1930s, advocates of a more vigorous approach to the antitrust laws had gained prominent positions within the Roosevelt administration. Consequently, some of the practices that had been encouraged, and even required, under the NRA codes became the target of new antitrust enforcement activities by the FTC and the Department of Justice. In 1939, for example, the government indicted four newsprint manufacturers on the West Coast for unlawfully combining and conspiring to fix and control prices for newsprint. The defendants pleaded nolo contendere and paid nominal fines. During the same year, the Department of Justice also investigated the Kraft paper branch of the industry for the first time, handing out indictments to the Kraft Paper Association and some thirty manufacturers for allegedly conspiring to restrain trade and eliminate competition by curtailing output and apportioning production among members. In 1940, the defendants agreed to a consent decree enjoining them from further control or apportionment of production. These were not serious crimes, but they illustrated the extent to which the business cooperation encouraged under the NRA became a liability during the Second New Deal, as practices that had been permitted under the codes were branded as unduly restrictive of competition.[61]

During World War II, government-sponsored price regulation returned. In the spring of 1940, pulp imports from Scandinavia suddenly came to a halt, leading to an immediate increase in paper prices. That summer, the government received assurances from leading paper firms that "unjustifiable" price increases would be avoided. The following year, the newly formed Office of Price Administration and Civilian Supply obtained an agreement from the nine leading producers of Kraft paper to hold prices on standard grades steady for the remainder of the year. Then, in 1942, the government issued its General Maximum Price Regulation, fixing all paper prices not yet under control. In doing so, the Office of Price Administration accepted the price zones and quoting methods that had been customary in the industry during the 1930s. The restrained competition of the NRA era had returned, lasting through the war years.[62]

More important to the industry, however, was the tremendous increase in demand that the war effort created. During the early 1940s, papermakers dedicated some 30 percent of U.S. paper production to defense. By 1943, the government recognized paper as an "essential industry" and reduced many of the previous restrictions on it in hopes of increasing production. Facing shortages of labor and raw materials, paper firms struggled to keep up with soaring demand. German and Italian prisoners of war began working in the woods. Blacks moved into mill jobs previously reserved for whites. Women joined the labor force in large numbers. As for raw materials, restrictions on consumer demand and a vigorous wastepaper recycling effort made up for part of the shortfall. By the end of the war, the future looked bright for the paper industry. Paper and paperboard products had found all sorts of new uses in the American economy. Pent-up consumer demand was poised to burst forth into a postwar consumption binge. Rebuilding the shattered economies of Europe promised large export markets.[63] Paper firms enjoyed a level of stability, cooperation, and profitability previously unknown in the industry.

In general, then, the period stretching from the beginning of the New Deal through World War II saw the development of strong patterns of cooperation within the industry and between the industry and the federal government. With the war effort, paper firms appeared to leave the old problems of overcapacity behind, entering a phase of enhanced stability and improved performance. During the late 1940s and 1950s, for example, the industry's rate of return consistently exceeded the all-manufacturing average—a striking contrast to its performance during the prewar decades.[64] Capacity utilization, which had averaged only about 75 percent between 1920 and 1940, reached 90 percent or better during every decade after 1940 (see figure 3.1).[65]

Although the industry had by no means vanquished its business cycle, it did seem to be settling into relative stability. Part of this stemmed from the more general postwar economic expansion. By mid-century, paper and paper products had become ubiquitous in American consumer culture. Paper manufacturers came to count on stable, growing markets in what turned out to be a golden age for the mass-production paradigm. In addition, most firms had diversified their product lines by this time, limiting their exposure to any one grade and allowing for more efficient utilization of equipment. Finally, the industry's new economic geography had crystallized by the end of the 1940s with the ongoing shift in production to the American South. This particular "spatial fix" represented a strategic restructuring that had important implications for industrial stability.

With Kraft pulping rapidly emerging as the standard process used in the industry, the large integrated production complexes in the American South provided much of the foundation for the sector's postwar growth. By virtue of their size and level of integration, moreover, these complexes created barriers to entry far more formidable than those of earlier years. While no one firm or group of firms could ever hope to control the raw material supply in the region, and although all employed the same basic production technology, those firms that established southern operations early on did enjoy substantial first-mover advantages. They were able to consolidate a land base quickly and cheaply. They could choose the best sites for plant location—those with deep-water ports and plentiful supplies of timber and other key raw materials such as water and energy. And they gained considerable experience in constructing a viable organizational form, particularly in wood procurement and forest management.

Because the decision to move south required a great deal of capital and organizational resources, larger firms, which typically had better access to such resources, tended to dominate. Smaller, single-plant firms, though they did exist, were often absorbed into the larger multiplant operations of the leaders. By the 1950s, with the southern expansion in full force, the large integrated production complex pioneered by firms such as International Paper and Union Bag provided the model for success.

Consequently, the southern pulp and paper industry grew at a very healthy rate of 11 percent a year during the 1950s. Between 1950 and 1959, seventeen new mills came on line in the region, two more were under construction, and some forty major expansion projects were completed. In addition to new capacity expansion, some firms bought their way into the region through mergers and acquisitions. Some of these mergers were vertical in nature, aimed at securing a low-cost production base to feed converting plants outside the region. Others derived from a desire to diversify operations across product lines. Some targeted both objectives. To take one example, the 1955 merger of Gaylord Container into Crown Zellerbach gave the West Coast giant access to a southern production complex (Gaylord operated a major pulp and paper complex at Bogalusa, Louisiana) and improved product diversification. According to president J. D. Zellerbach, the merger would allow his company "to achieve the broad geographic coverage, product diversification, and manufacturing and merchandising efficiencies which are essential in our dynamic and expanding economy."[66] By the 1950s, few companies could afford to forego a presence in the South.

The decision to establish southern operations, of course, did not end the many existing and ongoing challenges facing the industry. The massive scale and deep level of integration manifest in the new production complexes generated throughput imperatives that greatly exceeded those at the smaller northern mills. Running "full" was more important than ever. Barriers to entry, though certainly higher than before, were not high enough to keep new firms out, especially in the downstream converting sectors. Grade shifting, while less pervasive than in the past, continued to provide a powerful corrective to any effort to maintain "artificially" high prices. In short, the basic economic facts remained.

Moving South, then, did not banish the industry's business cycle, but it did contribute to a broader process of postwar rationalization in the sector. By establishing substantial southern operations, firms such as International Paper and Union Bag transformed themselves into low-cost producers, forcing the rest of the industry to follow. In doing so, they upped the ante for other firms in the industry. Without a southern production base, a company simply would not be a major player. Because the "optimal" size of these new mills was so large, moreover, the ongoing shift to the South meant that production was increasingly concentrated in larger units. By virtue of their size relative to the market, actions by individual firms, and even individual mills, could have a significant effect on the output and pricing decisions of competing firms. In the perennial struggle over market share, competitive advantage continued to derive largely from the ability to move materials through the production process as quickly as possible—to make paper faster than one's rivals. Any disruption could mean the difference between profit and loss—a basic fact that had significant implications for labor relations.

Race and Labor in the Southern Paper Mills

> It follows from these economic facts that pulp and paper mills must emphasize maximum utilization of equipment . . . The entire process is geared to continuous production much like a chemical plant or a petroleum factory. Thus, mill management is always anxious to avoid labor strife and to work out accommodations to prevent such strife. The threat of strikes over civil rights issues, for example, has been viewed most seriously by the industry in recent years, and undoubtedly contributed both to its reluctance to change basic hiring and employment practices, and to legal and industrial relations maneuvering to insure government support and approval when change could not be avoided.
>
> —*Herbert R. Northup, 1970*[67]

Labor had long been a prominent concern for paper mill managers. Given the substantial investments in fixed capital embodied in a pulp and paper complex, maintaining order and stability among mill workers was of paramount importance. As the scale of operations grew, managers were willing to trade high wages and union recognition for smooth and continuous operation. Pulp and paper unions were among the few industrial unions to achieve voluntary employer recognition prior to the enactment of the Wagner Act in 1935, and union membership climbed accordingly. By mid-century, four out of five pulp and paper workers enjoyed the benefits of union representation.[68]

When the industry moved into the South, it thus brought with it a model of industrial relations that was foreign to the region. High wages and union representation were hardly the norm in southern industries. In contrast to textiles and other major New South industries, pulp and paper seemed to represent a bright spot for those seeking to organize workers.

Generally, the two unions representing pulp and paper mill workers—the United Papermakers & Paperworkers and the International Brotherhood of Pulp, Sulfite & Paper Mill Workers—bargained jointly with individual mills or companies.[69] In some cases, agreements extended beyond the bounds of a single plant, particularly for companies with more than one mill, where the multiplant bargaining unit often prevailed. In 1938, for example, the unions signed a master agreement with International Paper's Southern Kraft division covering all of the company's mills in the South. This agreement then served as the basis for agreements with other companies in the region. By the 1950s, close to a fourth of the industry's workforce was covered under multiplant or multicompany agreements.[70]

Thus, although the unions did not develop a uniform wage policy for the industry as a whole, they did develop consistent policies for all the mills within each major producing region. As illustrated by the Southern Kraft agreement, the strategy typically involved negotiations with leading companies in each region to determine the key terms to be used as the basis for agreements with other mills and companies. Because paper industry managers were generally quite concerned about wage competition, they often favored such multicompany, quasi-regional approaches to bargaining.[71]

For most of the post–New Deal period, labor relations in the industry were relatively cordial. While labor-management disputes occurred, the industry avoided major strikes. When strikes did occur, they usually involved a small subset of the unions' membership and were quite short. As long as markets expanded, employers could grant improvements in wages and working conditions without seriously jeopardizing profitability. And since most mills were located in small, isolated communities, prolonged strikes also represented threats to the entire community's livelihood, not just the workers' earnings. This proved particularly true in the South, where paper mill jobs often provided both the best jobs in the area and the community's major source of income.[72]

The unions, however, did not represent all workers equally. When the industry and the unions moved into the South during the interwar and postwar decades, they typically followed regional practices of segregation. While new paper mills did hire black workers, sometimes in large numbers, they confined them to the dirtier, lower-paying jobs. Union representation, to the extent that it existed for black workers, usually meant token membership in segregated locals. Although such practices were certainly not exclusive to the South, they predominated there because of the higher proportion of black workers and the legacy of Jim Crow. As the labor economist (and future secretary of labor under President Jimmy Carter) Ray Marshall put it in 1965, "The race problem in the paper industry is restricted mainly to the South because there are relatively few Negroes in this industry outside the South."[73]

As they established their southern mills, pulp and paper firms hired blacks in large numbers, but they took great care to segregate them in accordance with regional norms. To be sure, these jobs typically provided black workers with significantly higher wages and a more stable employment situation when compared to their alternatives.[74] But Jim Crow cast a long shadow across the paper industry just as it did in virtually all other aspects of southern life. Blacks worked in the mill yard unloading and hauling pulpwood. They performed menial tasks in production and provided custodial services throughout the mill complex. Higher-paying jobs operating paper machines, pulp digesters, power systems, and heavy equipment remained "white men's preserves." All mill facilities—cafeterias, locker rooms, and recreation areas—were strictly segregated.[75]

During World War II, widespread labor shortages provided new opportunities for black workers in the industry. In some mills, blacks moved into jobs previously reserved for whites. Black women also entered the industry for the first time, but these gains proved temporary. By the end of the war, the proportion of all black workers in pulp and paper fell back to prewar levels; the war years failed to generate any permanent change in the industry's occupational structure. If anything, by 1950, segregation within the mills—in both jobs and facilities—was more rigid than it had been before the war. Black workers found themselves in a vicious cycle of inadequate training, job discrimination, and active exclusion from union governance.[76]

The unions excluded black workers in two ways. First, they created segregated locals, which effectively denied black workers any real power base. Whites had full control over all bargaining with management. Blacks did little more than pay dues.[77] Second, the unions adopted, developed, and systematized preexisting seniority systems at the mills to block promotion of black workers into more desirable jobs. These seniority systems were based on well-defined lines of worker progression in which each mill worker was tied to a specific line of job progression. Seniority status determined eligibility for promotions within that line. This plant-specific seniority model effectively insulated "skilled" and "semi-skilled" job categories from direct competition. Workers in one mill could not compete for jobs in another mill, nor could unionized employers seek to induce workers to move by manipulating wage rates. Like seniority rules in other industries, the system proved well suited to the needs of the paper industry. It provided on-the-job training, ensuring that only experienced workers would deal directly with the expensive machinery, and it created a system to govern movement of employees within a plant. Typically, employees worked their way up separate lines of progression for the paper mill, the pulp mill, and the woodyard. Such lines were "long and narrow"; an employee who started on one line could not transfer his or her seniority status to another.[78]

Viewed in the abstract, there was nothing inherently racist or exclusionist about the seniority system at pulp and paper mills. When combined with discriminatory hiring practices, however, the seniority system acted as a powerful tool for reinforcing segregation and denying opportunities to black workers. Hired into the lower-paying jobs, blacks found themselves facing a seniority system that made it virtually impossible for them to advance into the middle and upper echelons of the mills' occupational structure. By perverting seniority rules, creating certain nonpromotable jobs, and establishing separate "black" lines of progression, union locals denied black workers access to the higher-paying and more highly skilled jobs that whites held. Refashioned in the service of Jim Crow, the seniority system functioned to "curb the competition of Negro labor." When added to all of the other obstacles facing blacks—housing, education, transportation, politics—these practices deepened the already severe structural vulnerabilities of black workers.[79]

By 1950, segregated union locals and separate lines of progression for black and white workers had become standard practice at southern pulp and paper mills. Although the Congress of Industrial Organization's Operation Dixie, along with the tightened labor markets associated with World War II, briefly held out the promise of a more egalitarian approach to employment practices, the forces of reaction proved far too powerful.[80] Management, meanwhile, seemed to be going along for the ride. Rather than upset their white union locals, mill managers and company executives preferred to maintain an arm's-length relationship with the labor situation. As long as the mills were running smoothly, they hardly cared to change things. Equal employment opportunities took a back seat to the imperative of maintaining high-volume throughput and preserving industrial stability.

During the 1950s and 1960s, as pulp and paper firms worked to improve the efficiency of operations, they eliminated some jobs previously reserved for blacks. Mechanized handling of pulpwood and finished paper rolls replaced much of the traditionally black "hand" labor in the woodyard and the paper mill. Substituting continuous digesters for batch pulp digesters also eliminated many of the "black" jobs in the pulp mill.[81] Overall, labor productivity in the industry grew by 40 percent between 1950 and 1960, and by 55 percent between 1960 and 1970—far more than in previous decades.[82] Black workers, concentrated in the more labor-intensive jobs, tended to bear a disproportionate burden of the displacement caused by the ongoing trend to mechanize and automate mill operations.[83]

Under the combined impact of technological change and racial discrimination, therefore, black workers' share of employment in the industry declined throughout the two decades following World War II. During the 1950s, black employment in the entire southern pulp and paper industry slid from more than 20 percent to less than 14 percent.[84] In South Carolina, black workers' share of nonsalaried paper industry employment declined steadily after World War II, from a wartime high of 48 percent in 1945 to 14 percent in 1965–66.[85]

But the relative position of black workers did begin to improve, albeit slightly, during the second half of the 1960s—raising questions about the timing of the change and its relationship to broader developments within southern politics and the regional economy (figure 3.4). Recognizing the difficulty involved in taking South Carolina as reflective of the experience of black pulp and paper workers throughout the region, the data support the conclusion that federal civil rights legislation and litigation during the mid-to-late 1960s helped open up the occupational structure of the southern pulp and paper industry. Here, in short, is an illustration of what Gavin Wright has referred to as "the civil rights revolution as economic history."[86]

When civil rights issues began to heat up during the 1950s, however, it was not at all clear how quickly and to what extent the position of black workers in the industry would improve. With the *Brown* decision and the "rise of massive resistance" across the South, the racial attitudes of many white union locals hardened. Some of these locals became power bases for KKK and white citizens' council activities. In towns such as Savannah, for example, these organizations

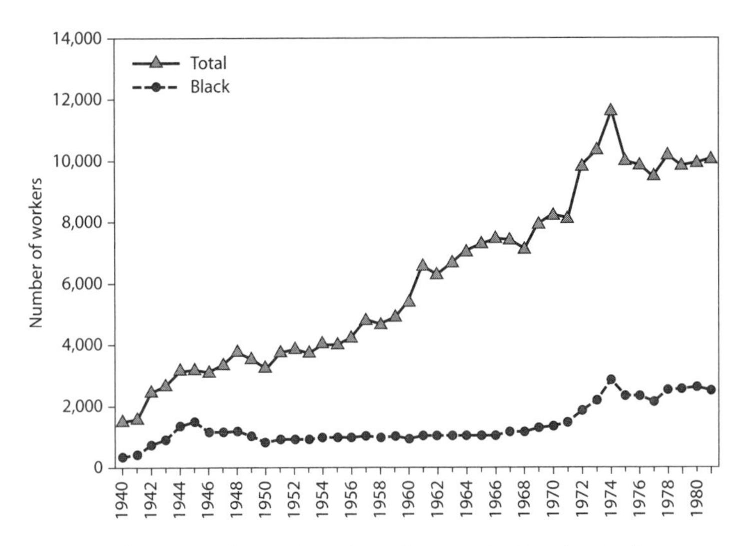

Figure 3.4. Employment of nonsalaried workers in the South Carolina paper and allied products industry, 1940–1981. SC Department of Labor, *Annual Report* (1941–1981).

often held meetings at union halls. In parts of Alabama and other states, the locals became "virtual extensions of segregationist organizations."[87]

Viewed from a broader perspective, however, the segregationist labor leaders were on a collision course with burgeoning civil rights activism and federal civil rights policies of the 1960s. In 1961, President Kennedy issued Executive Order 10925, prohibiting discriminatory hiring practices by federal contractors and creating the President's Committee on Equal Employment Opportunity (PCEEO) to investigate charges of discrimination. By this time, of course, sit-ins, marches, and boycotts were spreading across the South. At paper mills, black workers and their advocates began to demand equal job rights. National advocacy organizations such as the NAACP and CORE brought cases of employment discrimination in the paper industry to the attention of Kennedy's PCEEO. Formal investigations followed.[88]

The problem, though, could not be solved by simply integrating the formally segregated union locals and abolishing separate lines of job progression. While these were important goals, they would have to be approached carefully. In the absence of safeguards to protect the rights of black workers, the mere integration of white and black union locals would do little if anything to promote equal opportunity. Given that some white union locals had developed strong connections with the KKK and white citizens' councils, moreover, black workers could hardly be expected to be enthusiastic about such a plan. As for the abolition of separate lines of progression within the mills, this would do little to remedy the effects of past discrimination on a worker's place in the seniority system in the absence of other compensatory measures.[89] The challenges of integrating the pulp and paper industry were no less daunting than in other aspects of southern life.

Thus, when President Kennedy's PCEEO ordered the desegregation of paper industry employment in 1962, it elicited a diverse set of reactions from different groups in the industry. Black workers cheered the decision, but not without some trepidation given the resistance among whites. Company managers, fearing they would lose federal contracts yet hoping to avoid costly strikes, began to push very timidly for integration of union locals. For their part, the white union locals voiced bitter opposition, holding meetings and leading protest rallies sometimes in alliance with local KKK and white citizens' council groups. In Springhill, Louisiana, for example, nine white union locals from the massive International Paper mill convened a meeting in the fall of 1962 for "all those interested in maintaining our southern traditions" to discuss the effects of the executive order. Some 400 people attended, including union members, state legislators, and representatives from the local white citizens' council. Speaking to the group, one union leader charged that Executive Order 10925 was the "result of the Anti-God, Anti-Christ Kennedy's Inc. . . . They may cram it down our throats, but they can't make us agree and like it. Integration is now in Springhill. We don't want trouble and we won't agitate trouble, but we are against integration. Our local unions won't be responsible, but we'll stand behind our men."[90]

In the face of such reactions, black workers understandably lost some of their interest in an integrated union. While they clearly welcomed the opportunity to move into lines of progression that had previously been off-limits, they found progress to be painfully slow during the early 1960s. Company-sponsored tests, supposedly put in place to ensure that only qualified workers would get promotions, created serious obstacles. Because blacks had been denied access to the more lucrative lines of progression for so many years, they could not use the "seniority" status they had built up in the pulp mill or the woodyard as a basis for promotion in the paper mill. Consequently, for the next several years, few blacks moved into formally "white" jobs.[91] Facing a particularly virulent form of working-class racism and a timid industry management, black workers and their advocates pushed for stronger federal intervention.

It is important to remember in this respect that working-class racism was not an exclusively southern phenomenon. Such sentiments ran deep in organized labor. As Herbert Hill, labor secretary of the NAACP, noted in 1961: "Discriminatory racial practices by trade unions are not simply isolated or occasional expressions of local bias against colored workers, but rather, as the record indicates, a continuation of the institutionalized pattern of anti-Negro employment practices that is traditional with large sections of organized labor and industrial management. The pattern of union responsibility for job discrimination against Negroes is not limited to any one area of the country or to some few industries or union jurisdictions."[92] Indeed, despite the fact that the AFL and the CIO counted the elimination of racism as a major goal when the two organizations merged in 1955, union leadership failed to mount a systematic effort to counter discrimination and segregation within affiliated local unions during the late 1950s and early 1960s. Taking a "formal position" against segregated locals, the AFL-CIO refused to invoke sanctions against such locals or to establish a deadline for them to integrate.[93]

The exclusionary tendencies within organized labor took on a particular salience in the South during the post-*Brown* era. In the context of the broader civil rights movement, black workers' calls for equal opportunity represented a challenge not simply to industrial practices but also to the entire social structure. This was an issue that went far beyond segregated locals and separate lines of progression. Because many of these mills provided the major sources of employment and income for many small towns, job integration directly threatened the status quo. Highly paid white paper-machine operators made up the social and political elites in these company towns. If blacks began to move up the occupational ladder, they would gain an economic base for new political assertiveness. Integration of the mills thus constituted a deeper challenge to the dubious, increasingly tenuous, but still pervasive cultural hangover of an "organic society" wherein whites clung to a vision of their "superior" status as part of a natural order.[94]

But times were changing. During the early 1960s, some companies began to require the consolidation of segregated locals as a basis for contracts and to open up some jobs and job lines that had previously been off-limits for black workers. Blacks began "the slow process of working up the occupational hierarchies that were once white men's preserves."[95] By 1965, the tide had turned. In July of that year, Title VII of the Civil Rights Act of 1964 became effective, making it unlawful "to fail or refuse to hire or to discharge any individual, or otherwise to discriminate against any individual with respect to his compensation, terms, conditions, or privileges of employment, because of such individual's race, color, religion, sex, or national origin."[96] The newly created Equal Employment

Opportunity Commission (EEOC), charged with enforcing compliance with Title VII, was given broad authority to investigate employment discrimination complaints, facilitate voluntary compliance if possible, and bring civil actions if necessary. In the years that followed, the EEOC conducted investigations and initiated litigation against pulp and paper firms and specific union locals. By the end of the 1960s, the EEOC had brought charges or initiated litigation against virtually every paper mill in the South.[97]

Of course, the lasting effects of systematic discrimination could hardly be remedied overnight. As Herbert Northup put it, "Merely ending a discriminatory practice does not make whole those who have suffered discrimination nor does it alter for many years the pattern of employment which was created by years of discrimination."[98] Policies would need to be devised to compensate for the effects of past discrimination; the playing field would need to be leveled. The solution came largely through the litigation made possible by Title VII. The key case, *United States v. Local 189*, involved the Crown Zellerbach mill at Bogalusa, Louisiana, and the all-white union local of the United Papermakers and Paperworkers.[99]

Bogalusa was the quintessential company town. Founded in 1906 by the Goodyear brothers of Pennsylvania, it served as the headquarters for the Great Southern Lumber Company. During the 1910s, as lumber production peaked in the South, Great Southern's Bogalusa mill produced more lumber than any other mill in the world. In order to utilize the tremendous volume of sawmill residues, the company also built a pulp and paper mill. By the 1920s, Great Southern had initiated the South's first large-scale industrial timber plantation. A decade or so later, with the regional lumber industry in decline, the company closed its lumber mill but continued its pulp and paper operation.[100]

Throughout this time, Bogalusa bore the strong imprint of corporate paternalism. Not only did the company provide the vast majority of employment and income for the town's inhabitants, but it also built homes and schools for the workers and their children, provided for the local police force, owned the town's bank, supplied the residents with electricity, built hotels and a hospital, and operated a company store. Everything was rigidly segregated—from "cradle to coffin," as Judge John Minor Wisdom put it. Blacks, who accounted for as much as 40 percent of the town's population during the 1920s and 1930s, were expected to stay in their place.[101] In the view of company leaders, organized labor had no place in Bogalusa. Efforts to organize Great Southern in 1919 met with violent resistance. Later, when International Paper and other firms began recognizing the pulp and paper unions in the 1930s, Great Southern refused to follow. All of this changed in 1938, however, when Gaylord Container purchased Great Southern. Gaylord, like the other major companies, welcomed organized labor, and Bogalusa became a highly unionized town. Local 189 of the United Papermakers and Paperworkers Union was granted exclusive jurisdiction over bargaining at the mill. Black workers were confined to a segregated local. As was customary, separate job ladders were established for black and white workers, with the best jobs reserved exclusively for whites.[102]

In 1955, California-based Crown Zellerbach Corporation acquired Gaylord, continuing the previous company's cooperative stance regarding union relations and its predominantly hands-off approach to community affairs. Indeed, despite Crown's professed sympathies for the plight of black workers at the Bogalusa mill, company executives did very little to improve their position. Fearful of upsetting white workers and disrupting production, they considered the situation at their southern mill to be a local problem. Thus, when a new bag plant opened at the complex in 1957, whites received virtually all of the new jobs. Black men were confined to dirtier, lower-paying jobs in the woodyard and the pulp mill. And although the mill employed several hundred white women, not a single black woman worked there. Almost all of the "prosperity" that the Crown mill generated for Bogalusa went to the white population.[103]

By 1960, however, with the civil rights movement entering a more active phase, the racial climate in Bogalusa started to show signs of strain. When Crown initiated a mill modernization program that resulted in layoffs of both black and white workers, the union mounted a bitter, seven-month strike. Although the black local voted against the strike, they went along with it. In their view, such an action would only benefit the white workers. In reality, it did not benefit anyone and very likely worked to deepen the already substantial mistrust between the segregated locals. The company went ahead with its layoffs. Five hundred workers found themselves out of a job and were assigned to a reserve labor pool, known as the "Extra Board," which would be used to fill future vacancies. Suddenly, life in Bogalusa did not seem so secure.[104]

Meanwhile, black workers and their advocates began to issue louder and more sustained demands for an end to employment discrimination. President Kennedy's PCEEO pressured Crown and other southern mills to open up their job lines. The company, however, was reluctant to interfere with the situation in the wake of the strike. In 1963, it relented somewhat by adopting a policy of testing applicants (black or white) for job promotions, a move that ostensibly opened up previously segregated lines of progression to blacks with the appropriate qualifications. The following year, Crown also integrated the "Extra Board," allowing black workers to apply for job openings in the previously white lines of progression.[105]

As far as black workers were concerned, though, very little changed. Because job assignments were still based on job seniority in a particular line of progression rather than mill seniority, black workers had little chance of success in competing against white workers for job openings. By the end of 1964, more than a year after the changes had taken effect, only four blacks had moved into formerly "white" job lines.[106]

For white workers, even these rather timid efforts to integrate the mill represented a major threat. Laid-off and disaffected whites joined the burgeoning Klan movement in Bogalusa. According to one historian, the Crown mill "boasted a Klan unit of at least one hundred members and probably many more."[107] By 1965, the town as a whole apparently contained "the largest and most powerful Klan organization in Louisiana."[108] One writer from the *Nation* referred to Bogalusa as "Klantown, USA."[109]

Yet Bogalusa was also home to a courageous group of civil rights activists. Local men such as A. Z. Young and Robert Hicks, leaders of the black union local at Crown Zellerbach, emerged as formidable opponents of racial segregation and discrimination at the mill. Along with outside representatives from CORE and with protection from a local self-defense group known as the Deacons for Defense and Justice, these men mobilized resistance to the Klan's activities. Protests, demonstrations, and violence ensued. Some feared "a full-blown race war."[110]

With tension mounting, assistant attorney general John Doar visited Bogalusa and decided to make the town a test case for the 1964 Civil Rights Act. The Department of Justice quickly initiated separate lawsuits against several local police officials, three local restaurants, and the Klan. Though successful, the litigation proved slow and indecisive in leading to real results on the ground. Bogalusa, like so many other southern towns, was not going to change overnight.[111]

As for the situation at the Crown Zellerbach mill, local civil rights leaders finally succeeded in coaxing company officials to the negotiating table in July 1965. A. Z. Young and Robert Hicks, both longtime employees at the mill, and Bogalusa resident Gayle Jenkins urged the company to open up the mill to black men and women and institute promotion procedures based on length of service at the mill rather than in a particular job. The company agreed to merge all formerly segregated lines of progression but refused to replace the "job seniority" promotion system still in place, arguing that the union contracts then in force prohibited it from doing so.[112] Blacks who already worked at the mill could make no seniority claim in bidding for formerly white jobs. Their time at the mill counted for nothing as far as their chances of entering the previously "white" lines of progression were concerned.

In February 1967, the Office of Federal Contract Compliance began to pressure the company to amend the job seniority system, proposing in its place a system that took account of both job and mill seniority. Crown accepted the proposal and presented it to the union. Local 189 refused to go along. When the company went ahead with the new plan, the union voted to strike. At this point the Justice Department stepped in and filed a lawsuit against Local 189 in federal district court, asking the court to enjoin the strike because it was nothing more than an effort "to perpetuate a seniority and recall system which discriminates against Negro employees." The court granted the injunction the following day, one day before the strike was to begin. The Justice Department then asked the court to go beyond the system proposed by the Office of Federal Contract Compliance and prohibit any system based on job seniority. Such a system, the lawyers argued, should be replaced with a system based solely on mill seniority. The black local (189a) and two black employees then entered the case as plaintiffs on behalf of all black workers at the mill, charging employment discrimination under Title VII of the Civil Rights Act.[113] Acting on behalf of the black workers, the Justice Department brought suit against both the company and the union.[114]

After a three-day hearing, the court issued an order in March 1968 directing the defendants to abolish the "job seniority" system and replace it with one based on "mill seniority" for those black workers who had been hired prior to January 1966 (the date that the mill integrated all lines of progression). This proved to be a major breakthrough in bringing an end to the pervasive discrimination practiced in the southern pulp and paper industry. For the first time, black workers who had suffered all of the negative effects of past discrimination had a legal entitlement to their rightful place within the mill's employment structure. The jobs they would have occupied had they not been discriminated against were no longer out of reach.[115]

In July 1969, the U.S. Court of Appeals for the Fifth Circuit upheld the district court's ruling.[116] Writing for the court, Judge John Minor Wisdom noted that the problem raised by the case was

> one of the most perplexing issues troubling the courts under Title VII: how to reconcile equal employment opportunity *today* with seniority expectations based on *yesterday's* built-in racial discrimination. May an employer continue to award

> formerly "white jobs" on the basis of seniority attained in other formerly white jobs, or must the employer consider the employee's experience in formerly "Negro jobs" as an equivalent measure of seniority? We affirm the decision of the district court. We hold that Crown Zellerbach's job seniority system in effect at its Bogalusa Paper Mill prior to February 1, 1968, was unlawful because by carrying forward the effects of former discriminatory practices the system results in present and future discrimination. When a Negro applicant has the qualifications to handle a particular job, the Act requires that Negro seniority be equated with white seniority.[117]

The ruling went far beyond the Bogalusa mill. According to Herbert Northup, the Crown Zellerbach case proved instrumental in "altering the racial policies of the southern paper industry."[118]

Indeed, by the time of the decision, the OFCC and the EEOC had already been working to secure concessions from other major companies. Given its size and influence, International Paper was the obvious early target. In August 1968, the OFCC negotiated an agreement with the company and the paper unions in Jackson, Mississippi. Known as the Jackson Memorandum, the agreement applied to all of International's southern mills.[119] Now it was simply a matter of time. With *Local 189* affirming the need to institute measures to remedy the effects of past discrimination and using the Jackson Memorandum as a template, the federal government negotiated similar agreements with Union Camp, St. Regis, and Scott Paper Company.[120] For the more recalcitrant companies, and for those who dragged their feet in implementing the new system, litigation provided the persuasion needed to bring an end to discriminatory practices.[121]

Although it would take well into the 1970s to remedy the systematic discrimination of the past, black workers were finally able claim their "rightful place" in the occupational structure of southern pulp and paper industry.[122] It had been a long struggle, with local civil rights leaders such as A. Z. Young and Robert Hicks, national groups such as CORE, and government lawyers all playing key roles. From a structural perspective, the federal civil rights laws of the 1960s, particularly tools available under Title VII, proved instrumental in opening up opportunities for civil rights advocates to press their claims and transform their activism into mandatory protections sanctioned by law. Federal law, in this view, provided a crucial impetus for change in the industry.[123]

For black workers in the industry, the end of Jim Crow employment practices represented an important victory. But it was bittersweet. By the 1970s, the expansionary phase of job growth in the southern pulp and paper industry had passed. Total employment of nonsalaried production workers in the South Carolina paper and allied products sector peaked in 1974 (see figure 3.4). For the South as a whole, the number of production workers followed a similar path. The situation was especially bad at older plants such as Crown Zellerbach's Bogalusa mill, which reduced its total workforce by some two-thirds during the 1970s and 1980s.[124] Thus, while the employment share of black workers increased as blacks moved into jobs previously reserved for whites, their absolute numbers grew very little.

Part of the reason for this was that the southern pulp and paper industry had entered a more mature phase by the late 1960s. Although production continued to expand during the 1970s and 1980s, the rate of growth was far slower than in previous decades. Job creation suffered accordingly. A series of far-reaching organizational and technological changes in the mill labor process initiated during the 1970s and 1980s further reduced labor needs. These changes, which involved the widespread adoption of automated systems, dramatically altered the labor process, leading to the elimination of whole categories of "unskilled" and "semi-skilled" jobs, as well as to new occupational hierarchies in which process engineers gained significant authority and an increasing "abstraction" of work.[125]

Paper-machine operators, for example, could no longer use the craft knowledge they had acquired over many years on the job but instead had to adapt to the more abstract, technical requirements of automated systems. Meanwhile, the choices and decisions made by "non-expert" workers became increasingly standardized. Simply put, the shift to millwide information systems combined with widespread automation and extensive reliance on microelectronic controls transformed pulp and papermaking into "a major outpost of automated manufacturing."[126] In the process, the increased currency accorded to technical and engineering skills meant that black workers seeking to move up sometimes found themselves facing new barriers and obstacles that, at least formally, had less to do with their race than with their lack of previous education and training.

At the same time, the 1970s also witnessed the beginning of a decline in the strength of pulp and paper unions and the emergence of a significant nonunion sector in the industry. In the face of domestic stagflation, rising interest rates, rapidly increasing pollution control costs, and growing international competition, firms came under intense pressure to control costs and demonstrate a minimum level of financial performance. As the "golden years" of the postwar boom gave way to a much harsher and less stable macroeconomic environment, many of the industry's old problems returned. During the 1970s, for example, the return on net worth for the paper industry fell below the all-manufacturing average in every year but two. For companies seeking to placate the financial markets and avoid takeover attempts, cutting costs was critical.[127]

While many of the factors affecting costs during this time lay beyond the control of individual firms, labor costs did not. Taking aim at the wage increases built into union contracts, some of the major firms began to take a more aggressive, confrontational stance toward the labor unions.[128] Times were changing. The old Fordist model of labor relations that had dominated the industry during the post–New Deal period seemed to have lost its luster.

Pulp and paper unions thus found themselves under pressure on two fronts. First, at the handful of new greenfield mills built in the South after 1970, management typically pursued an explicit nonunion strategy. Part of the reason for this derived from the more general decline experienced by organized labor during this time. Unions were not as strong as they used to be. In addition, efforts within the industry to implement "flexible" team-oriented work regimes at their mills created a less-conducive environment for organized labor. The pulp and paper unions, which had spent decades establishing lines of progression within the context of a detailed, hierarchical division of labor, hardly welcomed the new principles of work organization that were sweeping through American manufacturing in the 1970s and 1980s. Flexibility and team work, in their view, had less to do with improved productivity than with job insecurity. Industry managers and executives, however, were eager to develop a less union-dominated work regime. As a result, not a single mill built in the South after 1977 was successfully organized by the unions. Major companies such as International Paper, Union Camp, Georgia-Pacific, Weyerhaeuser, and Mead, all of which had been completely organized at one time, each had by the 1990s at least one unorganized mill in the South.[129]

These new mills, of course, only accounted for a handful of pulp and paper workers in the region. Managers interested in extracting concessions from unions at existing mills would have to do so at the bargaining table. Starting in the late 1970s, some companies thus began taking a more confrontational "hard-bargaining" approach with the unions in which they demanded concessions, took strikes, and replaced workers. Labor-management relations deteriorated, reaching a low in 1987–88 when strikes and lockouts at several International Paper mills ended in union decertification. During the following decade, labor relations improved somewhat in the industry, but the unions never recovered the strength they had enjoyed in the past. Overall, between the early 1970s and the mid-1990s, union membership in the southern pulp and paper industry declined by more than 15 percent.[130]

When viewed against the general climate of labor relations in U.S. manufacturing at the end of the twentieth century, the general decline of union power and influence in the pulp and paper industry is not particularly surprising. The fortunes of workers and unions have typically reflected the relative stability of economic conditions in the sector, rising and falling with the industry's overall ability to grow in an orderly fashion. Given the industry's throughput imperatives, high wages and union recognition seemed a small price to pay to minimize disruptions during the expansionary years after World War II. Beginning in the 1970s, however, the saturation of domestic markets combined with increased macroeconomic volatility led to more vigorous (some might say destructive) competition within the industry and, as a result, increasing hostility toward the unions. Squeezing labor—through increased automation in the workplace and hard bargaining at the negotiating table—emerged as an important component of firms' efforts to maintain profitability.

That said, it would be a mistake to conclude that the pulp and paper industry, particularly in the South, simply abandoned its traditional "high-road" approach to labor relations. Southern pulp and paper workers continued to hold some of the highest-paying manufacturing jobs in the region as the twentieth century came to a close, and most enjoyed the benefits of union membership. Most importantly, black workers occupied a larger percentage of the workforce than at any time since World War II. The basic "economic facts" of the industry no longer dictated the exclusion of blacks from high-paying jobs as a means of maintaining order and stability. Notwithstanding the relative decline in job growth, the construction of a more integrated industrial order in the sector opened up well-deserved and long-delayed economic opportunities for black workers.

CHAPTER FOUR

Appropriating the Environment

> Until recently mankind has more or less taken for granted the . . . protective functions of self-maintaining ecosystems, chiefly because neither his numbers nor his environmental manipulations have been great enough to affect regional and global balances. Now, of course, it is painfully evident that such balances are being affected, often detrimentally. —*Eugene P. Odum, 1969*

In 1969, *Science* magazine published an important article by the esteemed American ecologist Eugene Odum.[1] Here, in only a few pages, was an encapsulation of Odum's ecology (his particular "idea of nature") with its emphasis on self-maintaining ecosystems as tending toward order, stability, and balance. Here also was a warning from one of the most influential ecologists of the postwar era that industrial society posed a serious threat to the "balance of nature." Marrying systems theory with ecology, Odum developed an ecological paradigm that provided part of the intellectual rationale for modern environmentalism and the system of environmental regulation that emerged in the 1970s, with its focus on restoring the proper balance between human activities and the environment.[2]

Odum, of course, also had a particular interest in the American South and the massive environmental changes that had transpired in the region during the twentieth century. As a native son (Odum grew up in North Carolina, where his father, Howard W. Odum, taught) and as a professor at the University of Georgia, he was keenly aware of the ecological disruption associated with the prevailing model of agricultural and industrial development in the postwar South. By the time he published his article, the environmental impacts of the southern pulp and paper industry had become a significant source of concern. During his research trips to the marshes and estuaries along Georgia's coast, Odum witnessed firsthand the disruptive effects the industry was having on these coastal ecosystems.[3] No doubt these experiences reinforced his distinctive ecological worldview.

Although Odum's ideas about stability and balance in ecosystems eventually gave way as ecologists rejected the equilibrium paradigm in favor of more dynamic models, his ecology exerted a powerful influence on the distinctive brand of environmentalism that emerged in the 1960s and 1970s and the ensuing debate over the environmental impacts of industrial and urban growth.[4] Indeed, the fact that many of the new environmental laws enacted during this time—all of which had broad, bipartisan support—made reference to the "balance of nature" suggested just how strongly many Americans of the day embraced the notion that industrial society was seriously at odds with the natural environment.[5]

In regions such as the South, where industrial boosterism ran thick, this new environmental ethos had to contend with powerful opposing forces. Eager for any sort of industrial development, many people growing up in the post–World War II South saw pollution as the inevitable by-product of industrial development—the "smell of prosperity" as Alabama governor George Wallace once described the odor from a local pulp and paper mill.[6] If catching up meant that the region's natural resources and environment would have to suffer, so be it. When pulp and paper firms moved into the South during the interwar and postwar decades, environmental pollution barely registered as a concern. The *Atlanta Constitution* captured the prevailing sentiment in 1945: "There are two objectionable factors about a paper mill. One is the smoke that bears an unpleasant odor . . . The other . . . is the charge that the chemical and other wastes dumped into the stream cause contamination . . . Communities in which paper mills are now established, however, consider these objections comparatively minor in view of the vastly stepped-up income the mill means to the community."[7] For the most part, southerners accepted the pollution as "the price of progress." Swept up in the ongoing "southern crusade for industrial development," few towns or states were willing or able to challenge an industry that promised so much prosperity.[8]

By the 1960s, however, the situation had begun to change. As pulping capacity increased, environmental disruption became so acute in some cases that even the industry's most ardent supporters could no longer deny the problems. Earlier fears of timber depletion and forest destruction were replaced by concerns over air and water pollution.[9] With citizens groups mobilizing around environmental issues, polluting industries received much more critical attention and pollution became a target for regulation. For the pulp and paper business, this had serious implications. In addition to being one of the most resource-intensive industries in the United States, it was also one of the nation's biggest polluters. In 1959, for example, the industry discharged more than 1.6 *trillion* gallons of wastewater into rivers and waterways. Taken as a whole, it accounted for about a fourth of the total organic wastewater pollution load generated by American industry.[10] In terms of air pollution, the industry emitted massive quantities of particulates and other pollutants. Few could forget the pervasive rotten-egg smell of a Kraft pulp and paper mill. Air and water pollution, in short, were as much a part of the industry's distinctive landscape as the carefully cultivated timberlands that fed the mills. Not surprisingly, the industry became a target of federal pollution control laws enacted in the 1970s. In the process, the environment emerged as a site of intense political, legal, and cultural conflict between the industry and its critics.

Much of the conflict revolved around the industry's access to and control over the capacity of local ecosystems to assimilate the wastes associated with pulp and paper production. Arguing that such access was necessary for economic survival, industry leaders fought hard against stringent pollution control standards. Viewed from a strategic economic perspective, their arguments made perfect sense. The firms rightly considered the absorptive capacity of the environment as important a resource as trees themselves. As one industry executive put it: "We're a company that depends on natural resources. We've got to have trees. We've got to have water. We've got to have rivers for our effluent. We've got to have land for the disposal of our by-products. And we've got to have the capability to put some of our release into the atmosphere. It's in our own best interest to protect the environment so that it is available for us to use in the future."[11] Like the industrial timberlands, ongoing access to the absorptive capacity of local and regional environments thus provided a vital component of the southern pulp and paper industry's postwar success. The difference, of course, was that air and water, unlike private timberlands, were public goods subject to multiple and often competing claims with fundamental importance to human health and well-being.

If appropriating the environment for waste assimilation represented a clear strategic objective for pulp and paper firms, the political means of achieving this appropriation seemed to lie in retaining state and local control over environmental regulation. During the debates over air and water pollution in the 1960s, pulp and paper representatives, along with their counterparts in other industries, mounted a well-orchestrated resistance to the push for stronger federal pollution control laws. "States' rights" quickly emerged as one of the most common refrains in their arguments that pollution control should be left to state and local governments. When one looks at the state pollution control laws of the day, particularly those in the South, it is not hard to understand the logic of this position.

As it turned out, they were swimming against the rising tide of American environmentalism. By the early 1970s, with the passage of tough federal pollution control laws, in part because of the abysmal record of previous state and local efforts,[12] the regulatory regime confronting pulp and paper firms shifted dramatically as power moved from the state house to the halls of Congress, the Environmental Protection Agency (EPA), and the federal courts. Older concerns with defining quality standards and allocating multiple uses soon gave way to across-the-board reductions of industrial emissions regardless of differences in the assimilative capacity of local environmental systems. The new end-of-pipe, command-and-control regulations focused directly on stopping or reducing the gross insults to the environment that had become an increasingly visible part of the American industrial landscape. As the new laws on air and water pollution forced many mills to adopt pollution control technologies to treat their wastes before releasing them into the environment, managing pollution loads became a much more expensive undertaking. During the early 1970s, for example, more than 30 percent of all capital spending in the industry was allocated to pollution abatement.[13] Over time, and after considerable political controversy, pollution loads were reduced and environmental quality improved.

Meanwhile, other problems emerged. By the late 1970s, growing concern over toxic compounds shifted attention away from earlier concerns with traditional forms of air and water pollution. For pulp and paper mills, this became a major issue during the mid-1980s when EPA researchers conducting a national study of dioxin contamination unexpectedly discovered dioxin in fish downstream from pulp and paper mills. These waters, which were supposed to be reference or background streams (i.e., uncontaminated), all had one feature in common: they were receiving wastewater discharges from paper mills that used chlorine bleaching to create white paper products. The EPA thus concluded correctly that the chlorine bleaching process used by some pulp and paper mills was a source of dioxin. This was a very different kind of environmental problem than more traditional types of air and water pollution. One of the most toxic compounds ever tested, dioxin persisted for long periods in the environment, and accumulated in increasing concentrations in food chains and living tissues.[14] As such, it posed immense challenges for regulation and came to occupy a prominent place in the increasingly politicized world of toxic substances during the 1990s, as the EPA, the paper industry, and environmental groups struggled to deal with the problem.

Other environmental problems associated with pulp and paper production have had less to do with pollution than with the industry's almost insatiable demand for key natural resources. In converting much of the southern landscape into intensively managed industrial timberlands, for example, firms have run into controversies involving endangered species and critical habitats such as wetlands. As noted in chapter 2, these issues at times have turned southern timberlands into political battlegrounds. Groundwater use by mills has also been a major source of political controversy in areas such as south Georgia,[15] where pumping rates by large mills exceeded natural rates of regeneration, leading to falling water tables and saline intrusion in outlying coastal areas.[16]

The unequal distribution of harms cut across all of these problems. As with other New South industries, the environmental impacts generated by the pulp and paper industry have often fallen disproportionately on poor and minority communities. Prevailing class structures, endemic practices of racial discrimination, and the distribution and exercise of political power have worked to limit the capacity of some people to avoid the environmental hazards associated with pulp and paper production. Pollution, poverty, and the political marginalization have long had a mutual affinity for one another.

This chapter explores the environmental implications of southern pulp and paper manufacturing. The focus is on the "downstream" impacts of mill operations and the ways such impacts have figured into the broader political economy of the industry. Rather than viewing such environmental pollution through the lens of externalities or spillovers, however, it considers these issues as normal, perhaps inevitable, aspects of "industrial metabolism."[17] In other words, the various types of pollution, environmental disruption, and attendant risks and vulnerabilities associated with industrial production are seen as part and parcel of the materiality of industrial production and, more importantly, of the social and political construction of industrial systems. To view them simply as externalities misses the crucial significance of the environment as a key component of firms' attempts to construct viable and competitive industrial forms.

Historicizing environmental disruption as part of the broader evolution of the industrial system shows the ways firms approached resource and environmental issues as strategic, political issues. The overall objective is to unpack how firms, in concert with other actors, attempted to maintain their access to the environment, manage the ecological disruptions associated with pulp and paper production in various political and legal arenas, and, in the process, determine the distribution of the associated risks and liabilities. At bottom, this reflects an effort to move beyond both traditional economic treatments of environmental harms as externalities and the Coasean approach to pollution as a question of correlative entitlements and bargaining. The aim here is to describe and analyze how politics at multiple levels shaped and gave meaning to "the problem of social cost" and how particular environmental harms were imposed on certain groups of people.[18] In the case of the southern pulp and paper industry, the challenge of securing access to the environment for waste assimilation and dealing with the impacts of the industry's pollution was saturated with politics at all levels—from early nuisance suits, to initial state laws, to the federalization of pollution control that transpired in the early 1970s. The dominant trend was a gradual diminishment of the industry's political influence and control and a migration of regulatory authority to the federal level, leaving firms to struggle with a more diverse and powerful set of stakeholders.

The first section focuses on the substantial challenge of water pollution, tracing the issue from some of the early nuisance cases brought by landowners whose property and livelihoods were directly impacted by the mills' massive pollution loads, to the efforts of various state governments to deal with the issue in the middle decades of the twentieth century, and then to the modern federal water pollution control regime established in the early 1970s. The second section discusses the challenge of controlling and regulating air pollution, focusing specifically on the struggle over federal versus state control during the 1960s, 1970s, and 1980s and some of the local political conflicts that ensued. Building on this, the penultimate section addresses the problem of dioxin contamination, which occupied significant attention during the 1980s and 1990s as the industry and the various communities in which it operated sought to respond to this new and unexpected hazard. Finally, the concluding section provides a more general discussion of the politics of environmental disruption and corresponding issues of distribution.

Living Downstream

> One of the necessary and legitimate functions of water in our urban and industrial economy is the utilization of the unique capability of running water to assimilate reasonable amounts of waste and to dispose of such wastes . . . Such use of the natural capability of water must be recognized as unavoidable.
>
> —*William H. Chisholm, 1964*[19]

Water is as important to modern pulp and paper manufacturing as any other input, including trees. The large southern production complexes used extraordinary volumes of water in the process of transforming wood fiber into paper.

By the mid-1950s, to take one example, Union Camp's Savannah mill was using close to 30 million gallons of water a day in its operations—all of which it pumped from the prolific Upper Floridan aquifer.[20] In North Carolina, Champion International's Canton mill, which was built in 1908, was using almost the entire flow of the Pigeon River for its operations during the 1980s, diverting some 46 million gallons of water every day and returning roughly 45 million gallons.[21]

As the Canton mill illustrated, the flip side of the massive water use involved in papermaking was the tremendous amount of wastewater that mills discharged into waterways. Although some water was consumed in the process of papermaking, the vast majority was returned to streams and waterways. As William Chisholm, a paper company executive representing the American Pulp and Paper Association, testified in Congress during 1964 hearings on water pollution, "The pulp and paper industry 'borrows' most of its water for its manufacturing processes . . . Approximately 95 percent of the total water intake of the paper industry is returned to the water supply."[22] In the early 1960s, the pulp and paper industry was discharging approximately 1.6 *trillion* gallons of water every year, accounting for about one-fourth of the total organic wastewater pollution load generated by all of American industry.[23] Reflected in this aggregate number, of course, were the substantial amounts of wastewater generated by particular mills in particular locations. One of International Paper's mills in Louisiana, for example, discharged wastewater into the surrounding bayou at a rate of 10,000 gallons per minute during the 1940s.[24] Forty years later, Champion's Canton mill was discharging more than three times this volume into the Pigeon River.[25]

The problem concerned not just the actual volume of discharge but also the quality of the water once it was "returned." In the absence of treatment, pulp and paper mill effluent contained large quantities of suspended solids and organic materials, not to mention pulping chemicals and other toxic compounds that often led to substantial degradation of water quality in receiving streams and waterways. Since most early pulp and paper mills made little or no provision for wastewater treatment, the environmental disruption associated with their effluent was difficult to miss. Along with their distinctive and pervasive odor, the mills' discharges of wastewater represented one of the most immediate and visible impacts of the industry. Understandably, some people did not appreciate their new industrial neighbors, filing lawsuits and publicly criticizing the industry. By the mid-1920s, articles on stream pollution had begun appearing in leading trade journals, and in 1926 leading firms created the National Waste Utilization and Stream Improvement Committee to explore the wastewater problem and provide assistance to individual mills involved in litigation.[26]

During the 1930s and 1940s, individuals from southern states began filing common law nuisance actions against pulp and paper firms in an effort to shut down the mills or recover damages incurred as a result of the wastewater discharges that threatened the livelihoods of landowners, fishermen, and others who depended on local waterways. The major problem was the quantities of organic matter (along with various chemicals and suspended solids) in mill effluent that, when dumped into a waterway, used up the dissolved oxygen in the water as bacteria worked to break down the material. As the organic load increased, more oxygen was needed for this process of oxidative decomposition.[27] When the oxygen demand of a particular waste stream exceeded the capability of the receiving waterway to replenish its supply of dissolved oxygen, aquatic life would suffocate and die. Once dissolved oxygen levels fell to zero, the remaining organic compounds would decompose anaerobically (without oxygen), discoloring the water and emitting foul odors, which is precisely what happened in many southern streams receiving untreated pulp mill wastes.

During the 1930s and 1940s, for example, Louisiana's Bodeau Bayou suffered substantial environmental degradation as a result of discharges from a local International Paper mill. In the late 1940s, these discharges became the subject of a series of lawsuits.[28] As described in one case from 1947:

> As a result of this deposit of waste matter, the fish and aquatic life in Bodeau Bayou have been entirely ruined . . . the water in said stream, instead of being free and sweet, fit for human and animal consumption as it was prior to the pollution, is heavily polluted with this black, poisonous substance that carries a vile odor and which . . . has rendered said stream and its tributaries entirely unfit and useless for agricultural purposes, cattle raising, human consumption, recreational purposes, or for any other useful purpose . . . [N]ot only has this condition ruined the usefulness of the stream, but [it] has devalued and practically ruined the valuation of all farms located on said stream or its tributaries, and particularly the place belonging to these plaintiffs.[29]

Like most other early pollution cases, this was a nuisance case in which local landowners were suing to enjoin the mill and recover damages incurred as a result of the pollution's negative impacts on the value of their land, livestock, and livelihoods.[30] Though eventually settled out of court, the case pointed to the difficulties inherent in using common law remedies to deal with large industrial pollution problems.[31] Even though these sorts of lawsuits did sometimes end with companies paying damages or building impoundments to allow wastes to settle and decompose before being released in more controlled fashion, the amount of damages was usually quite limited. In all of these cases, moreover, the burden of proof rested with the plaintiffs, and, except in the case of massive fish kills, the courts generally resisted plaintiffs' efforts to link declining land values directly to the wastewater discharges of the mills. Attempts to enjoin mill operations until the problems could be corrected always failed.

In fact, in contrast to prominent cases from other parts of the country where courts did issue injunctions against paper mills, there does not appear to be a single case in any southern state where a pulp and paper mill was forced to halt or suspend operations.[32] And although courts did award damages to plaintiffs who suffered harm from the mills' pollution, these same courts sometimes found creative ways to reverse or restrict such damages awards in a manner that reduced the mills' overall exposure to liability.

Two early cases from Louisiana, for example, denied any relief on the grounds that the plaintiffs failed to demonstrate that the pollution from a local International Paper mill had caused the alleged harms to their land.[33] As the court noted in one of the cases, "After reading and analyzing the evidence we feel far from satisfied that there were any chemicals in the waste waters from defendant's mills that were deleterious either to land or timber. They were of an alkaline nature, and should have been beneficial rather than deleterious. In fact, tests were made showing that the sludge from waters similar to these was beneficial to plant life."[34] Such language shows how uninformed and ill equipped these judges were to deal with the scale and scope of the pollution generated by the pulp and paper industry.[35]

As evidence of the harm associated with the mills' pollution accumulated, however, courts could not so easily dismiss cases. Indeed, a mere two years after denying relief to plaintiffs on grounds that they had failed to demonstrate any harm, the same justice from the Louisiana Supreme Court employed a new rationale to deny injunctive relief to a plaintiff who claimed timber loss as a result of pollution from the same International Paper mill at Bastrop. According to the court, the "plaintiff's land is hardly susceptible now to any damage that the water may cause it, and, if such should occur, it may be easily compensated in money. To enjoin defendant from using the stream to take off its waste water, and thereby deprive it of its only means of doing so, is virtually to close down mills costing several millions of dollars to prevent some possible damage, of no particular moment, on land which has but slight value, save possibly for mineral purposes."[36] In effect, as the links between the mill's pollution and the alleged harms became more obvious, the Louisiana court embraced the balancing approach favored by many courts throughout the country, weighing the social cost of enjoining the mill against the "slight" harm suffered by individual downstream landowners. By declaring that such a plaintiff had an adequate remedy at law (damages), the court effectively removed the possibility of injunctive relief from the available remedies. The resulting liability rule thus allowed polluting firms to purchase entitlements to pollute by compensating local landowners.

This basic approach received further elaboration eight years later. In a case that had been removed to a Louisiana federal district court at the request of the defendant company, the court refused to enjoin operation of the International Paper mill at Springhill, even though the court found that the plaintiff's fishing business had been "practically ruined" by the pollution from the mill. According to the court, "Even the officers of the defendant corporation admitted frankly that there was very little fishing at present, and for sometime past, on Bodeau Bayou, that the fish were killed, that the blackening of the water was taken by the public to mean that there was no fishing in this stream during those times, and that the fish, if caught, were not fit for human consumption." But the court denied any injunctive relief on the grounds that the plaintiff had an adequate remedy at law—$6,000 in this case.[37] Such an outcome accorded with the rule increasingly embraced by a majority of courts and represented an obvious and important example of how the common law could be refashioned to serve the needs of industrial development.[38]

Not all of the courts confronting similar claims embraced the balancing approach, however. In a 1943 case from North Carolina, brought on grounds that the wastewater discharges from a local mill constituted a public nuisance, the state supreme court reversed the lower court's dismissal of the complaint. In that case, the plaintiff, a local landowner who operated a commercial fishing operation, claimed that the pollution discharged into the Roanoke River had "destroyed or diverted" the seasonal runs of migrating herring, rockfish, and shad, thereby injuring the public in general and the plaintiff's business in particular. After finding that the plaintiff had made out a sufficient claim of special injury to proceed on his public nuisance claim, the court emphatically rejected any argument that the mill enjoyed an entitlement to the assimilative capacity of the local waterways: "We deal, therefore, not with a conflict of rights, but with the conflict between the right of the plaintiff to the security of established business and the wrongful conduct of the defendant in interfering with it."[39] The court then went on to reject the company's request that it "weigh the economic consequences" of a decision allowing the plaintiff's claim to go forward,[40] before closing with a brief admonishment questioning the wisdom of the position advocated by the company and its allies: "It is not amiss to say that a State which deals with its resources on the principle attributed to Louis XIV—'après moi le deluge'—is headed for economic ruin."[41] Bold as they might appear to be, however, such policy pronouncements fell on deaf ears.

Indeed, in the only other public nuisance case brought by private plaintiffs, the Florida Supreme Court took a decidedly different approach, finding a basis in Florida's Constitution to immunize the industry from any such suit in the state. In that case, ninety citizens of Duval County sought to enjoin the National Container Corporation from constructing a pulp and paper mill in the industrial section of Jacksonville on the grounds "that the mill will necessarily and unavoidably emit great quantities of smoke, fumes and gases of such powerful, foul, offensive, persistent smell as to cause material discomfort to the plaintiffs and to all the other people of the City of Jacksonville and the surrounding communities, both in their homes and in their places of business." The local residents further alleged that the wastewaters to be discharged into the St. John's River "will prove harmful, injurious and toxic to fish and aquatic life and that the supply of fish will be seriously reduced if not cut off entirely . . . [and] that a serious health menace will be created in that the waste waters will be of such great and additional oxygen demand that the sewerage of the City of Jacksonville now being emptied into the St. John's River will not be rendered sterile and harmless."[42]

But the court promptly rejected all of the plaintiffs' public nuisance claims, concluding that the people of Florida had already decided the issue. Specifically, the court cited a 1930 amendment to the state constitution that exempted from taxation all industrial plants established on or after July 1, 1929, engaged in the manufacture of, among other things, woodpulp, paper, paper bags, and fiberboard. According to the court, "By the adoption of this amendment industrial plants of the kind and character mentioned therein were not only authorized but were invited and importuned to take up their abode and operate in the State of Florida." The court elaborated: "When the people of the State of Florida by an overwhelming vote, wrote into the Constitution not only the permit for pulp and paper manufacturing plants to operate in the State but guaranteed to them as an inducement to come and operate in this State the exemption from all taxes from certain property used in such enterprise for a period of fifteen years, it did not mean for them to come in and bring their working capital and the facilities of employing hundreds of people in those plants but to leave outside the State its noises and offensive odors which would necessarily be produced when it began and continued operation." Thus, the court reasoned, as long as the National Container mill was "operated in such a manner as to, as far as is possible with presently known methods, reduce or eliminate the emission of such noxious and offensive odors, it cannot be held to constitute a public nuisance." In sum: "by a special and definite provision of our Constitution . . . the particular annoyance, a necessary incident to the manufacture of woodpulp has been immunized from having the status of a public nuisance."[43] In Florida, tax exemptions used to recruit industry carried the additional benefit of immunity from any public nuisance claim that might arise from the operation of a new facility built to take advantage of such favorable tax treatment.

Other states provided more direct statutory relief from public and private nuisance claims. An Alabama law held that "no manufacturing plant or other industrial plant or establishment, or any of its appurtenances, or the operation thereof, shall be or become a nuisance, private or public, by any changed conditions in and about the locality thereof after the same has been in operation for more than one year, when such plant or establishment or its appurtenances, or the operation thereof, was not a nuisance at the time the operation thereof began."[44] Thus, when a group of residents from a newly constructed subdivision in Mobile sought to enjoin a local Stone Container mill on the grounds that its wastewater discharges constituted a private nuisance, the state supreme court rejected the action, citing the statute and leaving the plaintiffs to pursue their action on the more difficult terrain of negligence.[45]

But even in cases where such statutory protections were not available, it was clear by the early 1950s that the common law would not protect property and livelihoods from the harms inflicted by industrial pollution. At most, individuals seeking relief would get damages. Simply put, the mills had come to be seen as far too valuable to the surrounding community to allow any equitable relief to interfere with their operations. A 1953 decision from Louisiana summed up the basic thinking: "To deprive [the International Paper mill] of its use of Bodcau Bayou for waste disposal would result in closing its plant and throwing out of employment thousands of well paid employees, and depriving the area of a very substantial revenue. If the slight annoyance and inconvenience to plaintiff is balanced against the loss to defendant and the community which would result from its having shut down its operations, the injunction must be denied."[46] With the threat of injunction removed, the mills simply had to negotiate and pay for the right to discharge their wastes. And so they did.

International Paper paid "permanent damages" to some forty landowners living along the Bodcau Bayou in Louisiana for "the perpetual right" to discharge its wastes into the bayou. According to a 1951 case, "The gross amounts paid for these servitudes represents an investment of several hundred thousand dollars and the instruments are so drawn to create, in favor of the defendant company, a servitude over and through such lands for the discharge of this effluent free from any other payment of damages." As the court remarked, "The present policy of the defendant company (and we have no disapproval of it personally or officially) is to own the right of flowage through Bodcau Bayou . . . [by] paying finally all the owners their legal claims for injuries." But the court also concluded by noting the irreparable harm that had been done to the bayou: "It is not questioned however but that the fishing of today will never be the wonderful . . . fishing that was enjoyed before the installation of the defendant's mill. Those days are gone . . . [T]he present waters of the Bayou Bodcau will never be what they were before the days of pollution. We visited the premises at a period when the discharge was not at its flux and one can tell that the waters are not natural. The bottom soil and the stumps and submerged branches show the permanent imprint of the black."[47] Though similar testimonials are scarce in the official court reports, there can be little doubt that such scenes repeated themselves across the region.

And yet, notwithstanding their success in fending off efforts to enjoin their mills, pulp and paper firms also recognized that the process of negotiating access and paying downstream landowners to secure their right to use waterways for waste assimilation was a messy, time-consuming, and unpredictable affair. Better to go to state legislatures and secure the entitlement through zoning or other means. Which is exactly what they did.[48] Through trade associations and through special organizations such as the National Waste Utilization and Stream Improvement Committee, which in 1944 became the National Council for Stream Improvement,[49] the industry explored technical options for treating mill effluent while orchestrating political and legal activities aimed at combating nuisance litigation and influencing new state and federal laws aimed at controlling water pollution.

During the 1940s, when several southern states passed new laws or amended existing laws to address the water pollution issue, the influence of the pulp and paper industry was readily apparent.[50] Alabama's Water Pollution Control Act provides a good example. Passed in 1947, the act did not establish any regulatory authority over water pollution, nor did it delegate any powers to a particular agency to control water pollution. Rather, it simply established a commission to study and report on water quality conditions in the state. In 1949, after the Alabama Water Improvement Commission delivered its report, the legislature gave it a limited degree of regulatory authority over water pollution control.[51] Fearing new regulations, industries in the state lobbied successfully for a provision that effectively exempted all industrial effluent existing as of August 25, 1949. In other words, those industries that were already polluting at the time of passage could continue to do so without any increased regulatory requirements. The commission had no authority over their effluent unless it could demonstrate a direct health effect.[52]

Most importantly, the 1949 amendments directed the commission "to designate streams currently used for industrial waste as *industrial streams* when not regarded by the commission as necessary, in the public interest, for sources of municipal or industrial water supply, or for the development of the commercial shellfish industry or which are not considered to be important by the commission for development as recreational areas."[53] In practice, this meant that these "industrial waters" would be off-limits as far as future regulation was concerned. The industry's entitlement to use these waterways for waste assimilation was now codified in the statute books.[54]

To say that Alabama's water pollution control laws circa 1950 were pro-industry would be an understatement. Not surprisingly, various interest groups in the state sought to amend the law. Beginning in 1953, a group of wildlife enthusiasts and outdoor sportsmen sought to remove the amendments protecting existing industries. They failed, as did many others in years to come. Indeed, every regular session of the Alabama legislature after 1953 saw another effort to remove the "grandfather clause." Even Governor Wallace, hardly a foe of industry, supported legislation in the 1960s that would have removed the 1949 protections.[55] Not until 1971 did the Alabama legislature remove the grandfather clause and strengthen its water pollution control laws.[56]

At the same time that Alabama and other southern states were enacting their first water pollution control laws, the federal government was taking its first tentative steps in the area since the 1899 Refuse Act.[57] In 1948, Congress passed the Water Pollution Control Act, authorizing expanded federal water pollution research and planning grants and low-interest loans to local governments (no funds were ever appropriated for these grants and loans).[58] For the first time, Congress had expressly recognized a national interest in controlling conventional forms of water pollution, signaling that water pollution might not remain the province of state and local government, but it refused to put any real muscle behind pollution control.[59]

As water pollution problems worsened and began to attract increasing national attention, Congress in 1956 amended the 1948 Act, authorizing states to establish water quality criteria and creating a new federal mechanism—enforcement conferences—to develop and enforce pollution control requirements on individual dischargers in the case of "serious" pollution problems in interstate waters.[60] Over the next fifteen years, the Department of Health, Education, and Welfare convened conferences throughout the country. The conferences' exclusive reliance on consensus and voluntary actions among dischargers severely limited their effectiveness in terms of actual pollution control.[61] But by bringing polluters, government representatives, and other interested parties together to discuss the issue, the conferences gave water pollution problems a heightened visibility.

Given the limitations of federal law and the highly uneven and largely ineffective efforts to control water pollution at the state level, it was clear by the early 1960s that more needed to be done. For southern states, the record of environmental disruption associated with water pollution was manifest in the growing incidence of fish kills, rising levels of water-borne bacteria, and increasing concentrations of organic synthetic pesticides and other chemicals. Between 1951 and 1963, Alabama officials recorded more than fifty significant fish kills in the state's waterways. By this time, the Mobile River above Mobile Bay had essentially been written off for lack of dissolved oxygen.[62] Along the lower Pearl River, which formed the border between Mississippi and Louisiana, pollution extended for some twenty-six miles below Bogalusa, home of the large pulp and paper mill operated by the Crown Zellerbach Corporation.[63] In Georgia, the cumulative effect of pollution loads on coastal ecosystems had reduced the state's annual oyster harvest from over 8 million pounds in 1908 to 160,000 pounds in 1961.[64] By 1964, according to a report issued by the Department of Health, Education, and Welfare, the lower Savannah River in the vicinity of Savannah had become so "grossly polluted" that it failed to support aquatic life.[65] And these were simply a few of the most visible instances of water pollution in the region.

With evidence of mounting pollution problems and in the face of increasing national concern, Senator Edmund Muskie of Maine, along with other members of Congress, convened a series of hearings on water pollution during the mid-1960s. Muskie, who grew up in the paper mill town of Rumford, Maine, along the Androscoggin River, was well aware of the tremendous pollution loads emanating from pulp and paper mills and their environmental impacts. He was also quite concerned about the existing regulatory framework concerning water pollution control. In opening the 1963 hearings before the Special Subcommittee on Air and Water Pollution, Muskie noted the piecemeal progress achieved under the existing laws but stated that "much more is indicated as urgently necessary to be accomplished."[66] To that effect, Muskie proposed legislation to enhance the federal role in water pollution control. The overall goal was "to establish a positive national water pollution control policy of keeping waters as clean as possible as opposed to the negative policy of attempting to use the full capacity of such waters for waste assimilation."[67] Here was a major shift in thinking about industrial pollution and its relationship to the environment. By emphasizing environmental quality rather than assimilative capacity, Muskie began to articulate a different set of goals for pollution control legislation.[68]

As Senator Muskie and his colleagues pressed the issue, they framed water pollution as an issue that went to the very core of American values. During House hearings in 1963, Rep. Robert E. Jones of Alabama, chairman of the House Subcommittee on Natural Resources and Power, emphasized this heightened sense of national mission: "The subject of water pollution control involves the very health and economic destiny of our country. Water pollution has become the Nation's most desperate single natural resources problem. Almost all of our major streams and rivers and lakes are suffering increasing pollution. This pollution jeopardizes our water supplies, menaces the public health, destroys aquatic life, and disgraces our environment."[69] Muskie echoed these sentiments at field hearings held in his home state during 1965: "Water pollution is an increasingly serious national problem affecting all fifty states. Our national health is at stake."[70] By the mid-1960s, these congressional leaders and their allies in the executive branch had begun to explore ways for the federal government to enter the field of water pollution control with real force.

Such efforts met with significant resistance from the business community. During the congressional hearings on water pollution in the mid-1960s, industrialists complained of the threat of overregulation and urged that the issue be left to the states. Prominent among them were the representatives of the pulp and paper industry. William H. Chisholm of International Paper argued that "based on the experience of our industry, we believe that the most effective unit for regulatory authority is the State. Water pollution control based upon State regulation can most nearly reflect the requirements imposed by . . . local factors." Chisholm also emphasized the "vital function [of water] in assimilating organic wastes" and urged that his industry not be burdened disproportionately with the costs of pollution control: "The welfare of our country is bound up with the economic well-being of our country's industries. If our industry, or any industry, is called upon to contribute an undue portion of the cost of water pollution control so that other water users may, without contribution to the cost, enjoy economic and other benefits, the economic welfare of the entire country will suffer."[71]

Other paper industry executives raised similar concerns. Paul C. Baldwin, executive vice president of Scott Paper, evoked the specter of overregulation and its negative implications for industrial competitiveness: "What can it benefit our industry and our national economy if in the pell-mell race for clean streams we lose out to domestic and international competition in the marketplace. And so it is that Scott and other companies with waste disposal challenges cannot correct their difficulties overnight. This economic fact of life must be better understood by all of those who would try to enforce clean streams into existence too quickly." Along these same lines, Richard M. Billings from Kimberly-Clark posited a direct link between local regulatory autonomy and the economic viability of the industry: "People in city, county, and State offices have the best overall picture of what is going on in their own locality. They know better than anyone else exactly where the pollution problems are, which ones under discussion are real, and how widespread is their effect. Knowing the economic needs of the local industries and communities, they are in a good position to decide how to approach the specific improvement required. Too much pressure, too soon, by arbitrary orders which do not take into account the local situation, could shut a mill down and disrupt the local economy." Government representatives certainly seemed to get the message, though some suggested that perhaps the industry was overreacting. In closing the hearings, Phineas Indritz, chief counsel to the House Natural Resources and Power Subcommittee, noted, "There seems to be a fear on the part of the paper and pulp industry representatives that the Federal Government is out to ruin the industry by forcing them to such extensive expenditures that they will have to go out of business." He assured them, however, that "we [the subcommittee] want pollution control, and we don't want to bankrupt the industry."[72]

During field hearings in 1964 and 1965, these issues of environmental federalism resurfaced and quickly came to dominate the discussion. At one set of hearings in Alabama, Governor George Wallace submitted a statement declaring, "Regulatory authority for the control of intrastate pollution is clearly a responsibility of the States."[73] Industry representatives reinforced this position. Speaking "on behalf of the water-using industries of Alabama," P. A. Bachelder, general manager of Kimberly-Clark's Coosa River newsprint mill and president of the Associated Industries of Alabama, strongly endorsed the "states' rights" position on pollution control:

> We claim neither that we have no problems nor that there is not water pollution. We do claim, however, that there is certainly not a pollution problem of such scope that additional Federal legislation is required. Though we do have problems to solve, we feel that they are being satisfactorily solved at the State level and will continue to be solved satisfactorily at the State level . . . We note with great concern the introduction in both Houses of Congress of bills that advocate the imposition of sweeping water quality standards applied nationwide, and which would require maximum treatment in all situations with little or no regard for local circumstances. The determination of satisfactory water quality standards must vary with local circumstances.[74]

Other participants offered a different perspective. Mrs. Atherton Hastings, president of Alabama's League of Women Voters noted, "State laws are inadequate . . . [T]he matter of pollution abatement—of having our waters clean—is not just a local community problem. It begins in the community, but it goes much further . . . Local, State, and Federal efforts must combine, and swiftly."[75] Ralph Richards, a fisherman representing the Alabama Fisheries Association, pleaded with the committee to recognize the plight of those whose livelihoods depended on the state's waterways: "We don't feel that the industrial people of this country have the right to concede the fishing industries or to steal the water resources that belong to the people of this country." As to the prospect of federal regulation, Richards stated, "The question arises as to whether we need Federal intervention on water pollution. We think we do, because these are national concerns doing most of the damage . . . Being national concerns, these industrialists can handle the legislature better on a State level than they can on a National level." In closing, he urged the committee to act: "Gentlemen, fish can't live in our waters and they are polluted. The fishing industry needs help . . . [I]f it is going to take 10 more years of research, . . . I would like for you gentlemen to go back and ask President Johnson to put us down for his program on poverty as the polluters are putting us out of business. Please start looking out for us so we don't need to go on the government relief."[76] Water pollution, from this perspective, went far beyond aesthetic concerns. For Ralph Richards and others like him, it was an issue of economic justice.

And their situation was hardly unique. Like Alabama, Georgia passed equally permissive water pollution legislation in 1964, declaring major sections of certain rivers to be "industrial zones."[77] This effectively guaranteed that industrial pollution in these areas would continue virtually unabated in the absence of new laws. By this time, water pollution problems had become so severe in the lower Savannah River that the Department of Health, Education, and Welfare decided to investigate. Their conclusions were not encouraging.[78] Finding the lower stretch of the river to be "grossly polluted" and unable to support fish and other "desirable" forms of aquatic life, the secretary of health, education, and welfare called on the U.S. Public Health Service to convene a water pollution enforcement conference for the lower Savannah River in February 1965. Representatives from industry, government, and the community met in Savannah to discuss the problem and develop a plan for dealing with it. Union Camp quickly emerged as the focus of attention. With a daily discharge of some 135,000 pounds of biochemical oxygen demand (BOD), this one mill produced 70 percent of the combined industrial and municipal pollution in the lower Savannah—the equivalent of a city of almost 800,000 people.[79] Speaking at the conference, mill manager James Lientz reminded the participants of his company's substantial "economic contributions" to the area, and he warned that any effort to increase the required amount of dissolved oxygen in the river "might be incompatible with any practical use of a major asset of the harbor—its ability to assimilate waste."[80]

Lientz also suggested that the organic wastes in the mill's effluent did not pose as much of a problem as some suggested: "The oxygen using materials in paper mill wastes are soluble materials from wood. They are not poisonous. They do not constitute a health hazard. In proper quantities they contribute food to the small organisms which provide food for fish."[81] Murray Stein, the federal official who convened the conference, called this a rather "curious statement" and suggested that Lientz was being somewhat disingenuous.[82] At the conclusion of the conference, the participants agreed that each of the industries on the river would reduce its BOD load by 25 percent and suspended solids by 90 percent by the end of 1967.[83]

Meanwhile, the push for a stronger federal presence in water pollution control gained momentum. In October 1965, President Johnson signed the Water Quality Act of 1965, giving the federal government new power to mandate state actions regarding water quality standards and pollution control in interstate and navigable waters.[84] By establishing the Federal Water Pollution Control Administration within the Department of the Interior, the act created an entity responsible for coordinating and enforcing state actions required under the law. Nonetheless, even though this represented an important step in building a national framework for environmental protection, the 1965 law contained shortcomings. In addition to the difficulties involved in establishing specific discharge reductions needed to meet quality standards, the act also left enforcement responsibilities largely with the states. Given that many states had neither the political will nor the institutional capacity for setting and enforcing new standards, the 1965 act turned out to be largely unworkable.[85] Progress in dealing with the water pollution problem was easier said than done. In states such as Georgia and Alabama, where pollution control often took a back seat to industrial recruitment and growth, the response was essentially business as usual.

Four years later, in October 1969, the Federal Water Pollution Control Administration convened a second enforcement conference to deal with the pollution problem on the lower Savannah River. Although some progress had been made in reducing pollution loads since the 1965 conference, things were not moving fast enough. By this time, environmental protection, and water pollution control in particular, had become a more pervasive concern for many Americans. In January of that year, for example, an offshore oil well near Santa Barbara, California, ruptured and spilled 200,000 gallons of oil into the ocean, blackening local waters and beaches. Seven months later, the Cuyahoga River near Cleveland, Ohio, burst into flames.[86] Feeling a new sense of urgency, federal and state regulators at the Savannah enforcement conference considered a greatly accelerated timetable for pollution control. As of 1965, Union Camp still contributed roughly 70 percent of the total daily BOD load going into the lower Savannah (figure 4.1).[87]

With Union Camp's Savannah mill discharging an amount of oxygen-consuming wastes equivalent to that produced by a city of 700,000 people every single day, even a mighty river such as the Savannah suffered. As a result of these massive pollution loads, a dead zone extended for several miles downstream from the mill where the dissolved oxygen often fell too low to support many forms of aquatic life.[88] For Eugene Odum the potential ecosystemic effects of such pollution were profound. Because coastal ecosystems, and particularly estuarine systems, operated at close to full capacity with regard to oxygen, even a slight reduction in dissolved oxygen could disrupt the delicate "balance" of the system and undermine its ability to support life.[89] This had long been clear to the shrimpers, oystermen, and others whose livelihoods depended on the health of the Savannah estuary. They knew all too well that the continued reductions in their catches derived directly from the stresses being put on the ecosystem by the massive pollution loads.[90] As one Savannah area shrimper put it: "Them

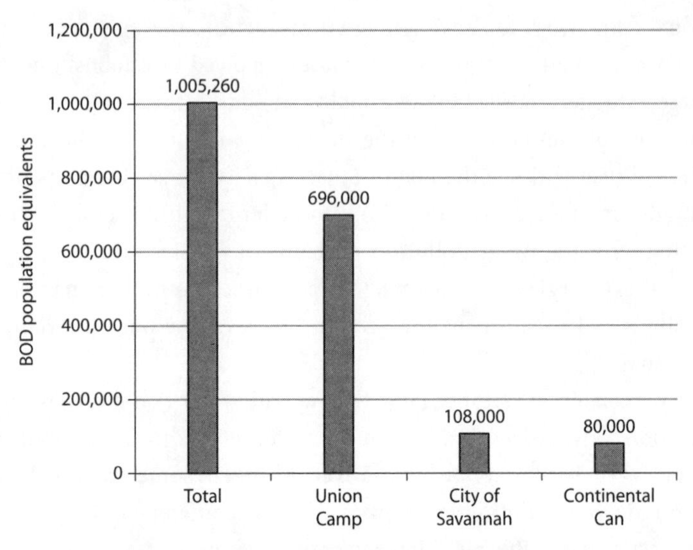

Figure 4.1. Daily BOD pollution loads in Lower Savannah River, 1969. Federal Water Pollution Control Administration, *Second Session of the Conference in the Matter of Pollution of the Interstate Waters of the Lower Savannah River and Its Estuaries, Tributaries and Connecting Waters in the States of Georgia and South Carolina, October 1969* (Washington, DC, 1970); Georgia Water Quality Control Board, *Water Quality Data Lower Savannah River* (Atlanta, 1969), 10–12.

paper bag people is fiddlin' with our livelihood . . . What's more important, a damn paper bag or us keepin' our livin'?"[91] In the face of these concerns and the mounting evidence of environmental disruption, the second enforcement conference raised the bar considerably for the region's polluters, requiring that all major industries and municipalities provide secondary treatment to remove at least 85 percent of the BOD from their effluent by the end of 1972.[92]

Union Camp responded with a major upgrade of its wastewater treatment facilities during the summer of 1970, the centerpiece of which was a large aerated lagoon on an island in the middle of the river. Wastewater would be pumped through a massive pipe for more than a thousand feet along the bottom of the river to the two-hundred acre lagoon, where the wastes would be aerated and allowed to decompose before being released into the river (figure 4.2). Completed in December 1972 (the deadline imposed by the second enforcement conference), the new $17 million treatment system reduced the mill's BOD load by some 115,000 pounds per day—equivalent to the 85 percent reduction mandated by the conference.[93] After more than thirty-five years of operation in Savannah, Union Camp had finally taken real steps to deal with the environmental impli-

Figure 4.2. The aerated waste ponds used to treat the effluent from Union Camp's Savannah mill, circa 1997. Photograph by the author.

cations of its wastes. Technologically, the solution to the BOD problem had always been relatively simple—the organic wastes just needed time and space to decompose. Politically, of course, solving the problem was far from simple, requiring, in the end, federal intervention. And this was merely the beginning.

By the time Union Camp installed its new water treatment system, the so-called decade of the environment was under way. With Earth Day and the creation of the EPA, it was increasingly clear that public opinion had shifted strongly in favor of increased federal regulation. Drawing sustenance from the civil rights movement of the previous decade, environmental advocates of the 1970s embraced a rights-based discourse that rejected any simple balancing of costs and benefits. During his 1970 State of the Union address, President Nixon acknowledged the power of environmentalism: "The great question of the seventies is, shall we surrender our surroundings or shall we make our peace with Nature and begin to make the reparations for the damage we have done to our air, our land, our water? . . . Clean air, clean water, open spaces—these would once again be the birthright of every American."[94] In addition to this new rights talk, environmentalism drew inspiration from the increasingly popular science of ecology, particularly the version espoused by Eugene Odum. As noted above, Odum's ecology exerted an important influence on modern environmentalism and many of the environmental laws that ensued. His conception of ecosystems as tending toward stability and balance suggested that in order to restore the so-called balance of nature (to "make our peace with Nature" in Nixon's words), pollution loads would have to be greatly reduced, perhaps even eliminated. Thus, when Congress enacted the Federal Water Pollution Control Act Amendments of 1972 (the modern Clean Water Act), it embraced the objective to "restore and maintain the chemical, physical, and biological integrity of the Nation's waters" and declared national goals of "fishable and swimmable" waters by 1983 and the elimination of pollutant discharges into navigable waters by 1985.[95] In retrospect, these goals turned out to be hopelessly unrealistic. Nonetheless, the clean water legislation of the 1970s marked a significant departure from past approaches, not only in its goals and objectives but also in the means outlined for achieving those goals.[96]

If the goals of the new water legislation paid homage to the concept of ecological balance, the tools for restoring this balance derived largely from engineering.[97] Any pretense of translating ecology into law, moreover, was quickly subsumed by the institutional challenges associated with establishing a technology-based approach to pollution control. In an innovative departure from the "ambient-based" standards approach of earlier water quality legislation, the 1972 water pollution control amendments mandated a system of technology-based effluent standards, which would define the maximum quantities of pollutants that each source would be allowed to discharge, regardless of the size or "assimilative capacity" of the receiving waters. Such an approach bypassed the extremely difficult task of translating quality standards for particular waterways into specific discharge reductions for specific users. It also avoided many of the intractable political controversies that ensued in the effort to establish such quality standards in the first place.

By federalizing the water pollution control effort and placing authority firmly in the hands of the EPA, the 1972 amendments also undermined efforts by state governments to use lax environmental standards as a tool for industrial recruitment. The law thus reflected a substantial diminishment of the political influence and control that pulp and paper (and other large industries) had enjoyed as they worked with state authorities to craft legislation that would protect their access to the absorptive capacity of the environment. The EPA was now charged with the responsibility of establishing effluent standards for classes and categories of industries, for issuing permits to individual dischargers, and for monitoring and enforcing compliance.[98] The focus was on "point sources" of pollution and dealt primarily with "conventional" pollutants (BOD, solids, color, etc.).[99] Eventually, all industrial facilities would be required to install the best available technology (BAT) for controlling these pollutants.

In practice, then, the clean water legislation embodied the top-down, command-and-control approach to environmental regulation that elicited so much criticism in later decades. At the same time, however, it opened the door to an onslaught of litigation from industry and environmental groups challenging the EPA's new technology standards. New rules and regulations would inevitably be subjected to a judicial "tug-of-war," as competing interest groups sought to shape the process and its outcome. But the end result was clear: regulatory authority had shifted dramatically to the federal level, and the industry no longer enjoyed the influence it previously exercised at state and local levels.

Given its significance as a point source and the largely conventional nature of its pollution load, the pulp and paper industry was an early target for the new water regulations. In discharging its responsibilities, the EPA promulgated a series of effluent guidelines for conventional pollutants for different subcategories of the pulp and paper industry in 1974 and 1977.[100] Lawsuits challenging the standards were quickly decided in the EPA's favor.[101] Those mills that had not already installed wastewater treatment facilities no longer had any choice in the matter. Many mills that had already installed treatment facilities were forced to upgrade. As a result, capital spending on water pollution abatement by pulp and paper firms grew to more than $300 million per year in 1976—approximately 16 percent of total capital spending in the industry.[102] Judged on the basis of per ton of product produced, the amount of wastes discharged into waterways by pulp and paper mills declined substantially. Many rivers and streams that had once suffered massive insults from pulp and paper wastes showed significant improvement.

This is not to suggest, however, that the pulp and paper industry's water pollution problems went away. Quite the contrary. Major controversies, such as that involving the Pigeon River in western North Carolina and eastern Tennessee, for example, continued to rage well into the 1980s and 1990s.[103] Toxic pollutants in pulp mill effluent, which had barely registered in the debates over water pollution in the 1950s and 1960s, also became an increasing source of concern beginning in the late 1970s and especially in the 1980s when dioxin was discovered in some pulp mill discharges. Even the success achieved in controlling conventional water pollution created a new set of problems since the process of allowing solids to settle and organic wastes to break down generated wastewater treatment sludge. This classic example of "cross-media" pollutant transfer became a significant source of environmental concern during the 1980s when trace concentrations of dioxin and other organochlorines were discovered in the sludges from pulp and paper mills. The single-media approach to pollution control, therefore, did not fully solve the water pollution problem but rather transferred a substantial part of it from water to land.

That said, it would be a mistake not to recognize the real progress achieved under the water pollution control laws of the 1970s. Stopping or diminishing the gross insults associated with pulp mill effluent was no small task. Although it may be easy to criticize the legislation of the 1970s from the perspectives of fairness or allocative efficiency, there can be little doubt that in the case of large industrial dischargers such as the pulp and paper industry, the new laws brought about substantial reductions.[104] Contemporary arguments that challenge the efficacy of these top-down, command-and-control approaches to pollution control as a basis for devolving environmental regulation back to state and local levels, moreover, would do well to recognize the historical reality of state inadequacies and the extent to which state regulatory efforts were captured by large industrial interests during the 1950s and 1960s.[105] Although the argument for maintaining state authority over pollution control was often premised on the need to promote industrial development in the South, it is clear that the industry survived the advent of federal pollution control laws and, more importantly, that meaningful pollution control in the region was not going to happen without federal intervention.

The Smell of Money

> On still days, when the air is heavy, Piedmont has the rotten-egg smell of a chemistry class. The acrid, sulfurous odor of the . . . paper mill drifts along the valley, penetrating walls and clothing, furnishings and skin. No perfume can fully mask it. It is as much a part of the valley as is the river, and the people who live there are not overtly disturbed by it. "Smells like money to me," we were taught to say in its defense, even as children. —*Henry Louis Gates Jr., 1994*[106]

In his memoir of his youth in Piedmont, West Virginia, Henry Louis Gates Jr. provides a compelling portrait of life in a paper mill town during the racially turbulent decades of 1950s and 1960s. His description of the powerful and pervasive odor emanating from the local Westvaco paper mill would doubtless ring true for anyone who has ever experienced the smell. For countless southerners growing up in the post–World War II period, the distinctive odor of the Kraft pulping process was as much a part of the landscape as pine trees or abandoned cotton fields. As industrial boosters were quick to remind them, this was the smell of money—an indicator of newly found prosperity, a welcome trade-off in a region desperate for jobs and investment. For much of the postwar period, local residents had little choice but to accept it or move away. Like polluted rivers and streams, air pollution was a fact of life in the industrializing South.

Air emissions from pulp and paper mills fell into three main categories: water vapor, accounting for most of the total volume; particulate matter; and trace gases and compounds formed during the production process. The specific mix of pollution from a mill depended on the type of pulping process used, the type of fuel burned, and the extent of pollution control. For Kraft or sulfate pulp and paper mills, the most common in the South, the major air pollution concerns included particulates and total reduced sulfur (TRS) compounds, particularly hydrogen sulfide and methyl mercaptons—the primary constituents of the distinctive odor. Combustion by-products such as carbon monoxide, nitrogen oxides, sulfur oxides, hydrocarbons, and volatile organic compounds are also present in virtually all pulp and paper mill emissions due primarily to the burning of fossil fuels for energy and chemical recovery systems.

All these pollutants posed dangers to human health and the environment. In addition to causing irritation of the lungs, respiratory tract, and eyes, small particulates could accumulate in the lungs and dissolve into the blood stream.[107] TRS compounds such as hydrogen sulfide blocked oxygen transfer in the blood in a manner similar to cyanide. Even at relatively low concentrations (between 22 and 42 parts per million) hydrogen sulfide exposure led to a variety of problems, including respiratory tract irritation, pulmonary edema, cardiovascular damage, disturbed equilibrium, nerve paralysis, loss of consciousness, and circulatory collapse. At high concentrations these compounds could be lethal.[108]

Yet, for much of the post–World War II period, there was a general consensus among government officials and industry leaders that air emissions from pulp and paper mills posed little danger to human health.[109] Part of this stemmed from the lack of good data on air pollution and its health effects. The prevailing view held that although some components of pulp mill emissions might be harmful at higher concentrations, the relatively low concentrations found in the air around mills did not pose a health threat. Witness the words of Dr. Harry W. Ghem, technical director of the pulp and paper industry's National Council for Stream Improvement: "Pulpmill emissions are not harmful to human health in the concentrations encountered in the ambient air. Concentrations of those gases which could be considered harmful to health, such as hydrogen sulfide and sulfur dioxide, are present at well below minimal adverse levels."[110]

The validity of such a statement depended, of course, on several assumptions: namely, that there were "threshold" levels or concentrations of these pollutants that could be identified as "safe" and that the pollutants were quickly dispersed once emitted. The first assumption was increasingly subject to question by a growing body of physiological and toxicological evidence suggesting that there may in fact be no safe levels for certain common air pollutants.[111] The second assumption failed to acknowledge that people living close to or directly downwind from an air pollution source typically received a much higher dose than that indicated by average, ambient concentrations.[112] Particulate concentrations in Savannah during the 1960s and 1970s varied considerably depending on how far away one lived from the Union Camp mill. In the primarily black, low-income neighborhood close to the mill, the Chatham County health department found that average particulate levels in 1970 were two to five times higher than in the more affluent white neighborhoods farther away.[113] Like the distribution of most other environmental hazards, the risks associated with air pollution tended to map hierarchies of race and class.

As with early water pollution problems, individuals whose property or well-being was harmed by the mills' air emissions could bring nuisance or trespass suits against the mill owners. In contrast to the relatively large number of nuisance actions in the water pollution context, however, there does not appear to have been a single nuisance action brought in state or federal court against a southern pulp and paper mill solely on the basis of the harms caused by its air emissions.[114] In part, this was because the direct environmental damage caused by the mills' air emissions was less visible than water pollution, even though the overall impact on public health and the environment was equally, if not more, serious. It also reflected the fact that air pollution was often viewed as an inevitable by-product of the productive use of natural resources, with large industrial facilities receiving as much right to use the air as anyone else. Finally, given the general embrace of balancing in modern nuisance law, most plaintiffs had no real chance of stopping or significantly altering the practices of large industrial concerns given the broad "societal benefits" that they delivered in the form of investment, tax revenues, and jobs.[115] As long as the courts provided the exclusive legal avenue for combating air pollution, industry had little to fear.

Early attempts to control air pollution through statutory means did not offer much of a corrective.[116] While some county governments began passing air pollution control laws in the early twentieth century, they were uneven and inadequately enforced. State air pollution control efforts stumbled along during the 1950s and 1960s in an ad hoc fashion, with only limited effect.[117] Most southern states did not take up the issue until the 1960s, and even though the federal government started to address air pollution in the 1950s, it followed much the same course as it did in the water pollution area, starting off with limited assistance for research and state pollution control efforts and only moving into a regulatory posture in 1970.[118]

By the 1960s, however, pressure for a stronger federal presence in air pollution control was mounting. In 1961, President Kennedy declared, "We need an effective Federal air pollution control program now."[119] Beginning in 1963, hearings were held in both houses of Congress to debate bills aimed at strengthening federal air pollution control efforts.[120] As with early water pollution hearings, some participants began to frame the problem in national terms. Rep. Kenneth Roberts of Alabama, whose Birmingham constituents had suffered an "air pollution episode" in 1962, opened the House hearings in March 1963 by declaring that the legislative goal was "to provide for a comprehensive national effort to control air pollution which presently jeopardizes the health and well-being of our Nation."[121]

The key question, of course, concerned the appropriate division of labor between federal, state, and local governments. Despite his concern for the national implications of air pollution, Rep. Roberts, like many of his colleagues, saw no compelling reason why the federal government should be involved in abatement and enforcement activities. At the Second National Conference on Air Pollution, in December 1962, he had stated in no uncertain terms that the federal government did not have "any business telling people of . . . Birmingham or Los Angeles how to proceed to meet their air pollution problems."[122]

Representatives from state and local governments, along with those from industry, agreed. Leading the charge were advocates for major polluting industries. In a statement submitted at the hearings, Robert E. O'Connor of the American Paper and Pulp Association outlined his industry's position: "Our industry believes that local and State rights and responsibilities in the prevention and control of air pollution should be preserved, protected, and enhanced, as these local agencies, through cooperative community action can most adequately fit control regulations and enforcement activities to local differences in topography, type, and degree of industrialization within the community." He urged, therefore, "that enforcement and control programs be kept at the local level where the problems can best be solved by cooperative community action" and that the federal role be limited to one of "research and advice."[123]

Others felt differently. Ivan A. Nestingen, undersecretary at the Department of Health, Education, and Welfare, argued that "unless there is an increased

Federal effort in air pollution control, the situation existing today is going to become much more aggravated." In his view, the "very nature" of the problem required that the federal government play a key role not only in "stimulating the establishment and improvement of State and local programs" but also "in reinforcing them in certain enforcement situations." Members of the nascent environmental community echoed these concerns. In a statement submitted on behalf of the Citizens Committee on Natural Resources, Dr. Spencer M. Smith summarized the dilemma facing advocates of stronger pollution control programs: "We suspect that local and State enforcement will not be successful in dealing with air pollution by an industry, national in character, possessing great economic power . . . [I]t seems most desirable that a more rapid bringing into play of Federal resources is necessary if effective enforcement is to be achieved . . . Local and State governments have not been too successful in the past in achieving the kind of pollution abatement that is necessary."[124] As for the actual record of state and local pollution control efforts during the early 1960s, it was abysmal. According to a 1962 report from the U.S. Public Health Service, some 60 percent of the nation's population (approximately 107 million people) lived in areas with air pollution problems, 43 million in areas with "major" problems. Although the number of people living in such problem areas had increased by some 23 million people since 1950, only seventeen states had air pollution programs with expenditures of $5,000 or more per year. Depending on how one interpreted "enforcement," only four to six states actually enforced pollution regulations.[125]

Pollution control efforts at the local level were even worse. Of 218 urban areas with more than 50,000 residents and major or moderate air pollution problems, only 119, or 45 percent, were served by an air pollution control agency. For the country as a whole, there were only eighty-six local air pollution control agencies spending at least $5,000 annually, despite the fact that there were roughly 7,300 communities with major air pollution problems. Almost 60 percent of all local agencies were one- or two-person operations. Taken as a whole, the median per capita expenditure for all local air pollution programs in the country was about eleven cents a year.[126] In the early 1960s, arguments that state and local regulation provided the most appropriate means of controlling air pollution faced some fairly stubborn facts.

After almost a year of debate, Congress passed air pollution control legislation in December 1963. President Johnson promptly signed it into law. Known officially as the Clean Air Act of 1963, the law provided for expanded financial and technical assistance to state and local air pollution control agencies and more research.[127] As for enforcement, the Clean Air Act followed the procedure laid out in earlier water pollution control legislation whereby the secretary of health, education, and welfare could convene a conference of relevant parties to resolve particular problems of interstate air pollution.[128]

Though clearly limited, the enforcement conference provision, and the Clean Air Act in general, did succeed in highlighting the national character of the air pollution problem. Air, like water, was coming to be seen as a limited resource that required careful attention to governance issues in the context of a rapidly expanding economy. As Senator Muskie noted in his subcommittee's report on the 1963 legislation:

> Air is probably the most important of all our natural resources. Everyone is aware that we need fresh air every few seconds in order to live. Less well known are the enormous demands upon our air supply—measurable in thousands of cubic miles annually—to sustain our modern technological way of life . . . Polluted air, like polluted water, is costly to our economy as well as a hazard to our health . . . [T]o provide safe air for its citizens, the country must curtail the discharge of pollutants into the air . . . Polluted air is not contained in a specific area but is carried from one political jurisdiction to another. It does not know State lines or city limits. Providing air of good quality to all of our people is a challenge and an obligation for Government operations on all levels.[129]

As Muskie traveled throughout the country in 1964 holding field hearings on air pollution, he witnessed firsthand the public's growing dissatisfaction with existing state and local air pollution programs.[130] Meanwhile, new research reinforced claims regarding the chronic health effects of exposure to particular types of air pollution.[131] After a four-day inversion in New York City in late 1966 left some eighty people dead, a consensus began to emerge both within the government and among the public that existing efforts were inadequate. In response, the Johnson administration proposed a major air pollution initiative, the Air Quality Act of 1967, which included national emission standards for major industrial sources and called for the establishment of regional air quality commissions to enforce pollution control in "regional airsheds" that included multiple state or local jurisdictions.[132] This represented a significant departure from previous legislation.

During the spring of 1967, Muskie's subcommittee held a series of hearings on the proposal, with the proposed national emissions standards occupying the heart of the debate. Those supporting such standards cited broad public concern about air pollution and its health effects, and asserted that state and local governments were unable (and in some cases unwilling) to impose emission standards on large industries. Dr. John T. Middleton, director of the National Center for Air Pollution Control, summed up the issue:

> Perhaps the most important of the factors that tend to discourage standard setting at the State and local levels is that such action seems inevitable to bring one major function of the State or local government—the protection of public health and welfare—into direct conflict with another—that of ensuring economic growth. No matter how often we remind ourselves that effective control of air pollution is not incompatible with economic progress, the history of air pollution control efforts in this country provides abundant evidence that State and local officials are unable to take decisive action to adopt and enforce effective standards for the control of sources unless the problems have become so obvious, so severe and obnoxious as a nuisance that they cannot be tolerated.[133]

One only had to look at Georgia's Air Quality Control Act adopted during this same year to see the logic of Middleton's position.

Like its water pollution control law, Georgia's air pollution law bore industry's unmistakable influence. During the state's 1966 legislative session, a relatively strong air pollution bill was introduced that aimed "to preserve, protect, and improve the air quality of [the] State so as to safeguard the public health, safety and welfare of the people of the State." The bill died in committee, whereupon Glenn Kimble of Union Camp offered a new proposal to the State Air Pollution Study Committee. Claiming to represent "all Georgia Industry," Kimble argued for a law that would "provide the safeguards and flexibility which are essential to prevent the crippling of segments of our economy by trying to reach Utopia too fast."[134] His proposal served as the basis for Georgia's 1967 Air Quality Control Act, which sought "to preserve, protect, and improve the air quality of [the] State so as to safeguard the public health, safety, and welfare of the people of the State consistent with providing for maximum employment and the full industrial development of the State."[135] The act further declared that standard setting by the state would have to be consistent with "the predominant character of development" in particular areas. For "highly-developed industrial areas," such as those occupied by Union Camp, standards could be adjusted to provide more leeway, regardless of the effects on nearby residents. Finally, the act sharply limited the financial and staff resources of the agency responsible for administering it.[136] As far as the State of Georgia was concerned, air pollution legislation was not about to get in the way of industrial progress.

And Georgia was hardly the exception. North Carolina adopted similar legislation, requiring that air pollution control efforts be consistent with the "maximum employment and full industrial development of the state."[137] In South Carolina, International Paper played a very active role in writing the state's air quality legislation. According to Ralph W. Kittle, a company vice president, such "coordination and participation" between the industry and state officials produced "effective, realistic air management programs . . . In the State of South Carolina, the pulp and paper industry . . . is taking an active part in drafting an enlightened state air pollution control program with which both industry and the citizenry can live and be proud. By such action, we can continue to help our state and local governments exercise responsible and effective control over their own environment."[138]

While southern state governments actively sought to accommodate industry in passing air pollution control legislation, the national trend in state and local air pollution control efforts across the country was not exactly promising. In Dr. John Middleton's view, and that of the Johnson administration, applying uniform national emission standards looked increasingly necessary in order "to insure that major industrial sources of air pollution, wherever they are located, would not, in themselves, result in the exposure of people and property to harmful or damaging levels of pollution."[139] At bottom, it was an issue of fairness. State and local governments, of course, would be free to adopt more stringent standards if they so desired.

Following much the same script that they had used in their opposition to national standards for water pollution control, industry representatives voiced unanimous opposition to national air quality standards. Prominent among them were representatives from the pulp and paper industry. Echoing testimony from previous hearings, J. O. Julson of the American Paper Institute reminded the Muskie subcommittee that his industry was "vitally concerned with the future air quality of [the] Nation," and urged that any new air pollution control program "be based on a clear delineation of responsibilities between the National and State Governments . . . New legislation should be consistent with the principles of our Federal system and be based on a recognition that a workable program depends primarily for its success on the State and local authorities acting in concert with the National Government. The task is too complex for the National Government to solve largely by itself."[140]

Julson then pledged his industry's support for the amendments offered by Senator Jennings Randolph of West Virginia, a strong supporter of the coal industry and the influential chairman of the Senate Public Works Committee, which oversaw Muskie's subcommittee.[141] Randolph's amendments removed the proposal for national emission standards, leaving standard-setting activities to state and local governments. According to Julson, this was the only way to build an effective air pollution control program. In his view, federal regulation of industrial emissions would actually be counterproductive in that it would "diminish the urgency, the incentive and the public demand for effective action by State and local governments." Given the differences in mill capacities, the age of machines, and pulping processes, moreover, Julson claimed that nationally uniform standards would "neither prevent some installations from being placed at a competitive disadvantage nor assure fairness to everyone." Instead of uniform standards, emission control requirements should be determined by the air quality needs of local areas as defined by state and local regulatory bodies. "Such an approach," Julson concluded, "would decentralize decisionmaking, assure reasonable attention to the economic and technological factors involved in carrying out the programs, and hopefully, avoid undue disturbance of the economy while protecting the public's health in every community or region."[142]

One area where the industry did reach out to the federal government, however, was in support for pollution control research and cost-sharing for pollution abatement. International Paper's Ralph W. Kittle argued that the costs of pollution control should be borne by society as a whole rather than solely by polluting industries: "As industrialization has passed along many benefits to society, sadly but truly, it has and must pass along to society some of its liabilities as well . . . Society . . . must share the burden of the basic and applied research for answers to problems, as well as the implementation of the technologies developed—all of which involve non-revenue producing capital and non-product oriented improvements . . . While clean air exacts its price from us all, each of us must be willing to provide the time, money, and effort required to assure that each next breath will be as pleasant as it is necessary."[143] A memorandum from the American Paper Institute reinforced the point: "Because of the non-profit-making nature of [air pollution control technologies] . . . and because their principal contribution is for the benefit of the public, the government should make a significant contribution to their cost."[144] Several prominent senators agreed, including Muskie and Randolph, urging that any new legislation include tax breaks and other financial incentives to stimulate the development and installation of pollution control equipment.[145]

When Congress passed the 1967 Air Quality Act in November, the tax breaks were there, while the Johnson administration's proposal for national emission standards had been removed. Instead, the law directed the Department of

Health, Education, and Welfare to conduct a two-year study of the standards issue in order to develop so-called Air Quality Criteria and Control Technology Documents, which would then serve as the basis for state standard setting and enforcement. As it turned out, the act proved to be as ineffective as previous legislation. For the most part, state and local air pollution control authorities continued to stagger along—underfunded, understaffed, and politically compromised. By 1970, only twenty-one states had actually developed implementation plans designed to achieve specific air quality goals, and none of these plans had been approved by the federal government.[146]

With air quality problems worsening and in the midst of the groundswell of national environmental concern then sweeping the country, President Nixon sent Congress a strong set of air pollution control amendments in early 1970. Among other things, Nixon's proposed amendments effectively resurrected Johnson's earlier proposal, calling for the establishment of national air quality standards and national emissions standards for new stationary sources and for sources of certain hazardous pollutants.[147] After some additional strengthening in the Senate at the hands of Senator Muskie (who at the time was seeking the Democratic presidential nomination), the Clean Air Amendments of 1970 passed the Senate without a single opposing vote; Nixon signed them into law on December 31, 1970.[148] Though state and local governments would still play important roles in air pollution control efforts (and would become key partners in the air pollution control effort under the model of cooperative federalism at the heart of the new law), the 1970 amendments contained substantial provisions for federal standard setting and enforcement.[149]

The major overhaul of federal air pollution policy contained in the 1970 amendments represented a significant change in the regulatory landscape. Despite their best efforts to retain state and local authority over air pollution control, pulp and paper companies now had to contend with the newly created Environmental Protection Agency and a whole series of requirements under the Clean Air Act. In general, the law charged the EPA with establishing National Ambient Air Quality Standards (NAAQS) for criteria pollutants such as lead, sulfur dioxide, and particulate matter, the last of which was an important constituent of paper mill emissions. These standards were to be uniform across the country, regardless of differences in the costs and benefits of controlling pollutants in specific areas.[150] As originally conceived, all areas of the country were to be in compliance with the NAAQS by 1975.[151] In order to ensure compliance with the NAAQS, states were required to develop state implementation plans (SIPs) to identify sources of air pollution and to determine what reductions were necessary to meet the federal air quality standards.[152] Other provisions aimed at regulating certain hazardous air pollutants and controlling tailpipe emissions from mobile sources such as cars and trucks.[153]

The Clean Air Act also authorized the EPA to establish nationally uniform emission standards for new or substantially modified stationary sources in particular industrial sectors.[154] Known as New Source Performance Standards (NSPS), these would be based on the "best system of emission reduction," with the somewhat ambiguous requirement that pollution control technologies be affordable by affected parties.[155] In 1978, the EPA promulgated New Source Performance Standards limiting emissions of particulate matter and total reduced sulfur (TRS) compounds from new or modified Kraft pulp and paper mills.[156] Any pulp and paper firm wishing to build a new mill or substantially modify an existing one after 1979 would have no choice but to install the best demonstrated technology to control these pollutants.[157]

As for existing facilities, the Clean Air Act left primary responsibility for establishing emission controls with the states, although the NSPS requirements, despite their name, did include provisions for "existing facilities," and any modification at an existing mill that went beyond routine maintenance or replacement would also trigger specific technology requirements tied to the air quality attainment status for the NAAQS.[158] Through their state implementation plans, which had to be approved by the EPA, states were to identify how existing sources would be controlled in order to achieve compliance with the NAAQS. If the EPA judged a state's plan to be inadequate, it reserved the right to step in and impose federal controls. By allowing state and local governments to retain substantial control over existing sources, therefore, the Clean Air Act gave them a certain political flexibility when it came to the sensitive issue of balancing the demands for pollution control with the imperatives of economic development.[159]

For older pulp and paper mills, such as Union Camp's Savannah complex, these provisions meant that the firm would continue to deal primarily with state and local governments in negotiating air pollution controls. Given the state's past practices of accommodating industrial concerns, Union Camp more than likely felt quite relieved that it did not have to deal directly with the EPA on air matters. In 1978, when the Georgia legislature amended the state's air pollution legislation, it effectively removed any doubt about where the state stood on air pollution control. Echoing its 1967 legislation, the new Georgia Air Quality Control Act declared that it was the "public policy of the State of Georgia to preserve, protect, and improve air quality and to control emissions to prevent the significant deterioration of air quality and to attain and maintain ambient air quality standards so as to safeguard the public health, safety, and welfare consistent with providing for maximum employment and full industrial development of the state."[160] As far as existing pulp and paper mills in Georgia were concerned, therefore, the new law turned out to be less of an immediate problem than previously expected—at least for a while.

For companies that continued to run the older, grandfathered mills, the real threat came from citizen activism. If the past was any guide, however, such grassroots politics posed little threat to large mills such as Union Camp's facility in Savannah, which provided good jobs and tax revenue to the local economy and had rarely faced much opposition. Yet times were changing. Beginning in the late 1970s, a group of environmental activists succeeded in making air pollution a major political issue in Savannah. These local residents were not satisfied with existing laws and did not particularly care that Union Camp was in "full compliance" with air pollution control regulations. They wanted clean air.

The campaign for clean air in Savannah began in 1978, when the Georgia Conservancy, a statewide environmental organization, secured funding from the EPA and several private groups to initiate the Chatham County Air Pollution Project. After a year of research, the Conservancy released its report, which provided detailed information on the air pollution problem in the area, an analysis of existing regulatory efforts, and recommendations for dealing with the problem. The report noted that Chatham County was the most industrialized area in the state of Georgia, with at least thirty major industrial sources of air pollution, and thus suffered from one of the most serious air pollution problems in the state.[161] As the largest facility in the county, Union Camp's complex accounted for a substantial share of the air pollution problem.[162] For major criteria pollutants such as particulates, the mill dwarfed all other sources, accounting for more than 60 percent of all industrial emissions (figure 4.3).[163]

As for the actual quality of the air in Chatham County, the Georgia Conservancy found existing information to be highly inadequate. Only a few pollutants were even monitored, and all measurements were based on twenty-four-hour averages, masking wide variations in pollution loads at particular times and places. As the report put it: "More is unknown than is known about the contents and effects of Savannah's air; more pollutants are unmeasured than are measured."[164] Still, ambient air monitoring indicated that particulate levels in the area surrounding the Union Camp mill were significantly higher than permitted by the National Ambient Air Quality Standards. Under the Clean Air Act,

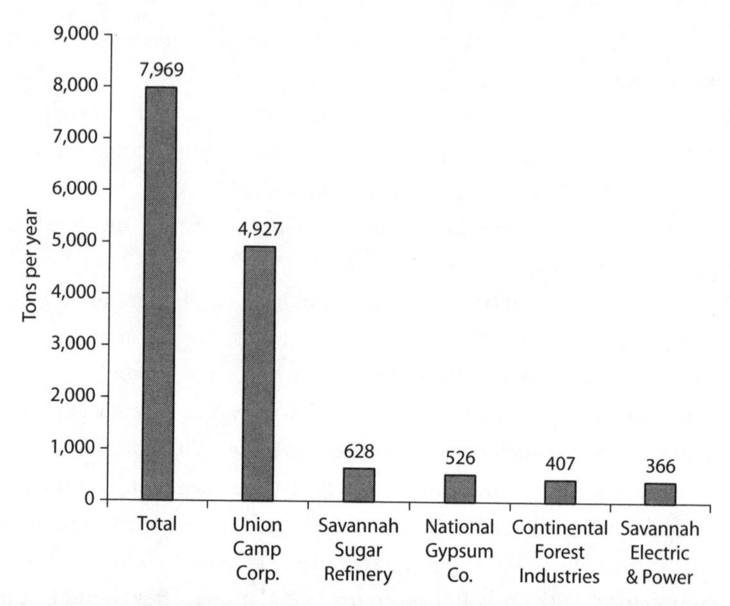

Figure 4.3. Major industrial sources of particulate pollution in Chatham County, Georgia, 1979. U.S. EPA, National Emission Data Systems, Region IV, January 1979, cited in Georgia Conservancy, *Air Pollution in Savannah: A Report on the State of the Air in Chatham County, Georgia, and What Can Be Done to Improve It* (Savannah, GA, 1979), 7.

such "non-attainment" status triggered additional requirements and controls to deal with the problem. Union Camp was subsequently required to install pollution control devices known as electrostatic precipitators to reduce its particulate load. By 1981, with particulate levels significantly reduced, the EPA took the Savannah area off of its "dirty air" list of nonattainment areas for particulate matter.[165]

Particulates, however, represented only one of several air pollution concerns raised by the Georgia Conservancy report. A more tangible concern for many local residents was the pervasive odor problem associated with Union Camp's total reduced sulfur (TRS) emissions. Because these compounds could be detected by the human sense of smell at concentrations of only a few parts per billion, solving the odor problem was an expensive proposition. Although EPA issued guidelines for the control of TRS emissions from existing Kraft pulp mills in 1979 under its NSPS authority, the agency gave states considerable flexibility in determining specific emission controls.[166] Indeed, despite the EPA's finding that "short-term effects (vomiting, headaches, shortness of breath, dizziness) occur[ed] in some individuals after prolonged exposure" to TRS emissions from Kraft pulp mills, the agency did not consider the ambient ground-level concentrations of TRS compounds near uncontrolled mills to be a public health hazard.[167] As far as the EPA was concerned, reducing TRS emissions from existing pulp mills was a state prerogative.

For residents of Savannah and other Georgia towns who looked to the state for action in reducing TRS emissions, the prospects at the end of the 1970s were not exactly promising. In 1979, in an effort aimed ostensibly at protecting the competitiveness of the state's pulp and paper industry, the governing board of the Georgia Department of Natural Resources (the agency responsible for controlling air pollution from existing sources in the state) declared that Georgia would take no action regarding TRS compounds until ten other states had submitted plans with TRS rules to the EPA.[168] As it turned out, the ten-state trigger effectively delayed state efforts to control TRS compounds for almost a decade.[169]

In the meantime, citizen activism in Savannah grew as did distrust for the state's environmental regulators. For its part, the Georgia Conservancy offered a serious indictment of the state's approach to air pollution control: "While people in Savannah are offended and concerned by industrial emissions, the Georgia Environmental Protection Division allows industry to do their own testing and gives them advance notice prior to inspection visits."[170] The air pollution problem in Savannah, according to the Georgia Conservancy, stemmed in part from the cozy relationship that existed between state regulators and the business community. Consequently, the report recommended stronger monitoring and enforcement of existing sources of air pollution, careful evaluation of new industries seeking to locate in the area, and improved public information. The overall aim was to make air pollution a major political issue—to stimulate greater citizen action in order to put pressure on elected officials.[171]

Local business leaders, of course, felt differently and attacked the report, suggesting that the authors were willing to trade jobs for environmental quality. As Tom Coleman of the Savannah Chamber of Commerce put it: "If we don't approach air quality problems with conclusive evidence, we won't get the opportunity to be selective about what industries we bring in here, because none will come . . . If you are going to clean the air, it is important to have someone here to breathe it."[172]

Given the state's position on TRS emissions, it looked as if the business community would carry the day. Efforts to persuade the Georgia Environmental Protection Division (EPD), which was under the state Department of Natural

Resources, to regulate air emissions from Union Camp and other large industries quickly ran into resistance. Because the governor appointed every member of the governing board of the Department of Natural Resources, as well as the director of EPD, finding a sympathetic ear among state regulators depended at least in part on where the governor came down on industrial pollution. Georgia's governor at the time, George Busbee, had always been a close friend of industry—the embodiment of the business-hungry New South politician. One did not have to look far for evidence of his pro-industry leanings. A mere two weeks after Busbee finished his second term in 1982 (he was barred from seeking a third term), he joined Union Camp's board of directors.[173] Not surprisingly, during the early 1980s, efforts by Georgia Conservancy members and other concerned citizens to persuade state regulators to do something about Savannah's air pollution problem met with little success. Queries to the EPA, which was then in the midst of Ronald Reagan's first-term assault on federal environmental regulation, proved equally futile.

In the face of such disappointments, several local air pollution activists decided to form a grassroots organization in Savannah to focus specifically on Union Camp's contribution to the air pollution problem in the area. Established in 1982 and known as the Citizens for Clean Air, the group's mission statement framed the air pollution issue in rights-based terms: "Citizens for Clean Air claims for the citizens of Chatham County the right to breathe clean air that does not smell bad." According to Dr. John Northup, a local physician and the group's leader, Citizens for Clean air sought to focus public pressure on county officials and, by pushing them to adopt a local air pollution ordinance, hoped to force the state and the industry to deal with the pollution problem. To that end, they quickly realized that the air pollution issue would have to be condensed into a marketable form that would stimulate public action. "Clean air that does not smell bad," a phrase borrowed from a young elementary school student, became the slogan for their campaign. People from all over Savannah—black and white, rich and poor—began to view the air pollution issue in a different light. Unwilling to accept air pollution as the "price of progress" or the "smell of money," concerned citizens started to assert their right to clean air. As local pressure grew, county commissioners and Union Camp officials found themselves in the increasingly uneasy position of defending the sanctity of industrial development in what was, at the time, the most polluted county in Georgia.[174]

By the mid-1980s, the conflict over Savannah's air had reached fever pitch. In 1986, state attorney general Mike Bowers cleared the way for the Chatham County commission to enact a local air quality ordinance that exceeded state and federal regulations if it so desired.[175] With legal obstacles out of the way, the pressure on the county commission increased considerably. John Northup and the Citizens for Clean Air began to take a more aggressive stance with Union Camp. "People in Chatham County," Northup argued, "are really deciding that Union Camp's image as a nice guy is nonsense. The fact of the matter is that they are putting toxic stuff in the air and people don't like it."[176]

In response to the growing controversy, the county commission appointed an Air Quality Task Force to consider an air pollution ordinance. Meanwhile, Union Camp began a $1 million study of the air pollution problem at the plant as part of a larger, long-term mill modernization program. Mill manager Jim Petite, however, cautioned local residents that the mill had actually lost money during three out of the previous four years and that costly air pollution controls might force it to shut down. Union Camp, he noted, was in full compliance with all state and federal air pollution control laws. Anything else, Petite suggested, might endanger the competitiveness of the mill and, by extension, the thousands of jobs it provided.[177]

Citizens for Clean Air refused to back down, stressing the potentially toxic effects of pulp mill emissions. Northup, a medical doctor, reminded his fellow citizens, "Hydrogen sulfide, of which Union Camp emits 15,000 pounds a day, works on your body like cyanide. Except that it is more potent."[178] Other members of the group raised the issue of corporate responsibility. "They [Union Camp] are going to stall and stall and stall until we put them in a position where they have no choice," member William Dickinson stated. "It boils down to one question. How do you convince a profit-motivated organization to spend non-revenue producing funds? I'm a businessman too. I don't like to spend money that's not going to make money."[179]

Meanwhile, the Air Quality Task Force that had been appointed by the county commission issued a report recommending unanimously that the county commissioners pass an air quality ordinance. Citizens for Clean Air then drew up a draft ordinance and presented it to the county commission, along with a petition of more than five thousand signatures. Time was running out for the county commissioners. The passage of a local air pollution ordinance looked like a real possibility. Yet just at this moment, perhaps by coincidence, the state Environmental Protection Division stepped into the controversy and activated its TRS regulations for pulp mills—nine years after passing its ten-state trigger rule.[180] Shortly thereafter, the head of Union Camp flew to Georgia to announce that the company would reduce its TRS emissions as part of a $375 million modernization program ($44 million of which would go to air pollution control).

According to Union Camp CEO Gene Cartledge, the modernization program would "eliminate 95 percent of the emissions of the mill's odor and [would] ensure that the mill [would] be an economically viable contributor to the local economy well into the next century."[181]

Representatives from the company emphasized that the firm was not bowing to local pressures but rather that the modernization program had been in the works for some time and was dictated purely by economics and the new state rules for TRS emissions. Citizens for Clean Air, however, argued that the threat of a county air pollution ordinance was the impetus behind the program, suggesting that the company had timed the announcement of its modernization program to head off a local ordinance.[182] Whatever the real reasons behind the modernization program, Citizens for Clean Air got the emissions reductions they were seeking, while the county commissioners quietly dropped the idea of a local air pollution ordinance.

But this hardly brought an end to public pressures on Union Camp to reduce its air emissions. Beginning in 1989, the EPA started releasing its annual Toxics Release Inventory (TRI) reports. Mandated by the 1986 Emergency Planning and Community Right-to-Know Act,[183] which had been enacted in the wake of the Union Carbide disaster in Bhopal, India, these reports provided information on toxic discharges (to air, water, and land) from industrial facilities throughout the country. When the first report came out for the state of Georgia (for the year 1987), Union Camp was quickly tagged as the state's number one polluter while Chatham County earned the dubious honor of most polluted in the state.[184] Headlines brought unwelcome publicity to the county and to Union Camp, putting added pressure on the firm to control its toxic emissions.[185]

Most of the compounds reported under TRI were previously unknown to the public.[186] More common regulated pollutants, such as particulates or TRS compounds, were not included. For pulp and paper mill towns such as Savannah, the major TRI air emissions of concern included methanol, acetone, vapors of sulfuric and hydrochloric acid, and toluene.[187] To the ears of average citizens, these were ominous sounding chemicals. Industry officials, however, raced to point out how small the concentrations of such pollutants actually were. Union Camp spokesman Bill Binns argued that his company's releases were "not in concentrations that cause a health hazard. Once they get into the atmosphere, the concentrations become very small."[188]

Concerned residents found little comfort in such pronouncements. Many expressed shock and deep concern over the new information. Even Dr. John Northup of Savannah's Citizens for Clean Air was caught off guard. "We used to be told it was just steam coming from the plants. Some of us were absolutely floored when we learned what was really coming out." He then went on to suggest a link between pollution levels and elevated disease in Chatham County: "We can't fully prove it, but I believe it is no coincidence that we have such high rates of cancer and heart disease, and we are the most polluted county in Georgia."[189] This was strong stuff, and the industry was going to have a hard time stemming the tide of public criticism if it did not demonstrate real progress in reducing its toxic emissions.

Consequently, subsequent TRI reports for Georgia indicated significant reductions in toxic emissions from the state's industrial facilities. Though some of these reductions occurred because certain chemicals were taken off the list (while others were added), the lion's share stemmed from real efforts to reduce toxic releases, primarily through process improvements and by switching to other chemicals. To take one example, total TRI releases (air, land, and water) from Union Camp's Savannah mill declined from more than 16 million pounds in 1988 to slightly more than 3 million pounds in 1997.[190] By 1994, the company no longer ranked as the state's number one polluter. Headlines in local papers began pointing to the progress being made, rather than the pollution being released.[191] Although TRI was hardly problem-free, its overall impact illustrated the power of information and public disclosure, rather than top-down regulation, in facilitating pollution reduction. In the words of one environmental activist from Savannah: "I'm a strong believer in the Toxics Release Inventory and community-right-to-know . . . If you've got to really go and get on the scales, then you might go on a diet, and I think that's why the Toxics Release Inventory is effective."[192] Even the industry admitted the power of TRI. As one pulp and paper executive put it, "TRI is a very powerful motivating tool."[193]

Dioxin

> The most alarming of all man's assaults upon the environment is the contamination of air, earth, rivers, and sea with dangerous and even lethal materials. This pollution is for the most part irrecoverable; the chain of evil it initiates not only in the world that must support life but in living tissues is for the most part irreversible. In this now universal contamination of the environment, chemicals are the sinister and little-recognized partners of radiation in changing the very nature of the world—the very nature of its life. —*Rachel Carson, 1962*[194]

Writing in the early 1960s, Rachel Carson concerned herself primarily with the impacts of DDT and other synthetic organic pesticides on human health and the environment. Challenging conventional views, she demonstrated the many interconnections between agro-industrial practices such as pesticide use and widespread environmental contamination. Her best-selling book, *Silent Spring*, was transformational, warning of dangers that were largely invisible to most people. As such it was both of the moment and ahead of its time. Indeed, many have suggested that it provided the initial spark that ignited the modern environmental movement—a "cry in the wilderness . . . that changed the course of history" according to Vice President Al Gore.[195] By forcing Americans to look anew at the *ecology* of certain industrial practices, Carson revealed the risks and vulnerabilities emerging from the very way modern industrial society produced its food. Hers was a story about connections and unintended consequences.

Silent Spring thus had a relevance that went far beyond the issue of pesticides and the environmental effects of chemically intensive agriculture. By pointing to the persistent nature of toxic substances, their movement upward through the food chain to inhabit living tissues in ever greater concentrations, and their ability to cause significant and lasting damage to human health and the environment at small concentrations, Carson challenged more traditional views of pollution. Microtoxicity mattered as much as, if not more than, gross insults to the environment. The challenges for regulation were immense. During the 1963 hearings on water pollution, Senator Muskie, no doubt aware of Carson's book, spoke of the difficulties posed by toxic substances: "The emerging problems involving chemical, pesticide, and radioactive wastes are filled with unknowns which need answers before adequate treatment methods can be developed. Every effort is needed, of course, to safeguard the public health and welfare through effective controls until research has found these answers."[196] These sorts of problems were different from traditional pollution concerns, such as the biochemical oxygen demand of wastewater or airborne particulates. Toxic substances released into the environment often persisted for long periods, were able to travel long distances, and tended to bioaccumulate in living systems. Once released, these substances could not simply be reconcentrated, isolated from the environment, and safely disposed. Efforts to regulate toxics (in air, water, land, and food) thus followed different trajectories, some proceeding independently from and others overlapping with traditional air and water regulation.[197] In the process, the science and practice of risk assessment came to occupy a central and highly contested place.

For the pulp and paper industry, toxics emerged as an increasing source of concern in the late 1970s and early 1980s, given the growing recognition that the industry's air and water pollution contained toxic substances.[198] But it was not until the mid-1980s, with the discovery of potential dioxin contamination in pulp and paper mill effluent, that the issue took center stage. Made famous by the controversies over Agent Orange in the 1970s and various toxic disasters such as Times Beach, Missouri, and Seveso, Italy, dioxin is the name of a family of some seventy-five related organochlorine compounds.[199] Like other chlorinated hydrocarbons, dioxins are extremely stable and persist in the environment for decades. These compounds are fat-soluble, and they bioaccumulate.[200] As a result, dioxins are found in living systems all over the world and are believed to inhabit the tissues of every person living in the United States. During the 1970s, researchers labeled dioxin "the most potent animal carcinogen ever tested"—the "Darth Vader" of chemicals, as one observer put it.[201] Both the World Health Organization and the U.S. Department of Health and Human Services have classified dioxin as a "known carcinogen." The EPA has labeled it "carcinogenic to humans."[202] More recently, attention has focused on the role of dioxin and dioxin-like substances in mimicking certain hormones and acting as endocrine disruptors—leading to a variety of environmental and health effects that have only just begun to be investigated.[203]

The issue of dioxin contamination from pulp and paper manufacturing arose rather unexpectedly in the mid-1980s during the EPA's National Dioxin Study.[204] The study, which had been undertaken after the chemical disaster at Times Beach, detected surprisingly high levels of dioxin concentrations in fish in several streams that were supposed to be uncontaminated and had been chosen to serve as reference or background streams for the study. On closer inspection, researchers determined that all of these contaminated reference streams were receiving wastewater effluent from bleaching pulp and paper mills. In the words of one EPA official, "A major unanticipated finding of the National Dioxin Study was the presence of dioxin in fish that were not downstream of already known dioxin sources. A review of the data indicated a close association between the elevated fish dioxin levels and woodpulping facilities."[205] In a follow-up study of twenty-two pulp and paper mills in Wisconsin, Minnesota, and Maine during late 1985, the EPA detected dioxin in wastewater sludge samples. Agency scientists hypothesized that dioxin was being formed as a byproduct during the bleaching of woodpulp with chlorine and chlorine derivatives.[206] To test this hypothesis, the EPA, in cooperation with the American Paper Institute and the National Council for Air and Stream Improvement (the renamed National Council for Stream Improvement), performed a study of five bleaching pulp mills during the second half of 1986.[207] The findings of this so-called five-mill study were not encouraging. The EPA found dioxins or furans, a set of related compounds, in the wastewater effluents of four of the five mills, the pulps produced at all five mills, and the residual wastewater treatment sludges from all five mills.[208] This provided compelling evidence in support of the conclusion that the chlorine bleaching process used in certain mills did in fact produce dioxins and dioxin-like substances.

Word of the EPA's findings soon reached members of the environmental community, some of whom suspected that the agency was not disclosing all that it knew about the issue. In December 1986, Greenpeace activist Carol Van Strum filed a Freedom of Information Act (FOIA) request for all EPA documents pertaining to dioxin contamination from pulp and paper mills. The following month, a letter from an EPA official to the American Paper Institute was leaked to the press. It suggested that the EPA would notify the industry in the event of any FOIA requests and would not release any results until publication of the five-mill study. The activists began to suspect collusion. In August 1987, Van Strum, along with Greenpeace colleague Paul Merrell, released a report on the issue, *No Margin of Safety*, which pointed to the connection between dioxin and pulp and paper production and, among other things, accused the EPA of attempting a cover-up.[209] One month later, the EPA officially released its two-year National Dioxin Study, confirming publicly that dioxin contamination from chlorine bleaching at pulp and paper mills was a potential problem.[210]

Assuming that the EPA's findings were correct, small levels of dioxin might turn up not only in wastewater discharges from bleaching mills (and the fish downstream) but also in the residual sludges left over from wastewater treatment (which were sometimes used to fertilize nearby croplands) and in the actual product itself. Indeed, one of the implications of the five-mill study, which found traces of dioxin in the pulps produced at all five mills, was that dioxin might also be present in the consumer products made from these pulps.

Already aware of the possibility and seeking to prevent a full-scale public relations disaster, the industry came out with its own findings that did indicate trace amounts of dioxin in certain consumer products. Not surprisingly, the prospect of dioxin-laced diapers and milk cartons created a media frenzy. In September 1987, the *New York Times* broke the story with a front-page article. Alarm bells were sounding. Here was a possible pathway for dioxin to migrate into the human body—part of the reason, perhaps, why virtually all Americans had detectable levels of dioxin in their bodies. As Dr. Ellen Silbergeld, a toxicologist from the Environmental Defense Fund, pointed out: "People clearly are getting exposed to and absorbing dioxin . . . I think we are facing a national health problem."[211]

In response, Red Cavaney, president of the American Paper Institute (API), pointed out that there was no evidence to support the claim that these "trace levels" of dioxin in paper products posed a threat to human health, but he conceded that the industry would continue to study the problem. He also emphasized, however, that "there [was] no readily apparent solution to the problem." The EPA effectively agreed, ruling out public health risks from using paper products. According to the agency, however, dioxin contamination in pulp mill effluent and wastewater sludge did represent a potential problem that would require more study and might entail significant regulatory action.[212]

The whole issue clearly had ominous implications for the industry and, more important, for the communities in which it operated. The problem was potentially quite serious in the South, moreover, not only because of the industry's massive presence there but also because southern pine, with its high resin content, requires more chlorine for bleaching than other, lighter types of pulpwood. Of course, not all mills in the South or elsewhere, produced bleached pulp and paper. Union Camp's Savannah complex, for example, never bleached. The Stone Container mill at Port Wentworth, Georgia (several miles away) and the Georgia-Pacific mill at Brunswick (about fifty miles down the coast), however, were both bleaching mills, as were some three other mills in the state and more than fifty throughout the region.[213] If dioxin contamination turned out to be a major hazard, the South would be disproportionately affected.

Reflecting the increasingly politicized and adversarial nature of environmental politics, industry leaders mounted a swift and vigorous campaign to shape the public's perception of the dioxin issue. According to a series of internal documents from API that Greenpeace made public in 1987, the institute developed an elaborate and sophisticated "Dioxin Public Affairs Plan" on learning of the EPA's "unexpected" findings regarding dioxin contamination in pulp mill effluent.[214] "This was not an issue we wanted to leave to chance," noted Carol Raulston, vice president for government affairs at API, when asked about the industry plan.[215] In pursuing its strategy, API hired a Washington public relations firm, trained corporate executives in how to address the issue, and conducted consumer surveys. The trade association also targeted the EPA for "intelligence gathering" and developed a three-tiered "crisis management" network for rapid deployment if and when the issue went public. API president Red Cavaney noted in one memo to his executive committee that the organization's reason for getting involved in the joint five-mill study with the EPA was in part to "forestall major regulatory and public relations difficulties." Overall, API sought to convince the EPA to "rethink" its dioxin risk assessment and to decouple pulp mills' dioxin contamination from any sort of threat to public health and the environment. By developing relationships with people in the agency, working through "selective" congressional contacts, and marshaling a small army of outside experts, API hoped to shape the EPA's approach to dioxin risk.[216]

For a major consumer products industry facing a potential public relations disaster, API's strategy made perfect sense. Whether the strategy actually did result in significantly shaping the EPA's approach to the dioxin issue is difficult to know. The federal judge who heard Carol Van Strum's FOIA action against the EPA in 1987 seemed to think so. In reviewing some of the EPA and API documents submitted to the court by Van Strum, the judge concluded, "The documents appear to support the existence of an agreement between EPA and the industry to suppress, modify or delay the results of the joint EPA/industry study or the manner in which they are publicly presented."[217] At the very least, the whole episode worked to further politicize the issue, creating even more distrust among certain environmental groups and the public for the industry and the EPA.

In the midst of the ongoing media fest and legal battles surrounding the revelations, the EPA began coordinating with API and the National Council for Air and Stream Improvement to conduct a larger study of all bleaching pulp mills in the country. The "104-mill study," as it was known, began in the summer of 1988 and was funded entirely by the industry.[218] This would, it was hoped, provide a definitive picture of the extent of the dioxin problem. As the study was getting under way, however, it also became wrapped up in a consent decree that the EPA was negotiating with the Environmental Defense Fund and the National Wildlife Federation. Seeking to end a five-year lawsuit brought by these environmental groups, the agency agreed in July 1988 to carry out a comprehensive assessment of dioxins and furans in wastewater effluent, residual sludges, and pulp and paper products from all bleaching pulp mills in the country. More important, the agency also agreed to propose dioxin regulations by April 1991. This specific regulatory proposal eventually became part of a larger set of rules regarding water effluent standards and national emissions standards for hazardous air pollutants from pulp and paper mills that has subsequently come to be known as the "cluster rule."[219] The consent decree not only increased pressure on the EPA (and the industry) to conduct a comprehensive study of dioxin contamination from pulp and paper mills, but it also set a timeline for developing appropriate regulations. The regulatory wheels were starting to turn.

And then the hearings began. Just as the EPA was working out its consent decree and launching the 104-mill study, the House Subcommittee on Water Resources convened hearings at the request of Tennessee representative James Quillen on dioxin contamination in the Pigeon River, which runs through western North Carolina and eastern Tennessee.[220] But instead of focusing on the source of the contamination—the Champion International mill in Canton, North Carolina—Quillen and others directed the bulk of their attention (and anger) during the hearings to the EPA and its supposed mishandling of the situation. Though Champion was certainly at fault, they felt that the EPA bore primary responsibility for failing to protect public health and the environment. This animosity toward the EPA stemmed both from years of frustration among Tennessee residents and politicians regarding the pollution of the Pigeon River and from allegations then coming to light regarding the EPA's relationship with the paper industry and its stance on the dioxin issue. Hopes for a balanced discussion were quickly dashed.

In his opening remarks at the hearings, Rep. Quillen charged that "the Environmental Protection Agency . . . has failed to discharge its duty . . . [and] failed in its responsibility to protect our citizens and our river . . . This is why I asked for today's hearings, to get some answers to serious questions which have been raised, and which continue to be raised, about EPA's handling of this distressing situation." Don Sundquist, another congressman from Tennessee, weighed in with equal vehemence: "I want to know . . . why EPA has allowed . . . [this] tragic environmental problem to escalate to obscenely hazardous proportions. It seems to me that EPA should have long ago been monitoring the water quality of the Pigeon River. It could have prevented the excessive, lethal, toxic pollution found in the Pigeon River today."[221] The EPA, of course, had only been aware of the *potential* for such a problem since 1985 and had only confirmed the presence of dioxin in the river in April 1988 (three months prior to the hearings). To suggest that it could have acted to *prevent* the problem reeked of political posturing.

Still, given the revelations of the agency's cooperation with the paper industry on the dioxin issue, suspicions of agency foot-dragging and a certain pro-industry bias may have been justified. Others who testified at the hearing, such as Carol Van Strum, provided damaging testimony regarding the EPA's relationship with the American Paper Institute, bolstering earlier claims of agency misconduct. In her view, the "EPA's handling of the pulp mill dioxin problem should be viewed in light of its traditional behavior. Leaked American Paper Institute documents . . . show that collusion remains the norm between EPA dioxin regulators and regulated industries . . . EPA's dioxin regulatory program is paralyzed and has been from the very beginning . . . More lives are at stake than just the lives of those who live near the Pigeon River."[222]

For their part, EPA officials attempted to remind participants at the hearing that the agency had only recently become aware of dioxin contamination from pulp and paper mills, that North Carolina and Tennessee still retained primary responsibility for issuing discharge permits to mills such as Champion's, and that the so-called collusion between the agency and API was little more than cooperation aimed at reducing the costs and time delays that would have occurred if the EPA had pushed ahead without such cooperation.[223] Representatives from the area were not persuaded. Jerry Clevenger, chairman of Tennessee's Cocke County Commission, left little doubt about his own feelings: "We are beginning to have serious doubts about the integrity of the Environmental Protection Agency . . . The EPA has a mandate to protect the environment, not to compromise the beauty and cleanliness of any waterway, including the Pigeon River . . . Several experts contend, and we believe, that allowing the continuous dumping of dioxin into the Pigeon River will constitute a death sentence for people who live along or nearby the river." Jerry Wilde, president of the local Dead Pigeon River Council, charged that the EPA's "compromise" with Champion represented one more example of the company's license to pollute: "Over 80 years ago, the Pigeon River was sold to the highest bidder. Are people's health and welfare now being sold to the highest bidder? I call on you, members of Congress and of this committee, to ensure that the EPA does, in fact, protect our environment for us and our children and to see to it that the truth can no longer be hidden from the people."[224]

While the bulk of the criticism at the hearing was directed at the EPA, Champion also received some harsh words. As Rep. Quillen pointed out:

> It needs to be understood that the pollution of the Pigeon River began when the Champion Paper Co. started up its mill operations in Canton, N.C., along the river 80 years ago. Since that time, down to the present, this once lovely and sparkling clean mountain river has been transformed into a foul-smelling, foaming, sludge-filled mixture that looks like oily coffee and stinks like rotten eggs. This is how Champion makes its money. This is not my opinion. This is fact . . . What we in Tennessee are . . . demanding is that the Champion Paper Co. begin at long last to operate its Canton mill in a manner which does not continue to grossly pollute the Pigeon River and in a manner which does not dump dioxin and other toxic compounds and chemicals into the river. We simply do not accept the destruction

of the river and the present public health crisis as a routine part of the mill's operation.[225]

Bob Seay of the Newport / Cocke County Tennessee Chamber of Commerce expressed similar outrage: "Champion International has long ignored Appalachia. Mr. Andrew C. Ziegler, president of Champion International Crop., your day has come. The people have spoken. No longer will Cocke County tolerate your corporate behavior . . . Don't attempt to hide behind the American Paper Institute or the EPA . . . Cocke County says be a credible corporate citizen. Let the Pigeon live."[226] Representatives from Champion and from the American Paper Institute did not respond to any of these charges. Instead, they merely stated that they would continue to cooperate with the EPA on the dioxin issue.[227] Soon they would be facing a class-action lawsuit. In the meantime, dioxin became a full-blown national obsession.

Less than six months after the Pigeon River hearings, Rep. Henry Waxman of California convened his own set of hearings on dioxin.[228] Referring to it as "one of the most toxic chemicals ever created," Waxman spoke of the fear and cynicism spreading among the American public regarding the unseen dangers in consumer products: "Every day, it seems, we learn that something else causes cancer or is detrimental to our health. Now we learn that the dread chemical of Agent Orange and the chemical that forced the abandonment of Times Beach, MO, is in baby diapers and milk cartons." Waxman also pointed to other pathways of contamination: "Valuable farm lands and the crops grown on them are contaminated with dioxin from the use of pulp mill sludge as a soil supplement. Fish downstream from pulp and paper mills are contaminated with dioxin flushed from mills during the manufacturing process. Concern is raised for those working in pulp and paper mills . . . While I understand there is continuing scientific disagreement over some characteristics of the dioxin risk, I do not believe anyone would argue that avoidable exposure to dioxin is acceptable. So the question is being asked: Can we avoid exposure to the dioxins formed in pulp making and paper bleaching? Can the process be changed?"[229]

Members of the environmental community argued that the process could in fact be changed, that alternative bleaching procedures existed that could greatly reduce or even eliminate dioxin formation during pulp and paper making. The problem, in their view, rested largely with the EPA's approach to the issue. According to Dr. Ellen Silbergeld of the Environmental Defense Fund, who was also a member of an expert panel convened by the EPA's own Scientific Advisory Board to reevaluate the carcinogenic risks of dioxins: "Fundamentally, we remain concerned that the reactive posture of regulatory agencies in this country perpetuates a situation which condemns us to catching up to situations of environmental food and product contamination instead of preventing these and other ongoing sources of dioxins and other compounds." It was not sufficient "to respond to [the] problem by developing a strategy that merely transfer[ed] risks from one vector to another." New processes needed to be developed that eliminated the problem altogether.[230]

In principle, the EPA seemed to agree with the goal of eliminating dioxin discharges. In September 1988, it issued an interim strategy calling for "aggressive action" on dioxin discharges from pulp and paper mills.[231] According to the strategy, such discharges would be eliminated by requiring states to adopt and enforce stringent water quality standards for dioxin. For those states that failed to act, the EPA would apply its own standards. At the same time, however, the EPA also noted that it was considering a reassessment of the cancer potency it used for dioxin, thereby opening the door for the states to employ different risk estimates when setting their own standards. As a result, the dioxin standards adopted by those states with chlorine bleaching pulp mills varied from the rest of the nation by some three orders of magnitude.[232]

Of those states with the most permissible standards, ten out of eleven were in the South. In late 1989, Georgia became the first southern state to relax its dioxin standard, proposing a new standard—7.2 parts per quadrillion—that was more than 500 times less stringent than the EPA's suggested standard of 0.013 parts per quadrillion.[233] Members of the environmental community expressed outrage, and local newspapers suggested that the state was bowing to industry pressure. In March 1990, the EPA notified Georgia that it would not approve the standard and would impose its own criteria if the state did not revise it.[234]

Meanwhile, Maryland and Virginia began revising their standards, settling on a level of 1.2 parts per quadrillion, which they arrived at by using the less stringent cancer potency estimates of the Food and Drug Administration and an "acceptable risk level" of 1 in 100,000, as opposed to 1 in 1 million, excess cancer deaths.[235] Recognizing the uncertainty involved in such risk assessments and ignoring protests from environmentalists, the EPA approved the Maryland and Virginia standard as within the range of "scientific defensibility," even though it was some ninety times less stringent than the federal recommendation.[236] Georgia and five other states (Alabama, Arkansas, Mississippi, South Carolina, and Texas) followed, adopting the "EPA approved" standard of 1.2 parts per quadrillion. New Hampshire and New York were the only other states with similarly permissible standards. At the other end of the spectrum, Minnesota adopted a standard that was one hundred times stricter than the EPA's criteria. Most other states simply adopted the EPA's criteria as their own standards.[237]

As the controversy over state standards raged on, the EPA also began releasing preliminary results from both the 104-mill study and another study of dioxin concentrations in fish and wildlife.[238] The findings were not encouraging, particularly for the South. In the fish study, which was part of a broader study on bioaccumulation of toxics in fish, EPA researchers found that dioxin concentrations in fish downstream from several southern pulp and paper mills greatly exceeded the FDA threshold of 25 parts per trillion. The highest level of contamination in the country, 180 parts per trillion, was found in a species of fish—the creek chubsucker—downstream from Weyerhaeuser's mill in Plymouth, North Carolina. The second highest concentration, 150 parts per trillion, occurred in carp living downstream from International Paper's Bastrop, Louisiana, mill.[239]

As for the 104-mill study, which measured dioxin concentrations in pulp mill effluent, the findings were even more troubling. Combining the study's preliminary results with those from the fish study, the EPA developed a list of twenty mills (sixteen of which were in the South) that presented significant cancer risks (more than 1 in 10,000) to those who regularly consumed fish from waters that received dioxin from the mills. In a 1990 press release summarizing the results, EPA ranked the mills by cancer risks. The top eleven were all southern mills.[240] International Paper's Georgetown, South Carolina, mill had the highest concentration of dioxin in its effluent, almost three times higher than the median for all mills where dioxin was found.[241] For those who ate fish from the Sampit River and Winyah Bay downstream from the mill, the risk was hardly trivial. According to the EPA, those who subsisted on a daily catch of fish from the Sampit faced an additional cancer risk of 1 in 50 over a seventy-year lifetime (the EPA's "acceptable" cancer risk from food was 1 in 1 million).[242] Because fish and shellfish from the Sampit and Winyah Bay had long been a major source of protein for many among Georgetown's poorer and predominantly black population, the implications were quite serious. As Dr. Carl Schulz, a toxicologist from the University of South Carolina put it, "I'm not a rabid environmentalist . . . but I felt for this particular group of people in Georgetown, it's a very serious situation." Most residents were confused and angry, particularly in light of the conflicting messages they were receiving regarding health risks. "I don't know which group of scientists is right," said Carol Winans, president of the local League of Women Voters, "but why should I take the risk." James Chandler, an environmental lawyer, put the issue in somewhat broader context: "We've won some cases lately, and you think you're making some progress, then something like this comes along and you realize you didn't even know what the problem really was."[243]

In response, the industry mounted a systematic and very public challenge to the EPA's findings, charging that the information released by the agency was misleading and confusing. Dr. Richard Phillips, staff vice president and director of process technology for International Paper, argued, "The EPA risk assessment needlessly alarms local communities by issuing recommendations based upon old data and faulty assumptions derived from incorrect and obsolete information. The suggestion that eating fish downstream from certain paper mills poses an increased cancer risk is not supported by what we consider to be the best available science and the most recent fish test data."[244] Phillips pointed to research done by Dr. John Squire, a scientist at Johns Hopkins University whose earlier work had served as the basis for EPA estimates of the carcinogenic risks associated with dioxins. Based on more recent research, Squire had concluded that "dioxin poses no cancer risk to humans at any anticipated levels of exposure."[245] The science wars had begun.[246]

In response to growing pressure from industry, and in the face of conflicting evidence regarding the mechanism of dioxin toxicity, EPA administrator William Reilly initiated a formal reassessment of dioxin risk in April 1991.[247] In calling for the reassessment, Reilly noted, "There has been much speculation about the effect of these new developments on our revised dioxin risk assessment. Some factors may decrease the level of concern. Others may result in estimates of increased risk."[248] Red Cavaney, president of the American Paper Institute, applauded the EPA's decision, charging that the existing assessment of dioxin risk was "too stringent in light of all the evolving science that has come out."[249] Adding fuel to the fire, Dr. Vernon Houk, the assistant surgeon general and director of the Center for Environmental Health and Injury Control at the Centers for Disease Control in Atlanta, concluded, "If it's a carcinogen, it's a very weak carcinogen and Federal policy needs to reflect this." Houk, who was the government scientist who made the original decision to evacuate Times Beach in the early 1980s, also suggested that, in hindsight, that decision was an error: "Given what we know now about this chemical's toxicity and its effects on human health, it looks as though evacuation was unnecessary."[250]

Members of the environmental community responded by arguing that the new studies supported the claim that dioxin was as dangerous as ever and suggested that perhaps the reassessment was politically motivated. Environmental Defense Fund scientist Dr. Ellen Silbergeld, who was also a professor at the University of Maryland School of Medicine in Baltimore, stated, "Nothing that has been learned about dioxin since 1985 when EPA first published its risk assessment finding on dioxin in the environment supports a revision of science-based policy or action." Speaking somewhat more directly to the issue of politics, Dr. Mary O'Brien, director of an Oregon nonprofit research group focusing on dioxin, claimed, "What's being protected here is not people or the environment but industries favored by the Government. The Government begins with the assumption that these industrial activities have to go on and they adjust the data to make the existing pollution practices acceptable."[251]

Reilly brushed off the criticisms and urged that all parties wait for the reassessment. Still, he more or less admitted to opening up a political can of worms: "I don't want to prejudge the issue, but we are seeing new information on dioxin that suggests a lower risk assessment for dioxin should be applied. I know the stakes and that I'm unraveling something here. There is not much precedence in the Federal establishment for pulling back from a judgment of toxicity. But we need to be prepared to adjust, to raise or lower standards, as new science becomes available."[252] Scheduled to take only a year, it took more than three to produce a draft. Meanwhile, the politics of dioxin raged on.

Part of the controversy stemmed from the difficulties of conducting quantitative risk assessments for toxic compounds such as dioxin. By the 1990s, quantitative risk assessment had come to occupy a prominent place in the EPA's approach to environmental protection, promising to provide a more rational and objective basis for efforts to reshape the agency's research agenda and allocate scarce regulatory resources more effectively.[253] In the view of its proponents, risk assessment would allow hazards to be reduced to some common calculus of expected harm or death, compared to one another, and ranked. Cost-benefit analysis could then be used to evaluate regulatory options. From the perspective of pure administrative efficiency, the goal seemed worthy enough.[254]

But quantitative risk assessment has never been able to provide a simple protocol for translating "good science" into risk predictions. Even former administrator William Ruckelshaus, a major proponent of risk assessment in the 1980s, realized its inherent limitations. Referring to risk assessment as "a shotgun wedding between science and law," he knew that it would always be plagued with uncertainties that required all sorts of assumptions, even guesses.[255] Not surprisingly, the practice of risk assessment has often elicited controversy, sometimes degenerating into a highly politicized process, with various interests lining up their own experts, their own "science," to support a particular approach or conclusion. As in other areas of environmental policy, this has led to charges of over- and underregulation, increasing distrust for regulatory agencies, and political balkanization.

Well before the EPA released the draft results of its reassessment, however, the dioxin controversy spilled over into the courtroom. During the early 1990s, plaintiffs' lawyers in several southern states rushed to file lawsuits against pulp and paper firms for polluting waterways with dioxin, which, the attorneys argued, endangered the health and diminished the property values of downstream residents.[256] The action began in southern Mississippi, with a 1991 case involving a mill in New August run by a Georgia-Pacific subsidiary, Leaf River Forest Products, which Georgia-Pacific had acquired as part of its 1990 takeover of Great Northern Nekoosa (Leaf River's parent company).[257] The EPA had listed the Leaf River mill as having the fifth highest dioxin level in its effluent in the country and had included it as one of the top twenty mills posing "significant" cancer risks to people who regularly ate fish downstream from the mill. As a result, the state of Mississippi had banned commercial fishing and catfish consumption along the river for some forty miles below New August.[258] Wesley M. Simmons, a retired commercial fisherman who owned a recreational camp on the Leaf River thirty-seven miles below the mill, decided to sue Georgia-Pacific for failing to warn him of the dioxin pollution, diminishing the value of his property, and causing him emotional distress because of the *fear* of future disease that came from years of eating contaminated fish.[259]

Simmons's lawyer, John Deakle of Hattiesburg, a man so tenacious "he'd fight a circle saw" according to one acquaintance, sought to turn the case into a moral crusade against Georgia-Pacific. Playing off local populist sentiments, Deakle used company and industry documents, including the American Paper Institute documents recovered by Greenpeace, to paint a picture of a greedy corporation willing to trade "pollution for profits." "Take the profit out of what they have done to the river," Deakle pleaded with the jury. "Your vote can make a difference." He then asked the jury to award his plaintiff compensatory damages for nuisance and trespass and some $70 million in punitive damages.[260] In reaching its verdict, the jury rejected Deakle's emotional distress argument, but it did award Simmons more than $40,000 in compensatory damages for nuisance, trespass, and loss of property value and some $1 million in punitive damages.[261] Although Georgia-Pacific immediately appealed, the verdict sent shock waves through the industry, as firms began to fear a potential avalanche of asbestos-style lawsuits seeking massive damages.[262] The highly uncertain, emotionally charged world of toxic torts was not a place these firms wanted to be.

By February 1991, more than 7,000 individual cases had been filed in four Mississippi counties known for their receptiveness to mass tort claims.[263] Deakle, who had started with only a handful of clients, was suddenly representing some 6,000 plaintiffs, including more than a thousand who had filed claims against a nearby International Paper mill.[264] Plaintiffs from other states soon began to file similar lawsuits. The floodgates had opened.

In Tennessee, J. A. and Joan Shults, who lived on the Pigeon River, filed a class-action lawsuit on behalf of several thousand property owners against Champion International.[265] They asked for $5 billion, alleging that Champion's pollution had diminished the value of their property, disrupted their lives, and threatened them with cancer. Coming on the heels of the 1988 Pigeon River hearings, this was a power play that had the potential to propel dioxin litigation into the big leagues. In their brief, the plaintiffs charged that Champion's conduct was "so outrageous in character and so extreme in degree as to go beyond all possible bounds of decency and must be regarded as atrocious and utterly intolerable in a civilized society." It would take another twenty-one months before the case went to trial.[266]

Meanwhile, John Deakle took his second Leaf River case against Georgia-Pacific to trial.[267] Initially, he was joined by another group of plaintiffs, though Georgia-Pacific quickly and quietly settled with this group for an undisclosed sum.[268] On his own, Deakle pursued the same basic strategy he used in the Simmons case. But this time he wanted to make the emotional distress argument stick. To that effect, one of his plaintiffs, Thomas Ferguson, offered some rather emotional testimony. Georgia-Pacific, Ferguson claimed, "took my rights away from me. I can't swim where I want to swim no more. I can't fish. I can't do lots of things. And got me worried about cancer. They've got a sign out there from the Mississippi Wildlife that says don't eat fish out of these waters, it's contaminated with dioxin and causes cancer. But I've ate hundreds of pounds of fish unbeknowing that it had dioxin . . . It gets you worried and shakes you up and makes you break down." Despite Ferguson's admission under cross-examination that he was not sick and had never been tested for dioxin, the jury awarded him and his wife $90,000 each for emotional distress, $10,000 for nuisance, and $3 million in punitive damages. This was the second judgment in favor of the plaintiffs in a dioxin case involving a pulp and paper mill in little more than a year. Again, Georgia-Pacific appealed.[269] The message seemed clear, however. There was money, perhaps a great deal, in these toxic tort cases.

The following March, less than three months after Deakle won his second trial, a group of lawyers filed a $100 billion class-action lawsuit in Texas seeking damages from thirty-three pulp and paper firms and the American Paper Institute for "dioxin poisoning." This was such an excessive claim that the suit was withdrawn the following month. Yet it suggested the extremes to which plaintiffs' attorneys might push the issue.[270] From the perspective of industry executives, the situation was threatening to spin out of control.

All of which gave added urgency to the Pigeon River lawsuit slowly working its way to trial in east Tennessee. In contrast to Deakle's cases, it would be tried in federal district court, giving it greater visibility and added weight. The financial stakes were also considerably higher, with some 2,600 plaintiffs seeking $5 billion in damages. The paper industry did not want to lose this one, and they worked hard during pretrial maneuvers to limit the scope of the trial.

Their efforts paid off. By the time the case went to trial in September 1992, the damage claims had been reduced to $2.9 million in compensatory damages and $365 million in punitive damages. More important, Champion's attorneys also succeeded in getting the judge to throw out all "emotional distress" claims and, by invoking the statute of limitations, to restrict any damages to the period after January 1988 (three years before the case was filed). Finally, they restricted the issue before the court to charges of trespass (by pollution rather than people) and nuisance.[271]

Referring to the company as "an arrogant power run amok," Don Barrett, the lead plaintiff's attorney, opened the trial by arguing, "This whole case can be summed up in a few words—pollution for profit." Champion's attorneys responded by arguing that pollution per se was not the issue. Rather, the question was whether the plaintiffs had suffered any diminution of property values because of Champion's actions. Champion attorneys also emphasized all of the scientific uncertainties associated with dioxin and the disagreements among scientists about its toxicity and cancer potency. Finally, Champion vice president Richard DiFlorio testified that the mill had complied with all rules and regulations, that, in his view at least, "the problem [was] aesthetic," and that the mill was in the midst of switching to a bleaching process that used less chlorine and therefore reduced the dioxin load in the mill's effluent.[272]

In closing arguments, Champion attorney Louis Woolf stated that the charges against Champion consisted of "rumors, gossip, and hearsay" and implied that the plaintiffs were simply after money they did not deserve. Plaintiffs' attorneys responded with an equally predictable argument, invoking the story of David and Goliath and asking the jury to "render a verdict that will cause Champion and all other Champions of this country to shudder in their boardrooms." After three and a half days of deliberation, the jury could not reach a verdict, and the judge declared a mistrial. For the time being at least, Champion had dodged a bullet. Rather than face another trial, the company decided to settle. In March 1993, Champion agreed to pay $6.5 million to the plaintiffs. In exchange, the court barred all further trespass, nuisance, and personal injury claims against Champion pertaining to any actions of the company prior to the date of the settlement. This case was closed.[273]

In the months and years that followed, there were other dioxin lawsuits and settlements throughout the South. In July 1993, for example, Kimberly-Clark paid $6.5 million to settle a class-action lawsuit involving dioxin contamination from its Coosa Pines mill in Alabama.[274] Three years later, Champion was back in the news, paying out $5 million to settle with a class of Alabama plaintiffs for damages associated with dioxin-contaminated effluent. As in Tennessee, under the terms of the Alabama settlement, Champion admitted no wrongdoing and was released from all further liability for property damages stemming from its mill's discharge prior to the date of the settlement.[275] While these were not exactly small settlements, they were clearly a far cry from the massive damages that plaintiffs had sought in the early 1990s.

Meanwhile, back in Mississippi, things were not going so well for John Deakle, the man who started the ball rolling in the first place. In 1995, the Mississippi Supreme Court reversed the $3.2 million judgment in the Ferguson case, refusing to recognize the emotional distress claims and rejecting the nuisance claims for lack of proof.[276] Then, in the summer of 1996, a Mississippi circuit court judge dismissed four dioxin suits brought by Deakle and others against International Paper on behalf of 1,800 plaintiffs. The following month the same judge dismissed more than 200 suits brought by Deakle and others against Georgia-Pacific's Leaf River mill on behalf of some 5,400 plaintiffs.[277] Finally, in January 1997, the Mississippi Supreme Court reversed the $1.047 million judgment that Deakle had won for his original plaintiff, Wesley Simmons, in 1990.[278] Based on the courts' holdings in these cases, the defendant pulp and paper companies won summary judgment in several other dioxin cases.[279]

In the end, the great toxic tort bonanza that John Deakle and others had hoped for turned out to be somewhat of a bust. Uncertainty over the nature of the harm associated with dioxin exposure, the concomitant difficulties of proving causation, and the absence of any illness severely limited the viability of personal injury claims under theories of negligence or strict liability. This left plaintiffs to pursue their claims in the tenuous world of emotional distress for fear of future illness, which the courts summarily rejected, or trespass and nuisance, legal doctrines that had inherent limits as vehicles for the kind of massive, almost self-sustaining tort litigation that characterized asbestos. As a result, dioxin pollution from pulp and paper mills failed to ripen into the kind of toxic harm that could get traction in American tort law. Although dioxin tort litigation continued in Mississippi and elsewhere, it never regained the urgency it once had. Paper company executives who once feared a litigation explosion that might threaten the solvency of their firms moved on to other concerns.

Still, the *potential* of substantial asbestos-like liability, particularly in the wake of the initial victories during the early 1990s, had two important effects on pulp and paper firms. First, it reinforced incentives to implement procedures that used less chlorine in the bleaching process in order to demonstrate progress in dealing with the problem. Simply put, if corporations could show that they were not only meeting state and federal standards for dioxin but also modernizing their mills to reduce their dioxin loads, they might have a stronger case.[280] For the most part, this involved substituting chlorine dioxide for elemental chlorine in the beaching process.[281]

Beginning in the late 1980s and accelerating during the early 1990s, firms began voluntarily switching to chlorine dioxide in their bleaching process. As a result, dioxin levels in their effluent began to decline. Between 1988 and 1993, the discharge of dioxin in effluent from bleached pulp and paper mills declined from 201 grams per year to 71 grams per year. By 1994, according to the National Council for Air and Stream Improvement, bleaching mills had further reduced these discharges to less than 15 grams per year.[282] Those firms involved in dioxin litigation were quick to use this as evidence of how promptly they had responded to the problem.

The second major effect of these tort cases had to do with dioxin science. At the same time that they were making process improvement to reduce their chlorine use, pulp and paper firms involved in dioxin litigation also sought to emphasize the scientific uncertainty surrounding dioxin. During the court battles, industry advocates constantly pointed to the many disagreements among reputable scientists about dioxin's cancer potency. By emphasizing such uncertainty, they hoped to cast doubt on some of the arguments being made by the opposition. Scientific uncertainty, in short, could be as effective in the courtroom as it was in the regulatory arena.

By the mid-1990s, with dioxin litigation waning, attention had shifted back to the regulatory arena. In September 1993, the Natural Resources Defense Council (NRDC) and fifty-five other environmental groups filed a petition under the Clean Water Act demanding that the EPA ban all dioxin discharges by the pulp and paper industry by prohibiting the use of chlorine in the bleaching process.

Three months later, the EPA proposed effluent rules for dioxin discharges as part of a larger "cluster rule" containing air and water standards. The proposed dioxin rules required total chlorine-free (TCF) bleaching only at certain types of sulfite mills, a small subset of the industry. For the rest of the mills, the new rules called for complete substitution of chlorine dioxide for chlorine combined with oxygen delignification—a process that further reduced chlorine use in the bleaching process. According to EPA estimates, compliance with the proposed rules (including those for air emissions) would cost the industry $4 billion, result in closure of eleven to thirteen mills, and lead to a loss of between 2,880 and 10,700 jobs. Dioxin (and related compounds) in the industry's wastewater discharges would subsequently be reduced by more than 85 percent, from an estimated 410 grams per year to around 56 grams.[283]

The industry quickly responded with its own estimates of compliance costs that were far more drastic. According to the American Forest & Paper Association, the successor to the American Paper Institute, compliance with the new rules would shut down some thirty mills, cost $10 billion in capital expenditures, and eliminate 275,000 jobs, including 19,000 at the mill level.[284] Luke Popovich, a spokesman for the association, claimed that the new cluster rule was "likely to be the costliest regulatory program ever imposed on a single industry by the EPA."[285] Not surprisingly, during the comment period on the rules, industry advocates flooded the EPA with data supporting their contention that any move to eliminate chlorine altogether from the bleaching process could not be justified on an economic basis. Environmental groups, by contrast, criticized the proposed rules as too lenient and urged that the government require total chlorine-free bleaching.[286] The EPA went back to the drawing board, and the debate raged on for the better part of four years.

Meanwhile, in 1994, the EPA released the first external draft of its highly anticipated dioxin risk reassessment.[287] Based on three years of work by more than a hundred scientists from inside and outside of the agency, several public meetings, and an extensive peer-review process, the 2,400 page report was widely considered to be the most thorough assessment of dioxin risks to date. As for the report's major findings, they were not encouraging, particularly for those who had previously advocated the reassessment on the presumption that dioxin's risks had been overstated. Indeed, not only did the draft report reaffirm dioxin's status as a "probable" human carcinogen, but it also pointed to a variety of other noncancer health effects that were far greater than previously expected. These included various reproductive and developmental effects, disruption of the endocrine system, immunotoxicity, and several other potential metabolic and hormonal disturbances. Some of these effects were thought to happen at or close to "background" exposure levels.[288]

Regarding sources and exposures, the report noted that hazardous and municipal waste incinerators accounted for the majority of releases to the environment. Pulp and paper bleaching was listed as a source, though not of the same magnitude as incinerators. Atmospheric deposition was considered to be the major vector of environmental transport, while food intake (particularly fish, beef, pork, and chicken) was hypothesized as the predominant pathway of human exposure. Background levels in the general population were estimated to be between forty and sixty parts per trillion when all dioxins, furans, and PCBs were included.[289] High-end estimates of the body burdens for the top 10 percent of the general population were possibly three times higher than the average. Overall, the EPA estimated that the average "body burden" of dioxins and dioxin-like compounds in the general population resulted in an additional cancer risk of 1 in 10,000 to 1 in 1,000 per year.[290] In other words, based on these estimates, current "background" levels of exposure were already posing additional cancer risks that were considerably higher than the 1 in 1 million threshold the EPA considered "acceptable."

Although the draft report did not point to any significant differences in exposure rates due to ethnicity, socioeconomic status, or geographic location, it did identify several potential highly exposed populations. Breast-feeding infants constituted one such group. Due to the levels of dioxins and dioxin-like compounds in mother's milk, nursing infants received between 4 and 12 percent of their total lifetime intake during the first year of life. This finding was particularly disturbing as developing infants are considered to be far more sensitive to dioxins and dioxin-like chemicals than older people. Moreover, given the cumulative effects of these substances, early exposure might also have a considerable influence on later susceptibility to dioxin toxicity. In other words, the timing of the dose may turn out to be as important as the dose itself in determining the risks of adverse health effects over a normal lifetime. Other highly exposed subpopulations identified explicitly by the report included workers exposed in occupational settings and during industrial accidents, people who lived near discrete local sources, and subsistence fishers who consumed large amounts of fish from areas where dioxin concentrations in the fish were high. Although the data were spotty, several studies of subsistence fishermen around the world reported levels of dioxins and dioxin-like compounds in their blood that were three to twenty times higher than the average "background" levels for the general population.[291]

On the whole, the reassessment confirmed dioxin's status as one of the most potent toxicants ever studied and suggested that current levels of exposure constituted a human health hazard. This was obviously not what industry representatives had hoped for when they asked for the reassessment back in 1991. Not surprisingly, industry advocates criticized the findings. One group of scientists assembled by the American Forest & Paper Association argued that the EPA did not have "sufficient scientific evidence" to reach its "alarming conclusion" regarding the existence of adverse health effects at or near current background body burden levels.[292]

As expected, members of the environmental community generally praised the reassessment. Some also clearly relished that the results had not turned out as the original advocates of the reassessment had hoped. Peter deFur, a toxicologist with the Environmental Defense Fund, suggested that the industry was "about to be bitten by the snake it loosed."[293] Others pointed to research contained within the report that reinforced earlier arguments claiming that there was no "safe" level of exposure to dioxin and that even if there were, background exposure levels were already high enough to generate "unacceptable" risks by the EPA's own standards. Consequently, they argued, the "EPA must rigorously eliminate sources as well as exposures, recognizing that each exposure threatens human health and the environment."[294]

As for the review conducted by the EPA's Science Advisory Board (SAB), the general impression was positive. Only minor changes were suggested for the vast majority of the report. The SAB specifically commended the EPA for including a broader class of dioxin-like compounds and for focusing on noncancer effects in addition to carcinogenic ones. It also praised the way in which the EPA had conducted the process. However, the SAB review did note "three major weaknesses" in the EPA's characterization of dioxin risks. First, most, though not all, members of the SAB review committee concluded that the report exhibited "a tendency to overstate the possibility for danger" associated with dioxin exposure. Second, the EPA had not been as thorough as it could have been in identifying and analyzing some of the "important uncertainties associated with the Agency's conclusions." Finally, the characterization of noncancer risks had not been performed in a way that could "facilitate meaningful analysis of the incremental benefits of risk management alternatives." On the basis of these suggestions, the SAB asked the EPA to rework the sections of the report dealing with dose-response modeling and overall risk assessment before submitting it to them for a final review.[295] In September 2000, the EPA formally submitted the reworked portions of its draft dioxin reassessment and the results of the peer review process. In addition to confirming the conclusions found in the 1995 reassessment, the 2000 draft reclassified mixtures of dioxins and related compounds as "likely human carcinogens," while characterizing 2,3,7,8 TCDD (the most potent and thoroughly studied of the dioxin compounds) as "carcinogenic to humans."[296] After another decade of back and forth between the SAB, the EPA, and the National Research Council, the final completed dioxin reassessment for noncancer health effects was released in 2012. The dioxin reassessment for cancer health effects was still pending as of early 2015, almost twenty-five years after the process was begun.[297]

While the EPA was working on its dioxin reassessment in the 1990s, it was also reworking the so-called cluster rule for the pulp and paper industry, considering whether to go forward with a full ban on all chlorine or something less.[298] Representatives of the environmental community argued that the EPA had a mandate to require the "best available technology" to control dioxin pollution and that total chlorine-free bleaching was available—a point that was hard to argue with since several mills in Europe had converted to TCF bleaching. Even in the United States, Louisiana-Pacific's Samoa, California, mill had switched to TCF bleaching as part of a settlement with the local Surfrider Foundation. Given the right incentives, there was no reason why pulp and paper production in the United States could not be chlorine free.[299] For its part, the industry continued to issue dire predictions of the economic impact associated with the more restrictive approach. Labor unions soon joined in fighting the more expensive regulations. Here, in their view at least, was a classic example of the jobs versus environment tradeoff. The EPA, they argued, should be careful to focus on cost-effectiveness and economic impacts in promulgating the new standards.[300]

In November 1997, the EPA issued the final version of the cluster rule.[301] The industry got what it wanted: total substitution of chlorine dioxide for chlorine with no additional requirement for oxygen delignification. To reward those mills that had already installed such processes as well as those that might do so in the future, the rule also included the voluntary incentives program. In total, the EPA estimated that the rule would cost some $1.8 billion in capital expenditures. Dioxin levels in pulp mill effluent would be reduced by 96 percent, a reduction that would eventually eliminate all dioxin-based fish advisories attributed to pulp and paper mills. In announcing the rule, EPA administrator Carol Browner stated that the agency was "taking significant steps to protect the health of millions of American families from contaminated air and water from pulp and paper mills. This action puts us well on our way to cleaning up more than 70 rivers and streams throughout the nation."[302]

Environmental advocates expressed dismay. Jessica Landman, an attorney with the Natural Resources Defense Fund, charged that the standards would "allow the pulp and paper industry to continue their routine contamination of our waterways. We have cost-effective technologies readily available, but you would never know it from these standards." Rick Hind of Greenpeace referred to the rule as "low-lead instead of no-lead gas." In defending the rule, EPA staff noted that the original 1993 plan would have cost an extra $1.2 billion in order to achieve an additional 1 percent reduction in dioxin discharges.[303] Clearly the agency did not consider this extra margin of protection economically justified.

The industry was generally pleased with the final rule. By the time it was issued, many firms were well on their way toward compliance. Most firms expressed relief that they would not be required to meet the more stringent standards. Industry giant Georgia-Pacific referred to the rule as a "reasonable" and "right" choice for protecting the environment. The American Forest & Paper Association applauded the EPA's conclusion "that a totally chlorine-free alternative was not a viable option."[304] Meanwhile, the United Paperworkers International Union praised the rule for "protect[ing] the environment while minimizing job loss in the pulp and paper industry."[305] In short, the industry and its allies succeeded in convincing the EPA that the more stringent requirements contained in the original 1993 cluster rule proposal were regulatory overkill. By taking voluntary steps to reduce their dioxin discharges, they effectively got out in front of the "regulatory curve," demonstrating that they were willing and able to make changes.[306]

Such efforts clearly reflected a more proactive approach by pulp and paper firms to environmental issues. During the 1990s and 2000s, some firms along with industry trade associations went to great lengths to emphasize their concern for "environmental stewardship." Sustainability and a commitment to improved environmental performance became standard industry rhetoric. Some firms even began to discuss the possibility of moving to zero-discharge mills as way of avoiding problems like the dioxin issue. Technological innovation aimed at preventing pollution rather than simply controlling it had become an objective for managers and executives. Jerry Ballangee, an executive from Union Camp, noted in the mid-1990s that "a no-discharge plant is a very attractive proposition and at some point we will get there." R. L. Erickson, vice president for manufacturing and technology at Weyerhaeuser made a similar point: "We had dioxin in 1987. What will it be in 2005? By closing the loop, you can prevent future issues."[307]

In retrospect, the dioxin controversy clearly affected the industry's approach to environmental regulation, marking a departure from its more recalcitrant stance on environmental issues in the past. There can be little doubt that the industry's initial cooperative approach on dioxin gave it considerable input into the regulatory process.[308] By working with the EPA rather than fighting it at every turn, the industry gained access to the regulatory process in a manner that allowed it to challenge the scientific uncertainties regarding dioxin's health effects and the costs and benefits associated with various regulatory options. This is not to suggest that the industry succeeded in getting everything it wanted or that the EPA was somehow co-opted. Indeed, environmental groups, though perhaps unhappy with the final results, were also successful in shaping the process from the beginning. By making dioxin a major public issue in the late 1980s, they effectively put the industry on the defensive. By forcing the EPA to accept a legally imposed timeline for developing dioxin regulations for pulp and paper discharges, they set the regulatory train in motion. Throughout it all, moreover, these environmental groups, like the industry, pushed for a particular interpretation of dioxin science that bolstered their own agenda.

What often got left out of all of these deliberations and debates, however, were the actual impacts of dioxin contamination on real people living real lives in real places. Of course, no one will ever really know how many people ended up with cancer or some other chronic disease because of exposure to dioxin from pulp and paper production. The various risk estimates of excess cancers and other illnesses associated with eating contaminated fish or with some other exposure pathway were always going to be unable to make visible the actual harms imposed by industrial pollution on individuals and the people who cared about them. Their histories will never be told. What can be said with some confidence, however, is that for decades before and after the pulp and paper industry adopted cleaner production methods, dioxin would and will continue to persist in the environment, bioaccumulate up the food chain, and contribute to a diverse and complex set of latent harms—all the while enacting a quiet, slow, largely unseen violence on environments and communities across the rural South.

Pollution, Politics, and Environmental Equity

> No segment of the population should have to bear the brunt of the nation's industrial pollution problem. Yet, institutional barriers still limit residential choices and mobility options for millions of minorities, working-class persons, and poor community residents. While much progress has been made in bringing blacks

> and other minority groups into the mainstream, all Americans do not have the same opportunities to escape the ravages of environmental toxins.
>
> —*Robert D. Bullard, 1990*[309]

That many of the hazards generated by industrial society are distributed unevenly is incontestable. Prevailing class structures, racial discrimination, the distribution and exercise of political power—all of these work to reinforce the fact that "all Americans do not have the same opportunities to escape the ravages of environmental toxins." And although some of the hazards associated with industrial society may transcend the lines of race and class, many others do not. This has been the basic insight and rallying cry of the environmental justice movement for more than two decades.[310] Its relevance for the South would be difficult to overstate.

Given the region's long history of poverty and racial discrimination, its headlong rush to industrialize, and the accommodating stance of business-friendly state and local governments, one would expect to find more than a few cases of environmental inequality. This is not to imply, of course, that such inequality is unique to the South. Nor is it to suggest that it is somehow new. In fact, one would not have to look far for evidence of environmental inequities in previous periods of southern history. Tenancy and sharecropping, for example, created vulnerabilities and dependencies that made it far more difficult for blacks and poor whites to escape the impacts of land degradation, resource depletion, and the ecological liabilities of row-crop agriculture. Extractive industries such as lumber and mining often left behind denuded lands and depleted mines, further exacerbating the poverty of local residents. Mill towns and logging camps subjected workers to industrial pollution and environmental diseases that wealthier communities could easily evade. Basic sanitation and other public services enjoyed by so many in the urbanizing New South did not always reach poor and minority communities, particularly in rural areas. Many of the working environments of New South industries were replete with environmental hazards easily avoided by managers and owners. Black lung, brown lung, silicosis, and occupational injuries of various sorts all offered poignant reminders of environmental inequality. Within the workplace, moreover, blacks often performed the dirtiest and most dangerous jobs. Outside of it, they tended to live in more polluted and less sanitary areas. In short, as the New South industrialized and urbanized, the politically disenfranchised and economically deprived found it more difficult to avoid the environmental hazards associated with these processes. Jim Crow made it all too easy to displace such hazards onto African

American communities; while the prevalence of low-wage industries further limited the mobility of poor working-class people, white and black, creating any number of "uneven places," to use George Tindall's evocative phrase, that bore a disproportionate burden of the region's environmental hazards.[311]

Simply put, environmental hazards, whatever their origin, have almost always fallen more heavily on those southerners occupying positions of structural vulnerability. This is a point so obvious it hardly needs repeating. Yet if the positive correlation between environmental risk and structural vulnerability is widely accepted, there has been considerable and ongoing controversy regarding the underlying causes. Where some see outright racial discrimination, others see market dynamics at work.[312]

To be sure, there are often multiple factors behind environmental inequality. Any particular explanation of why certain types of environmental risks have ended up in certain poor or minority communities would require detailed investigation of specific cases. Identifying a pattern, in other words, should not be confused with efforts to understand the processes that have generated the pattern and its impacts on particular people and places. The detailed local-level research required to understand these processes—to explain why polluting industries, toxic waste dumps, and other undesirable land uses end up in poor and minority communities; why some enjoy the benefits of environmental protection laws more than others; and what kinds of remedies might be appropriate for these forms of inequality—has been quite rare, in part because it is so difficult.

Such issues are obviously a long way from Eugene Odum's "balance of nature," and some have suggested that they provide a healthy corrective to the white, upper-middle-class bias of mainstream American environmentalism. Indeed, despite (or perhaps because of) more than four decades of progress in environmental protection, perennial questions of equity and fairness continue to plague efforts to further control and reduce environmental risks. Distribution, of course, has always been a central issue in environmental law and regulation. Most of the concern in the past, however, has turned on the distribution of the various costs or risks associated with environmental disruption between private firms and society at large. Far too little attention has been directed toward the differential and sometimes regressive impacts of the risks that remain on particular classes and groups of people *within* society.[313]

In the case of the southern pulp and paper industry, the many and varied efforts to control pollution during the twentieth century, whether through the courts, the legislature, administrative agencies, or social movements, all had distributional consequences. If there was a master trend in the evolution of pollution control during this period, it was federalization, progressing largely in response to the abysmal record of state and local governments in mitigating environmental insults and aimed ostensibly at leveling the regulatory landscape. It may well be, moreover, that many of the federal environmental regulations intended to control pollution from pulp and paper mills have done as much or more to protect politically and economically vulnerable populations than those who have had greater opportunities to escape the harmful effects of such pollution. People in positions of structural vulnerability, who were disproportionately affected in the first place, had more to gain from pollution control.

That said, the fact that some of the more important federal pollution control laws of the 1970s contained various "grandfather clauses" exempting existing facilities, covered only certain types of pollutants, and delegated considerable authority to the states in the areas of standard setting, monitoring, and enforcement ensured a certain unevenness in environmental protection. While information-based strategies such as the Toxics Release Inventory have worked to enhance fairness in certain areas of environmental protection by leveraging a broader public to sanction individual polluters in poor and minority communities, such gains have never been guaranteed, especially in a world of imperfect information. Ultimately, for such information-based strategies to succeed in reducing environmental inequalities, they need to be backed by processes to ensure accountability and by an administrative state willing and able to compensate for the differential results achieved by those who wield such information effectively and those who do not.

It almost goes without saying that the growth and success of the southern pulp and paper industry, like other New South industries, generated significant environmental costs or risks that too often fell disproportionately on poor and minority communities. Air pollution obviously affected those who lived closer to the mills. The massive discharges of untreated pulp mill effluent into surrounding waterways destroyed aquatic life and undermined the livelihoods of local fishermen. Elevated dioxin levels found in fish downstream from certain mills meant that those who subsisted on these fish faced higher cancer risks than the population at large.

None of this is intended to suggest that pulp and paper firms explicitly discriminated against particular southern communities in terms of where they put their pollution. As suggested, economic vulnerability, political marginalization, and environmental pollution have a certain mutual affinity for one another. This appears no less true today than in the past. Indeed, although pollution loads clearly have been reduced, they certainly have not been eliminated and continue to fall disproportionately on poor and minority communities.

The situation surrounding the Union Camp mill in Savannah illustrates some of the challenges of understanding how the distribution of environmental hazards affected a particular community living around a particular mill. Out of all of the households living within three miles of the Union Camp mill in 1990, roughly two-thirds were African American and more than 42 percent had incomes of less than $15,000 per year. For the city of Savannah, some 45 percent of all households were African American and 35 percent made less than $15,000. For Chatham County, about a third of all households were African American and almost 29 percent had incomes of less than $15,000. Finally, for the state of Georgia, about a fourth of all households were African American and about a fourth made less than $15,000.[314] Do these numbers prove environmental discrimination? Hardly. Do they indicate that certain African American and poor families have been disproportionately exposed to the environmental hazards (particularly toxic air emissions) generated by the mill? Indeed they do. Do they do anything more than confirm what many people likely already suspected? Probably not.

Ultimately, the distributional impacts of pollution and environmental disruption can never be understood fully, much less resolved, if they are treated exclusively as economic issues. Nor can environmental problems be dealt with effectively if they are seen only through an ecological lens. In the real world, the "problem of social cost" resolves into the micro- and macro-politics of social relations. Questions of politics and distribution do not arise as mere afterthoughts in the quest for improved environmental protection. They have been present, though perhaps not acknowledged, since the beginning.

CHAPTER FIVE

New South, New Nature

> Here is nature and there stands the folk. Behind the folk stands a tragic history. What we need to know is that, in spite of its tragic history, the mold in which the South is to be fashioned is only now being laid. —*Rupert Vance, 1932*

When Rupert Vance concluded his magisterial portrait of the depression-era South, *Human Geography of the South*, the pulp and paper industry was just beginning to move into the region on a significant scale. Along with many of his contemporaries, Vance welcomed this development. For him it represented the possibility of utilizing southern resources in a manner that comported with his ideals of conservation, scientific management, and regional planning. In contrast to the irrationality and waste associated with cotton tenancy and lumbering, pulp and paper promised a new regime of progressive resource management. As Vance noted on repeated occasions, the South held great potential for a dominant industry based on scientific forestry.[1] All it needed was the capital, technology, and institutional changes to realize this potential—none of which would come easy.

Today, no one can deny the success of the southern pulp and paper industry. From the perspective of economic theory, such success seems easy to understand. As New Deal agricultural and labor policies worked to undermine the farm tenancy system and open up the southern economy to new investment, outside capital moved into the region to take advantage of cheap land and labor, plentiful resources, and a hospitable regulatory climate. Based on this comparative advantage, the South came to provide a new cost structure for the industry. As leading firms rushed to build southern production complexes, the South emerged as a national and international leader in the production of pulp and paper based on intensive cultivation of timber.

Such an explanation, however, misses many of the most important questions about how this actual process of regional industrialization played out on the ground. Indeed, on closer examination, it appears that the success of the southern pulp and paper industry was contingent on a complex set of interactions between various actors, few of which could be predicted ex ante based on a logic of economic self-interest. Rather than deriving simply from regional comparative advantage in key factors of production and the attraction of branch plants controlled largely by northern corporations, the growth of the southern pulp and paper industry was a historically situated process of social, political, and ecological construction.

Attending to the ways in which pulp and paper firms, in concert with other actors, mobilized and organized the productive capacities of southern land and labor and integrated them into a highly competitive industrial system also departs from more traditional top-heavy narratives of American industry that privilege national patterns of industrialization and the large mass-production enterprise.[2] To be sure, pulp and paper embodied many of the principles of the mass-production paradigm. But it was also an industry that reached deep into the rural South and fashioned its competitive advantage out of a complex and changing ensemble of regional institutions and resources. By following these firms into the rural communities and environments in which they operated and uncovering the various forms of social coordination and conflict that constituted the larger industrial system, this book has sought to develop a deeper appreciation for the actual process of regional industrialization. Likewise, by focusing specifically on how biophysical systems get incorporated into the larger industrial system—whether in the context of the intensive cultivation of timber or in the struggle over pollution from pulp and paper mills—the story told here engages directly with the industrialization of nature and its impact on people and places.

In many respects, then, this book represents an effort to join three different kinds of histories—agrarian, industrial, and environmental—in explaining the construction of a new regional industrial order. Given its dependence on rural land and labor and the central importance of intensive cultivation of timber, the southern pulp and paper industry cannot be understood without attending to the South's distinctive post–New Deal agrarian transition. At the same time, given the large investments in fixed capital embodied in a pulp and paper mill and the fact that some of these firms were among the biggest corporations in America, the development of the industry cannot be adequately explained without attention to corporate strategy, organizational capabilities, and the requirements of keeping fixed capital in motion. Finally, given the vast biophysical requirements of the industry, one cannot ignore the central importance of nature in the making of this industrial system.

Conceived as the weaving together of these three historical trajectories, the regional industrial order concept provides a vehicle for telling a larger story about the deep interpenetration of social and environmental change. The lessons here go well beyond the regional context. Although this book can be read as contributing primarily to the history of southern industrialization and, by extension, to the story of national industrialization in twentieth-century America, it is also directed at a larger set of questions about industrialization and economic development that gives more attention to noncorporate actors and institutions and takes the problem of nature, from multiple perspectives, as a central object of concern. This chapter elaborates on some of these larger questions, situating the study of southern pulp and paper in a broader set of concerns relevant to understanding regional industrialization and environmental change. It also provides a brief discussion of some of the challenges facing the industry in the early twenty-first century.

Industrial Metabolism and The Problem of Nature

> *Industry* is the *real* historical relationship of nature to . . . man.
>
> —*Karl Marx, 1844*[3]

Investigating how the varied and variable biophysical processes that are often grouped under the term "nature" get incorporated into the process of regional industrialization requires moving beyond standard views of nature as a source of raw materials or as a set of obstacles and constraints to be overcome. By examining how nature has been appropriated and (re)made to operate as a productive force—a process intimately bound up with broader political economic changes and one with multiple social and ecological consequences—the story told here has sought to highlight the ways in which industry reveals and embodies distinctive aspects of the relationship between nature and society. Focusing on the *industrialization* of nature, in other words, uncovers some of the connections between social and environmental change that might not be as apparent through an examination of the extension of markets and attendant processes of commodification that have occupied other environmental histories.[4]

This is particularly important in those sectors that are engaged directly in the production of food, fiber, and raw materials, where nature operates not simply as a feature of the firm's external environment but as part and parcel of the basic problem of organizing production.[5] To fully understand these industries, one must investigate how specific biophysical processes get refashioned and pressed into service for purposes of industrial production. In the case of pulp and paper, this involved substantial efforts that were at the very center of the industrial complex aimed at rationalizing the southern landscape, intensifying the biology of tree growth, managing the spatial and temporal challenges of wood procurement, and externalizing the massive wastes associated with pulp and paper production.[6] In each of these areas, it makes little sense to view nature as something external to industrialization. Rather, it was a vital part of the industrial process itself.

Thus, in contrast to the logic of extraction that prevailed in the early twentieth century southern lumber industry, where the vast forests of the region were treated largely as a resource to be mined, pulp and paper firms embraced a logic of cultivation that treated timber as a crop. In doing so, they had to develop strategies to manage the long biological time lags (measured in decades) and extensive spatial requirements (measured in square miles) involved in timber growth and forest regeneration in a manner that comported with the requirements of a high-throughput, mass-production enterprise. Given that much of the timberland in the region was owned by thousands of small landowners, coordinating timber growth, harvest, and delivery to ensure an adequate and continuous supply of fiber for the mills proved to be one of the more difficult challenges facing firms as they established their southern production complexes.

Efforts to meet this challenge proceeded along two distinct but complementary paths. The first entailed the creation of a thoroughly industrialized forest out of a severely degraded landscape, a task that required extensive cooperation between competing firms and with other public and private entities. Indeed, before timberlands could even be approached as objects of intensification, the landscape of investment had to be rationalized via systematic and ongoing inventories of standing forests, an expansive regional program of fire control, tax and credit reform, and substantial changes in land use policies. By midcentury, with forest regeneration programs moving into high gear, efforts to accelerate tree growth and maximize desirable characteristics had become economic imperatives. Tree improvement emerged as the principal vehicle for overcoming the spatial and temporal challenges of timber growth and consolidating the biological foundations of the industry's comparative advantage.

But making trees grow faster and better was only part of the solution. The mills also had to coordinate the harvest and delivery of wood in sufficient quantities and at the appropriate times to meet the vast and changing fiber requirements of their pulp and paper machines. Through careful construction of local pulpwood markets and through the use of wood dealers to manage the uneven labor demands associated with timber harvesting, mill managers developed a flexible, low-cost procurement system that refashioned the social and biophysical landscape of the rural South to meet the needs of the industry. In effect, the geography of local pulpwood markets together with the wood dealer system provided an effective way of overcoming the temporal lags associated with timber growth in order to maintain a continuous flow of wood to the mills.

Together these strategies of tree improvement and a flexible wood procurement system worked to rationalize and discipline the spatial and temporal requirements of industrial timber production, allowing more timber to be grown faster on less land across a carefully ordered procurement area designed to deliver continuous wood flow while adjusting rapidly to the changing fiber requirements of the mills. As the industry grew, this new bio-geography spread across the South, inscribed not only in the physical landscape but also in the functioning of local markets for land and labor.

By allowing for increased throughput to meet the expanding needs of pulp and paper production, these strategies also translated directly into increasing volumes of waste and pollution. As paper machines grew bigger and faster, the mills operated as huge concentrating agents, drawing pulpwood from across hundreds of square miles and funneling it through the industrial complex. Delivered by truck, train, and barge, pulpwood flowed to the mills around the clock, feeding the massive digesters and Fourdrinier machines that turned trees into paper at incredible speeds. Together with the vast amounts of water, energy, and chemicals used in the process, this relentless throughput of fiber resulted in discharges of wastewater and air pollution that dwarfed anything the region had ever seen before. Managing these pollution loads and ensuring ongoing access to the assimilative capacity of local environments thus became as important to the industry as the effort to industrialize the forest and move wood cheaply to the mills. As the industry's metabolism accelerated in response to growing demand and the imperative to speed up the process, firms struggled to maintain such access, facing growing opposition from local people and communities who saw their homes and livelihoods threatened and, in some cases, destroyed by the resulting environmental degradation.

In searching for a vocabulary to make sense of the various social and environmental aspects of this process, the concept of "industrial metabolism" seems particularly apt, if a bit old fashioned. The term draws on conceptions of the labor process as the metabolic interaction between nature and society and provides a way of engaging the problem of nature as an integral part of the effort to organize and maintain industrial production without lapsing into the tired dualism of nature and society on the one hand or the particularisms of local environmental histories on the other.[7] Thus, in building a regional industrial order, pulp and paper firms constructed a distinctive form of industrial metabolism that married the biological intensification of the forest to efforts to appropriate the absorptive capacity of ecological systems for waste assimilation. At the heart of this industrial order, in other words, was a voracious metabolic process that combined nature and technology in new ways, in the process pulling different people and places into a new set of socio-ecological arrangements. In many respects, these new arrangements thoroughly disrupted existing social and environmental patterns, illustrating the deep contradictions (the metabolic rift) between capitalist economic imperatives and socio-ecological systems.

The results of these disruptions were readily apparent to even the most casual observer. Large segments of the landscape were effectively transformed into a vast machine to serve the mills' growing appetites, resulting in a radically simplified version of the southern forest. To be sure, much of the enthusiasm for this new regime of industrial forestry emerged out of, and drew sustenance from, the Progressive conservation movement of the early twentieth century. What better illustration of the "gospel of efficiency" than a rational, scientific program aimed at transforming a degraded landscape into a vast "organic machine"?[8] But to suggest that this process was simply another aspect of a larger set of government projects aimed at legibility and simplification would be a mistake.[9] Indeed, while government programs played important roles in establishing many of the conditions necessary for remaking the southern forest, they were hardly the prime movers. This was (and is) ultimately a story about how capital seeks to harness nature as a productive force.

Harnessing nature to the imperatives of large-scale industry, of course, also created new risks and vulnerabilities. By ramping up the biological productivity of the southern forest, pulp and paper firms increased the scale and scope of simplification and, as a result, the potential for ecological disruptions. In effect, the industrial tree improvement programs that came to provide the biological basis of the industry's competitiveness brought with them new forms of ecological vulnerability, manifest in increased susceptibility to pathogens, loss of habitat, and diminished biodiversity. The seemingly imminent commercial introduction of genetically engineered trees promises to push this process of intensification, with its attendant risks and contradictions, even farther.[10]

Moving downstream through the pulp and paper complex reveals more disruption. As the mills worked to digest the increasing throughput of fiber, water, and chemicals, they generated massive waste flows that severely disrupted local ecosystems, further challenging the long-term viability of the industry. With little or no incentive to treat their wastes and in the absence of regulatory controls, the mills externalized their wastes onto surrounding environments and communities, sacrificing local livelihoods and places to the needs of industrial production. But as the scale of pulp and paper production grew, this strategy of externalizing pollution ran up against an alternative set of values and priorities. By the 1970s, as the environmental movement gained steam, the ecological disruptions caused by the industry's accelerating metabolism had become sources of intense political and legal conflict. Federal environmental regulation emerged as an increasingly important force in shaping the development of the southern pulp and paper complex by restricting access to the absorptive capacity of local ecosystems and forcing the industry to adjust and manage its pollution loads.

In all of this, the incorporation of nature into the pulp and paper industry was less about commodification than it was about the appropriation and subsumption of biophysical properties and processes. As trees were remade to work harder, faster, better—subordinated to the dictates of industrial production—a thoroughly industrialized forest was created, looking nothing like the forests of the past and displacing older ways of relating to the land. As local environments were appropriated for waste assimilation, substantial degradation and disruption of receiving waterways and airsheds brought unprecedented and, in some cases, irreversible changes to the ecology of the region. In both cases—that is, in the care and attention that firms directed at the industrialization of the southern forest and in the near total disregard (at least in the early decades) for the disruptions caused by widespread air and water pollution—firms created a distinctive and lasting industrial landscape that marked a definitive end to the older agrarian ways, with significant impacts on people and places across the region. Fields and farms; bayous, streams, rivers, and estuaries; towns and cities; people of different classes and races—all were swept up in various ways by the efforts of pulp and paper firms to bend and shape natural systems and local environments to satisfy the needs of a vibrant and growing industry.

Constructing a Regional Industrial Order

> Economic agents . . . do not maximize so much as they strategize . . . [T]hey are at least as much concerned with determining, in all senses, the context they are in as they are in pursuing what they take to be their advantage within any context. Self-interested adjustment to conditions taken as given thus proceeds paradoxically yet as a matter of course together with efforts to find or create a more advantageous set of constraints. —*Charles F. Sabel and Jonathan Zeitlin, 1996*[11]

Economic actors construct, rather than simply adapt to, the social and environmental contexts within which they operate. Building an industrial system is an active process taking shape out of the many interactions (some planned some not) between economic agents operating in a dynamic institutional environment. In natural resource industries, understanding this process of construction requires attention to the interactions between these agents and the various biophysical properties and processes they engage as they attempt to subordinate resources and environments to the dictates of industrial production. Thus, as pulp and paper firms moved into the South, they developed specific strategies that fit with, but at the same time refashioned, a set of regional institutions governing access to land, labor, and the environment.[12] They had to do this, moreover, in the context of profound shifts in federal and state policies regarding southern agriculture, land use, and labor markets initiated during the New Deal—shifts that they and their advocates sought to shape in various political and legal arenas. Finally, they had to do it in a manner adapted to the biophysical challenges involved in growing and harvesting large amounts of timber in concentrated areas and managing tremendous pollution loads.

In short, the various actors involved in the pulp and paper industry had to construct new "worlds of production" capable of straddling the flexible labor markets and informal institutions of the rural South, the biological rhythms of tree growth and the dispersed nature of regional timberland ownership, the imperatives of heavy industry and the need for a stable mill workforce, and the substantial pollution loads associated with pulp and paper production.[13] That they did so in the context of massive inequalities is undeniable. But this does not mean that pulp and paper firms could simply impose an optimal organizational form on a prostrate region. Working both with and against other actors, they had to develop solutions that recombined and re-elaborated vernacular institutions.[14] The forms of industrial organization and economic life that emerged thus bore the deep imprint of regional institutions and dynamics, illustrating Karl Polanyi's notion of the economy as instituted process.[15]

This view of economic activity as embedded within larger social and political practices where economic agents operate as strategic actors seeking to construct the worlds in which they act departs from the traditional neoclassical conception of economic actors as maximizing agents following their endogenous preferences through a world of exogenous constraints. To be fair, important work in economic history and the new institutional economics has adopted more elaborate notions of economic activity, but too often these "thicker" conceptions of economic behavior are based on neoclassical foundations and, consequently, continue to dis-embed economic actors from their institutional contexts.[16] Institutions, in these frameworks, tend to emerge out of the interactions between individual actors (often viewed in principal-agent terms) or are simply taken as given background conditions and constraints. Such a view of institutions leaves aside or assumes for its purposes the difficult and important questions about how actors respond to and refashion existing institutions as part of their economic activity and how, in the process, particular forms of economic life become embedded within these institutions.[17]

Viewing economic agents as strategizers intent on constructing the institutional environment in which they act, in contrast, directs attention to contingency, opportunism, and creativity in the face of uncertainty, thereby creating a space for understanding the role of politics, social norms, and law in the organization of economic activity.[18] By refusing to isolate these strategic agents from the institutions that they are simultaneously responding to as objective constraints and refashioning as they attempt to construct new forms of economic organization, such a perspective also recognizes that although actors do in fact make their own history, they can never do so, as Marx long ago observed, under the conditions of their choosing.[19]

In the context of regional industrialization, such an approach leads to a different kind of inquiry that results in a different sort of story from Alfred Chandler's canonical history of American industry. Chandler's work, drawing on Ronald Coase's early insights regarding the nature of the firm, explained the rise of the large vertically integrated mass-production enterprise as the product of an underlying logic intent on solving the coordination problems that emerged from increased economies of throughput.[20] In rejecting the heavy functionalism of this interpretation, the constructivist perspective views industrialization as a historically specific process that is heterogeneous and open to alternatives.[21] Organizational forms and strategies do not emerge out of some pre-institutional state of nature as a response to transactions costs; rather, they are made out of existing institutions and practices as firms struggle to deal with economic imperatives in the context of a dynamic institutional environment—a process that can only be fully understood in specific historical and social contexts.

The value of the Coasean perspective, of course, is that it provides a relatively simple explanation for the emergence of institutions (efficient frameworks for economizing transactions costs)—a sort of Occam's razor for economic history. In doing so, however, it glosses over the ways economic agents fashion new forms of economic organization out of existing ensembles of social practices, norms, and power relations. Industrial orders, in short, are made rather than discovered.[22] And individual firms are only one element (albeit an important one) in these larger industrial orders.

In the case of southern pulp and paper, individual firms worked with various actors to build a competitive industrial complex that both reflected and elaborated the institutions and the biophysical conditions particular to the region. In doing so, they engaged opportunistically with regional social norms, law, and politics at multiple levels in an effort to take advantage of and shape the distinctive institutions governing land, labor, and the environment.

Thus, the prevailing pattern of land tenure in the rural South—widely dispersed private ownership—required that firms simultaneously negotiate local solutions that fit with existing practices and norms, most prominently through the use of wood dealers to purchase timber from local landowners, while also working to shape larger, macro-level policy changes that would create incentive structures for landowners to participate in the industry. Efforts to rationalize the landscape to promote long-term investments in timber growing required changes to a range of local, state, and regional institutions, from the criminalization of customary woods burning practices to changes in tax laws and credit systems. Federal agricultural programs, such as those initiated during the New Deal, the Soil Bank Program of the 1950s, and the Conservation Reserve Program of later years, refashioned the property held by landowners and worked to tie their economic fortunes more closely to those of the pulp and paper industry. Important regional initiatives in fire control and tree improvement, which drew together actors from government, universities, and the private sector, provided a basis for technological innovation and regional collective learning.

As with efforts to industrialize the southern forest, the wood procurement system that served the mills so well emerged out of a complex articulation between regional norms, politics, and formal law. Here, however, leading firms approached regulation differently than they did in the context of land use. In building a flexible procurement system, they found ways to take advantage of the South's highly localized, informal labor markets, resisting efforts to bring logging contractors and their crews under federal labor laws. Invoking the hybrid status of these workers, pulp and paper companies and their allies blocked the extension of New Deal labor protections to woods workers for more than twenty-five years. At the same time, industry executives and managers employed basic principles of contract and agency law to refashion the dense network of local relationships between wood dealers, landowners, and thousands of itinerant woods workers into a system that delivered wood to the mills at low cost and with limited risks for the pulp and paper companies. By maintaining an arm's-length relationship with those working in the woods, they effectively insulated themselves from the costs and liabilities associated with full-time employees, while also making it much more difficult for loggers to organize for the purpose of collective bargaining. Law, in this instance, played a fundamental role in constraining the political capacity of woods workers and, in the process, determined to a significant extent those workers' economic relations to the mills.

In other areas, however, pulp and paper firms were less successful in using law and social norms to achieve their objectives. Seeking to maintain a stable mill labor force, managers and executives acquiesced to the endemic racial discrimination of the Jim Crow South and allowed the union locals to retain the best jobs for whites. After passage of the Civil Rights Act of 1964, litigation brought under Title VII by the federal government on behalf of black workers forced the companies and their unions to dismantle the discriminatory job progression lines that had denied so many black workers the opportunity to compete for the best mill jobs. As one of the most litigated industries in the South, the pulp and paper industry found itself in a defensive posture during the civil rights era, responding to rather than shaping larger legal and social changes. In this instance, at least, the industry and its unions could hardly be considered agents of progressive change in the region.

The federalization of pollution control laws during the 1970s also represented a fundamental legal and political shift that dramatically affected the industry. Prior to this time, pulp and paper firms succeeded in maintaining, through both the common law and state legislation, largely uninhibited access to the absorptive capacity of local ecosystems. But the federal statutory schemes enacted in the early 1970s to regulate air and water pollution constrained the industry's access to the environment, creating significant legal obligations in the process. In mandating substantial investments in pollution control technologies, moreover, the laws also reshaped the industry's competitive landscape. Given the costs involved, it is no surprise that environmental regulation occupied a great deal of the industry's attention during the 1980s and 1990s.

In all of these areas, politics, law, and social norms were woven into the very fabric of the industrial system, giving it a distinctive southern accent and determining in large part the terms of access to land, labor, and the environment. For much of the post–World War II period, pulp and paper firms successfully adapted to and shaped regional institutions to fit with the imperatives of mass production. In doing so, they created a regional industry that came to dominate national markets and dwarfed its international counterparts. By the 1970s, however, the stable prosperity of the postwar American economy was coming to an end, and the industry no longer wielded the influence in state and local politics that it once enjoyed. Firms found themselves facing a new set of challenges and constraints not only within the region but also in the larger world beyond the South.

Globalization, Industry Restructuring, and the Challenges Ahead

> Basically, International Paper needs to become more international. Right now, we do 75 percent of our business within North America. That needs to be more of a 50-50 split. We're going to make more choices and some of them will be tough. Not every [International Paper] facility you see today will be around 10 years from now. Our job is to make sure we figure out which ones to grow aggressively. I think you'll see a company more willing to change because drastic change is needed in today's marketplace. —*John Faraci, 2004*[23]

By the early 1970s, Union Camp's Savannah mill had produced enough paper to manufacture over 230 *billion* paper bags. According to the company, if laid end-to-end, these bags would have reached from Earth to Venus and back with enough left over to encircle both Earth and Venus almost one hundred times. As ground zero for the brown paper bag revolution in postwar America, Union Camp's Savannah complex produced the paper for roughly 20 percent of all bags used in the United States during this period. Reflecting on his industry's contributions to American prosperity, John Ray, an executive vice president at Union Camp, told the Savannah Rotary Club in 1970, "One of the surest measures of a country's economic development and well-being is the amount of paper it consumes in a year. In the United States, that amount is 576 pounds per person. In the Soviet Union, it is 57 Pounds. In India, just a little under four pounds."[24]

What Ray's numbers really illustrated was the maturity of the American paper industry and the growing saturation of domestic markets. While per capita paper consumption had more than doubled during the twenty-five years preceding 1970, over the next twenty-five years it increased by less than 30 percent.[25] Having come of age in an era of expanding markets and general macroeconomic stability, paper industry executives found themselves confronting a very different economic landscape in the decades after 1970. As domestic markets saturated and the economy slowed, the older strategy of "running full" no longer worked, leading directly to problems of overcapacity, declining prices, and poor returns.[26]

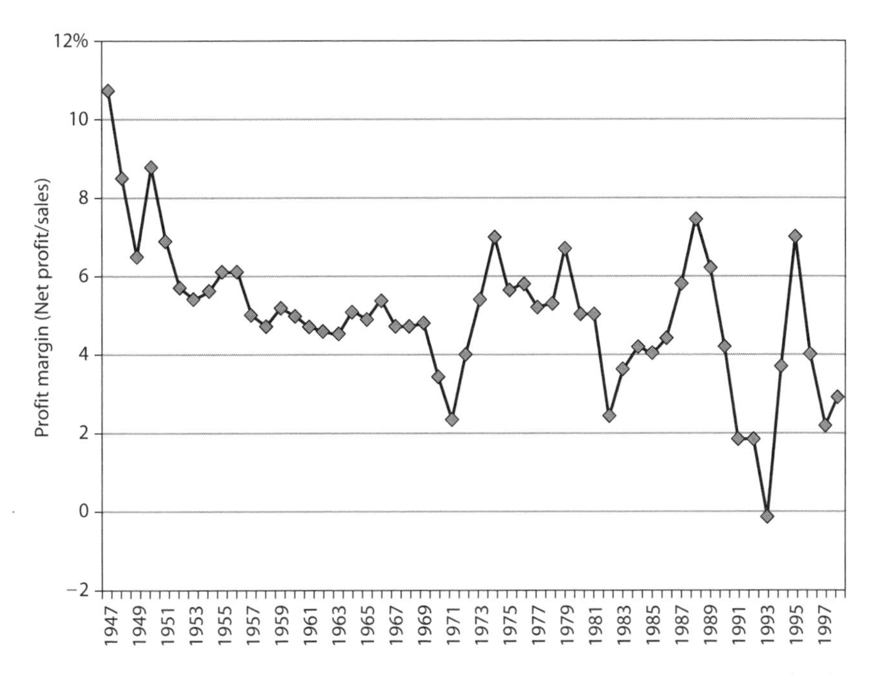

Figure 5.1. U.S. paper industry profit margins, 1947–1998. American Paper and Pulp Association, *Statistics of Paper* (1947–1972); American Paper Institute, *Statistics of Paper and Paperboard* (1973–1980); and the American Forest and Paper Association, *Statistics of Paper, Paperboard, and Wood Pulp* (1981–1997).

Indeed, the changing macroeconomic environment of the 1960s and 1970s—reflected principally in rising domestic inflation, the abandonment of Bretton Woods for a system of floating exchange rates, the concomitant proliferation of new financial instruments and offshore financial markets, and, as the 1970s progressed, oil shocks, rising interest rates, and "stagflation"—confronted large mass-production industries with a host of new challenges that were not amenable to the industrial strategies of previous decades.[27] With growing competition from abroad for both domestic and international markets, profitability for pulp and paper firms was increasingly linked to the relative strength of the U.S. dollar and fluctuations in international markets rather than the ability to "run full."[28]

Overall, the profound instabilities that marked the post-1970 macroeconomic landscape made consistent profitability much more elusive for a cyclical industry such as pulp and paper. Profit margins fluctuated dramatically after 1970 (figure 5.1). One 1998 survey of nineteen major U.S. paper producers found that their return on capital averaged *negative* 1.7 percent between 1975 and 1997. Not surprisingly, the relative stock market performance of these companies lagged the overall market between 1970 and 1999 during every year except 1983 to 1986.[29]

Obviously, the world had changed—quite radically in some respects—in a manner that challenged the basic assumptions of the mass-production paradigm. The relative stability of the post–World War II economy, which had provided such a hospitable operating environment for large vertically integrated firms, had given way to an era of increased economic uncertainty and competition. Efforts to adapt to the new business landscape were further constrained by the substantial investments in large pulp and paper mills, a relatively unconcentrated industry structure, and a homogenous commodity-grade product—all of which made the collective problem of managing capacity at a sectoral level very difficult. And even though these large investments in fixed capital were essentially sunk costs, they often created significant barriers to *exit*. As long as individual firms could cover their variable costs, they kept producing. Having either already paid for their machines or committed to a stream of debt payments that would come due regardless of whether the machines were actually operating, there was little incentive to shut them down over the short term. When prices fell low enough, firms began to cut back on capacity. But as soon as they began to rise, firms tended to crank their idle machines back up as quickly as possible, collectively canceling out any potential gains from price increases.

This was precisely the kind of "industrial insanity" that *Fortune* magazine and other industry trade journals had pointed to in the 1930s, and it led some to question the intelligence of those in charge.[30] As one industry manager put it in 1997: "We're the dumbest industry in the world. Here we've gone on for three years with really sorry markets. It's been horrible. We get a little bit of a ray of hope with a possible price increase or a little bit of inventory adjustment that puts things in balance so that we should be able to be in a profitable situation. And what do we do? We crank up every damn paper mill and every machine we've had shut down and we're going to flood the market again. And why these brilliant people that are running these companies can't figure that out is beyond me."[31] Another industry executive made a similar point: "The great bane of this industry is overbuilding. We get into a cycle when the prices are up and you're making a lot of money and you've got wonderful cash flow, and so you go out and buy new paper machines. It's like kids in Toys-R-Us."[32] In effect, individual firms behaved in a manner that translated directly into a collective failure to adjust to changing market conditions. The symptoms—overcapacity, "ruinous" competition, and poor returns—brought renewed calls for corrective action throughout the 1990s.

Squeezing labor emerged as one part of the effort to cut costs and maintain profitability (see chapter 3). But this was not really a viable strategy over the long term. Although it might improve the competitive position of certain firms in the short run, it could do little to solve the larger structural problems of overcapacity and macroeconomic instability. At best, such an approach might reduce some of the pain associated with the industry's business cycle by transferring some of the costs of adjustment from capital to labor. In the end, though, any relative devaluation of labor did nothing to alter the basic stock of fixed capital in the industry and the imperative to run full.

Two other potential solutions to the problems of overcapacity and excessive competition were product differentiation and consolidation. Product differentiation proved limited, given the commodity characteristics of paper, although some firms did push into niche markets with specialty papers and other new products. As for the consolidation strategy, it had been tried in the past but failed to achieve any lasting effects. Despite perennial calls for consolidation, the overall level of concentration in the primary-processing segments of the industry remained fairly stable from the 1960s through the mid-1980s. Starting in the mid-1980s, however, concentration among the largest firms in the sector did increase as a result of mergers.[33]

By the second half of the 1990s, as the industry turned in one of its most dismal performances on record,[34] industry leaders pointed with increasing urgency to the capacity problem and the corresponding need to consolidate. John Simley of Chicago-based Stone Container lamented his industry's "prolonged state of negative profit [stemming from] rampant overcapacity . . . with too many machines pushing too much paper chasing too few buyers."[35] For Pete Correll, chairman and chief executive of Georgia-Pacific, the solution was simple: "The industry needs to consolidate."[36] Otherwise, he warned, it would be in danger of "sinking like the *Titanic*."[37]

During the late 1990s and early 2000s, merger activity picked up considerably. International Paper's acquisitions of Union Camp and Champion International, along with other high-profile mergers, suggested to some that the long-awaited era of consolidation had finally arrived.[38] By 2002, the four largest firms in the industry accounted for 50 percent of the value of all shipments, up from only 22 percent in 1982.[39] The key question, of course, was whether consolidation would be followed by actual reductions in capacity; that is, whether the remaining firms would have the discipline to shut down older machines and temper their appetites for expansion. Looking back over the period since this most recent merger wave, the evidence suggests that the answer is yes. During the ten-year period between 2000 and 2009, U.S. paper and paperboard capacity actually *declined* by an annual average of 0.9 percent.[40]

To some extent this newfound discipline together with the ongoing process of consolidation reflected the increased global exposure of North American firms. The accelerating globalization of the industry, manifest in increased integration of product markets, cross-border mergers and acquisitions, and the emergence of a handful of truly global companies, had a significant impact on industry organization and structure during the 2000s.[41] In particular, the success of low-cost production complexes in the southern hemisphere since the 1990s looks very much like another round of spatial restructuring akin to what happened in the United States during the 1930s and 1940s when the industry moved into the American South. With fiber costs accounting for a substantial part of competitive advantage in an increasingly global industry, the development of pulp and paper complexes based on intensively managed short-rotation pulpwood plantations in Asia and South America created a new cost structure for the industry, leading to a further "rationalization" of the North American and European sectors.[42] Such a geographically based and, to a large extent, biologically driven process of creative destruction appears to have produced the kind of devaluation that industry leaders had long been incapable of achieving through more traditional methods of consolidation and corporate restructuring.

Likewise, China's rapid rise to become the world's leading paper producer in 2009 put additional pressures on the North American industry.[43] As a result of generous subsidies from the Chinese government (one estimate put the total value of such subsidies at more than $33 billion between 2002 and 2009), the paper industry expanded dramatically in China during the late 1990s and 2000s.[44] Paper production quickly outstripped domestic demand, notwithstanding dramatic growth in per capita paper consumption, forcing Chinese paper producers to export an increasingly large portion of their product. In the process, they undercut other global producers and captured a significant share of global markets.[45] Given the importance of government subsidies in fueling China's dramatic growth, and the fact that China has no natural competitive advantage in paper production, the ability of the Chinese paper industry to continue taking global market share is not clear. But there can be little question that the rise of China combined with its voracious appetite for imports of pulp and wastepaper to feed its mills marked a dramatic shift in the competitive landscape of an increasingly global pulp and paper industry.

None of which is to suggest that the industry is going to pack up and leave North America or the South anytime soon. Pulp and paper is hardly a footloose industry capable of rapid migration. More importantly, the southern pulp and paper industry is still quite cost competitive. Given its proximity to the U.S. market, a vast and intensively managed biological production base, and a set of organizational capabilities that has been refined over more than three quarters of a century, the industry appears to be well positioned going forward. But in order for the industry to thrive, the pressure to extract higher biological returns from timberlands will likely increase, with biotechnology and intensive management playing increasingly important roles in refashioning the southern forest yet again.

In adjusting to this new competitive landscape, the industry will also have to confront ongoing challenges inside the region. The tensions embedded in the wood procurement system are likely to persist as firms attempt to further control and even reduce the cost of fiber delivered to the mills. Changes in local labor markets and the broader regional economy also threaten to undermine the foundations of the southern wood procurement system, as people continue to leave the rural South. The sale of significant acreages of company timberlands to large timber investment management organizations and real estate investment trusts during the first decade of the twenty-first century also raises questions about the industry's resource base.[46] As timberland becomes a more liquid asset, subject to multiple market opportunities rather than the singular needs of the forest products industry, mill procurement managers may lose some of the traditional advantages they have enjoyed in local timber markets.

Environmental concerns also pose ongoing challenges for the industry. The proliferation of satellite chip mills and the heightened awareness of the impact of industrial forestry on biological diversity have turned parts of the southern forest into deeply contested landscapes. Climate change and the attendant problems of prolonged drought; increased incidence of fire, pests, and disease; and intense storms and other extreme weather pose threats to timberlands and to the stability necessary for rational, long-term investment. Further downstream, despite the vast progress that has been made in reducing pollution from pulp and paper mills, the industry is still a major source of air and water pollution in certain areas, with some environmental groups pushing for totally closed-loop, zero-discharge mills—a very expensive proposition from the industry's point of view. As the region witnesses increased growth in tourism and suburban development, particularly in coastal areas, mills are also facing opposition from boosters seeking to protect and enhance the aesthetics of local landscapes. In short, although still influential, the industry no longer enjoys the privileged position it once held in state and local politics and no longer has access to the environment on the terms and conditions that it enjoyed during the expansionary post–World War II decades.

New South, New Nature

> The South must choose between tradition and progress. It must choose because . . . Southern tradition has essential elements . . . which are irreconcilably at war with regional economic progress. —*William H. Nicholls, 1960*[47]

William Nicholls saw the antinomies of tradition and progress as the fateful choice confronting the post–New Deal American South.[48] For him, as for so many of his contemporaries, the pathologies of racism, endemic poverty, undemocratic politics, and a stunted sense of social responsibility all grew out of the South's agrarian heritage, creating formidable barriers to lasting economic progress.[49] "The South," he wrote, "cannot have it both ways"; it would have to make a fundamental choice "between the values of agrarianism and industrialism."[50] And it was only by embracing industrialism that the region could finally shed the burdens of its history and achieve the progress that had eluded it for so long. Only then would the South be able to join the rest of the country. Only then would it become truly American.

This conflict between the values of a premodern, agrarian way of life and those of industrial capitalism had preoccupied southern intellectuals and political leaders for decades.[51] William Faulkner saw in the postbellum South a region struggling to come to terms with its deeply tragic past as it was swept up in a new economic order. For him, the destruction of the landscape was a symptom of a larger regional tragedy; the contradictions and violence at the heart of southern society mirrored in the "doomed wilderness" that was fast disappearing under the relentless assault of logging and plantation agriculture. Forcing his readers to confront the substantial human and environmental costs incurred in creating the "tamed and ordered" landscape of the modern South, Faulkner showed how the social order was intimately connected to the land and to the particular places that people made out of that land.[52]

For the Nashville agrarians, a group of intellectuals writing during the 1930s, industrialism threatened to destroy the last vestiges of a fragile humanism that might provide an antidote to the materialism of modern, twentieth-century America.[53] Rightly criticized as elitist, even reactionary, with no real program for the region, the agrarians wanted to find a way for the South to manage its engagement with industrialism.[54] In their effort to articulate a counternarrative to the sometimes shrill New South boosterism of the chamber of commerce, the agrarians saw the instrumental rationality of industrial capitalism as a source of profound alienation.[55] Put simply, the question they posed was whether the South could industrialize on terms and conditions that would allow it to hold on to some of the values that came with older ways of life.

But, of course, it had always been so much harder than that. The region—its many people and places—faced contradictions, burdens, and constraints that the rest of America never really confronted. The idea that it could somehow forge its own path to a modern industrial future—that it could join the rest of America on its own terms—was always a fantasy. Equally mistaken was the view that the industrialization of the South could be reduced to a simple narrative of modernization and progress in which something old and backward was replaced with something new.[56] Without question, the vernacular institutions and landscapes of the old agrarian South underwent profound transformations during the transition to an industrial economy. They were taken hold of, sometimes violently, refashioned, and pulled into an emerging industrial order that, on the surface at least, bore little resemblance to what had come before. Yet in order to really understand this process of industrialization we have to look beneath the surface, following the connections and linkages that underwrote this new industrial order back into the rural South to uncover the networks of informal relationships and practices that provided so much of the foundation for this industrial transition.[57]

For it was mainly in the process of recombining and elaborating existing institutions and practices and in the appropriation of existing landscapes and resources that a new industrial order was constructed. Southern industry, in this sense, had always embodied a distinctive combination of tradition and progress, and too much of southern economic history seems to have missed the importance of understanding how tradition and progress have been made to work together in the process of industrialization. Instead of a linear, singular process of modernization, in other words, the story of southern industrialization, like that in other developing countries, represented a combination of seemingly contradictory elements, drawing from both old and new—always uneven, messy, and contingent.

As one of the twentieth-century South's most successful and promising industries, pulp and paper embodied this mix of contradictory elements. It combined one of the world's leading programs in scientific forestry with a repressive and backward approach to logging and wood procurement that drew on, and even nurtured, the inequalities of the rural South to fashion a low-cost, flexible system for delivering wood to the mills. It brought with it huge investments in industrial facilities with high-paying union jobs but acquiesced to prevailing racial norms and allowed the best of those jobs to be reserved for whites in order to keep the mills running. It showed great care in its management of industrial timberlands while demonstrating an almost complete disregard for the local environments and communities affected by its massive pollution loads.

Viewed in the aggregate, the development of this industry surely represents a success story of immense proportions—a realization of the ambitions of those keen on building an industrialized South out of a backward agrarian economy. Whether measured in jobs, wages, taxes, the utilization of regional resources, or value added, this was exactly the kind of industry that so many southern boosters had sought for so long. But while some of the changes wrought by the industry were undeniably positive when set against the backdrop of the 1930s, others were not. The industrial timberlands that took shape out of the wreckage of the 1920s and 1930s seemed sterile, artificial, and monochromatic when compared to earlier landscapes—devoid of people, "tamed and ordered" more thoroughly than perhaps even Faulkner could have foreseen. Small contract loggers of both races who worked to deliver wood to the mills faced a degree of political and economic marginalization as severe as that facing any other workers in America. Until the civil rights revolution of the 1960s, black mill workers were denied access to higher-paying jobs and the opportunities that came with them. And poor and minority communities ended up bearing the brunt of the pollution associated with the expansion of the industry in the region. This sort of uneven development is hardly atypical of regional industrialization. Indeed, it is an intimate part of any such process.

Epigraph. Raymond Williams, "Ideas of Nature," *Problems in Materialism and Culture* (London: Verso, 1980), 83–84.

Introduction

Epigraph. Interview with Elwood R. Mauder, in Elwood R. Mauder, *Voices from the South: Recollections of Four Foresters* (Santa Cruz, CA: Forest History Society, 1977), 38.

1. The colonial analogy has a long history in southern economic thought. Among the historians and social scientists who employed the model to explain the persistence of southern "underdevelopment" during the half century after Reconstruction, see especially Rupert B. Vance, *Human Geography of the South: A Study in Regional Resources and Human Adequacy*, 2nd ed. (Chapel Hill: University of North Carolina Press, 1935); Howard W. Odum, *Southern Regions of the United States* (Chapel Hill: University of North Carolina Press, 1936); C. Vann Woodward, *The Burden of Southern History*, 3rd ed. (1960; reprint, Baton Rouge: Louisiana State University Press, 1993), 190–91; C. Vann Woodward, *Origins of the New South, 1877–1913* (Baton Rouge: Louisiana State University Press, 1951); and George B. Tindall, "The 'Colonial Economy' and the Growth Psychology: The South in the 1930s," *South Atlantic Quarterly* 64, no. 4 (1965). For an insightful overview of the subject that traces the intellectual lineages of the colonial analogy from the revolutionary period through the mid-twentieth century, see Joseph J. Persky, *The Burden of Dependency* (Baltimore, MD: Johns Hopkins University Press, 1992). For a sustained critique, see Gavin Wright, *Old South, New South: Revolutions in the Southern Economy since the Civil War* (New York: Basic Books, 1986), 156–97.

2. Vance, *Human Geography of the South*, 474, 489. Vance's colleague and teacher Howard Odum put it this way: "In fine and in sum, the agrarian problem is *the region*, for better or for worse, and the agrarian statecraft which is involved." Odum, *Southern Regions*, 57.

3. Vance, *Human Geography of the South*, 489.

4. That the South was a "low-wage region in a high-wage economy" meant that its policy options were sharply constrained. There was no regional government, no control over foreign trade and factor flows, and little control over the banking system. The proximity of a vast high-wage economy to the north also created a vested interest on the part of southern landowners and other employers to maintain regional isolation. For an elaboration of these points, see Wright, *Old South, New South*.

5. For general discussions of underdevelopment in the postbellum South and extensive documentation of regional trends and indicators, see Odum, *Southern*

Regions; Vance, *Human Geography of the South*; and Rupert B. Vance, *All These People: The Nation's Human Resources in the South* (Chapel Hill: University of North Carolina Press, 1945).

6. Howard Odum referred to it as the "the poverty-breeding system of tenant and landlord." Odum, *Southern Regions*, 55.

7. As a region, the South accounted for 61% of the nation's total acreage of eroded lands. Odum, *Southern Regions*, 38–39. Land degradation, which peaked in the 1920s and 1930s, greatly reduced the tax base of many state and local governments—further straining the already limited fiscal ability of southern states to respond to the crisis. See also Vance, *Human Geography of the South*, 99–106, and Gerald W. Johnson, *The Wasted Land* (Chapel Hill: University of North Carolina Press, 1937).

8. Odum, *Southern Regions*, 341.

9. For discussions of federal land policy and its effects, see Paul W. Gates, "Federal Land Policy in the South, 1866–1888," *Journal of Southern History* 6, no. 3 (1940), and "Federal Land Policies in the Southern Public Land States," *Agricultural History* 53, no. 1 (1979).

10. Detailed research by Paul Gates found that for "sales of 3,728,000 acres of southern land sold between 1881 and 1888, . . . 68 percent were acquired by lumbermen and dealers in timberland from Chicago, Michigan, and Wisconsin, including Frederick Weyerhaeuser who was rapidly becoming king of them all. Probably a considerable portion of the remaining 33 percent acquired by southerners was partly made possible by northern capital" ("Federal Land Policies in the Southern Public Land States," 220). See also Gates, "Federal Land Policy in the South," 315 n45.

11. In May 1888, a joint resolution of Congress suspended all further entries of land except under the Homestead Act for Alabama, Arkansas, and Mississippi. In July this was extended to Florida and Louisiana—effectively reinstating the Homestead Act of 1866. In March 1889, land sales for private entry were halted in every public lands state except Missouri. The following year, Congress established a limit of 320 acres of public land that any person could acquire by any combination of measures. Gates, "Federal Land Policies in the Southern Public Land States," 220; Woodward, *Origins of the New South*, 115–18.

12. Gates, "Federal Land Policy in the South," 330. Gates noted in his 1979 article "Federal Land Policies in the Southern Public Land States" that there was still much research to be done on the southern land question, "that there are enough unanswered questions about southern land history to keep historians busy for years" (227).

13. Woodward, *Origins of the New South*, 117–18. Woodward goes on to note that three Florida legislatures in the early 1880s had actually granted or attempted to grant more than 22 million acres of the public domain to projected railroads. Yet, the so-called public domain in Florida had never at any time exceeded 14,831,739 acres! "The Carpetbaggers in all their glories could hardly match such deeds."

14. See U.S. Bureau of Corporations, *The Lumber Industry* (Washington, DC: Government Printing Office, 1913), xvii, xxii. Initiated in 1911 at the behest of both houses of Congress and based on a two-year investigation in all of the major timber-producing regions of the country, the report compiled a wealth of evidence on tim-

berland ownership and offered some hard-hitting conclusions on the role of public land policies in inviting land speculation on a massive scale.

15. Within the southern pine region, the Bureau of Corporations found significant concentration in the higher value species, particularly longleaf pine and cypress. For the region as a whole, 67 "holders" owned 39% of the longleaf pine, 29% of the cypress, and 19% of the shortleaf and loblolly pine. The 925 holders who owned more than 60 million board feet of standing timber controlled 73% of the longleaf pine, 72% of the cypress, and 46% of the shortleaf and loblolly pine. The report also pointed out that throughout the southern pine region, the larger holdings, which were typically of higher quality than smaller holdings, could be logged at less cost per unit, thus giving them a greater importance in the market. According to the investigators, larger holdings could also be used to "block in" smaller holdings, thus providing the larger holders with a greater area of control. Bureau of Corporations, *Lumber Industry*, 21, 96–97.

16. Ibid., xvii.

17. Vance, *Human Geography of the South*, 124–25. See also George B. Tindall, *Emergence of the New South* (Baton Rouge: Louisiana State University Press, 1967), 82.

18. Howard Odum noted, "Of the 125 million acres in virgin pine, one hundred million have been cut-over, ten and a half million absorbed into agriculture, and thirty three million remain without new growth." Odum, *Southern Regions*, 576. See the table on p. 80 for numbers and a map on the cutting in the pine belt. The table reports that of 115 million acres of total pine area in the South, only 23.46 million acres of old growth remained in 1919 and only 12.65 million in 1927. The same table also reports that for the state of Georgia, the area of old-growth pine was reduced by half between 1919 and 1927, from 700,000 acres to 350,000 acres (ibid., 80).

19. R. D. Forbes, "The Passing of the Piney Woods," *American Forestry* 29, no. 351 (1923).

20. Cited in James I. Pikl Jr., *A History of Georgia Forestry*, Research Monograph no. 2 (Athens, GA: University of Georgia Bureau of Business and Economic Research, 1966), 17.

21. Faulkner wrote powerfully about the devastation left behind by the lumber industry. In *Light in August*, he described a scene that doubtless replayed itself throughout the region: "All the men worked in the mill or for it. It was cutting pine. It had been there seven years and in seven years more it would destroy all the timber within its reach. Then some of the machinery and most of the men who existed because of and for it would be loaded onto freight cars and moved away. But some of the machinery would be left since new pieces could always be bought on the installment plan—gaunt, staring, motionless wheels rising from the mounds of brick rubble and ragged weeds with a quality profoundly astonishing, and gutted boilers lifting their rusting and unsmoking stacks with an air stubborn, baffled and bemused upon a stump-pocked scene of profound and peaceful desolation, unplowed, untilled, gutting slowly into red and choked ravines beneath the long quiet rains of autumn and the galloping fury of vernal equinoxes." William Faulkner, *Light in August* (1932; reprint, New York: Vintage, 1987), 5. In *The Hills Beyond* (1935; reprint,

New York: Sun Dial, 1943), Thomas Wolfe described a scene of similar destruction: "The great mountain slopes and forests of the section had been ruinously detimbered, the farm soil on hill sides had eroded and washed down; high up, upon the hills, one saw the raw scars of old mica pits, the dump heaps of deserted mines. Some vast destructive 'Suck' had been at work here; and a visitor, had he returned after one hundred years, would have been compelled to note the ruin of the change. It was evident that a huge compulsive greed had been at work: the whole region had been sucked and gutted, milked dry, denuded of its rich primeval treasures: something blind and ruthless had been here, grasped and gone" (236–37).

22. The phrase is from C. Vann Woodward: "Cut off from the better-paying jobs and the higher opportunities, the great majority of Southerners were confined to the worn grooves of a tributary economy. Some emigrated to other sections, but the mass of them stuck to farming, mining, forestry, or some low-wage industry whether they liked it or not. The inevitable result was further intensification of the old problems of worn-out soil, cut-over timberlands, and worked-out mines" (*Origins of the New South*, 319–20).

23. C. Vann Woodward, again, said it best: "for a long time to come race consciousness would divide, more than class consciousness would unite, Southern labor" (ibid., 222). V. O. Key Jr. offered similar reflections. His summation of "the politics of color" in South Carolina was particularly apt: "South Carolina's preoccupation with the Negro stifles political conflict. Over offices there is conflict aplenty, but the race question muffles conflict over issues latent in the economy of South Carolina. Mill worker and plantation owner alike want to keep the Negro in his place. In part, issues are deliberately repressed, for, at least in the long run, concern with genuine issues would bring an end to the consensus by which the Negro is kept out of politics. One crowd or another would be tempted to seek his vote. In part, the race issue provides in itself a tool for the diversion of attention from issues. When the going gets rough, when a glimmer of informed political self-interest begins to well up from the masses, the issue of white supremacy may be raised to whip them back into line." V. O. Key Jr., *Southern Politics in State and Nation* (1949; reprint, Knoxville: University of Tennessee Press, 1984), 131.

24. National Emergency Council, *Report on Economic Conditions of the South* (Washington, DC: Government Printing Office, 1938), 8.

25. See Albert O. Hirschman, "A Generalized Linkage Approach to Development, with Special Reference to Staples," reprinted in *Essays in Trespassing: Economics to Politics and Beyond* (Cambridge: Cambridge University Press, 1981), 59–97.

26. The Kraft or sulfate process was invented in Germany at the end of the nineteenth century and employs sodium sulfate as the pulping agent. Paper produced by this process is very strong, hence the word "Kraft," which is German for "strength."

27. Beginning in 1925, for example, industry leader International Paper established a foothold in the region with its purchase of the Bastrop Pulp and Paper Company of Louisiana. Two years later, the company bought another Louisiana firm (the Louisiana Pulp and Paper Company) and created a new subsidiary to manage its southern operations. Known initially as the Southern International Paper Com-

pany and later as the Southern Kraft Corporation, the division was the leading pulp producer in the region by 1930—earning profits even during the depths of the depression and making up for International Paper's unprofitable northern operations. See "International Paper & Power," *Fortune* 16, no. 6 (1937): 229; and Jack P. Oden, "Development of the Southern Pulp and Paper Industry, 1900–1970" (PhD diss., Mississippi State University, 1973), 578–79.

28. Thomas Clark, *The Greening of the South: The Recovery of land and Forest* (Lexington: University Press of Kentucky, 1984), chap. 9.

29. "Union Bag & Paper Corp," *Fortune* 16, no. 2 (1937): 49, 126.

30. For a discussion of early Kraft pulp manufacture in the South, see Oden, "Development of the Southern Pulp and Paper Industry," and Jack P. Oden, "Origins of the Southern Kraft Paper Industry, 1903–1930," *Mississippi Quarterly* 30, no. 4 (1977). Oden's dissertation represents the most comprehensive treatment of the southern pulp and paper industry to date. See also William T. Hicks, "Recent Expansion in the Southern Pulp and Paper Industry," *Southern Economic Journal* 6, no. 4 (1940); and Helen Hunter, "Innovation, Competition, and Locational Changes in the Pulp and Paper Industry: 1880–1950," *Land Economics* 31, no. 4 (1955).

31. The Paper and Allied Products industry (SIC Code 26) contains seventeen distinct industries. The basic organization can be broken down into three major tiers of activity: woodpulp production; paper and paperboard production; and finished products conversion. Integrated pulp and paper mills, which consist of large pulp mills and paper/paperboard mills on the same site, allow for production to run continuously from raw materials to bulk paper and paperboard products. Large-scale integrated mills are necessarily located close to the sources of raw materials and represent the foundation of the modern pulp and paper industry. Maureen Smith, *The U.S. Paper Industry and Sustainable Production: An Argument for Restructuring* (Cambridge, MA: MIT Press, 1997), 19–27.

32. U.S. Forest Service, *The South's Fourth Forest: Alternatives for the Future*, Forest Resources Report no. 24 (Washington, DC: Government Printing Office, 1988), 329 table 2.27; U.S. Forest Service, *An Analysis of the Timber Situation in the United States: 1952–2030*, Forest Resource Report no. 23 (Washington, DC, 1982), 298–99; American Forest & Paper Association, *Paper, Paperboard, and Woodpulp Statistics* (Washington, DC, 1995), 49. The South, according to these government reports, includes Delaware, Maryland, Virginia, West Virginia, North Carolina, South Carolina, Georgia, Florida, Tennessee, Kentucky, Alabama, Mississippi, Texas, Oklahoma, Arkansas, and Louisiana.

33. *Pulp and Paper International*, January 1997; American Forest & Paper Association, *Paper, Paperboard, and Woodpulp Statistics* (1995).

34. F. Cubbage, R. Abt, W. Dvorak, and G. Pacheco, "World Timber Supply and Prospects: Models, Projections, Plantations and Implications" (presentation at Central American and Mexican Coniferous Resource Cooperative Meeting, Bali, Indonesia, October 15–24, 1996). Figures for the United States and the South are based on linear projections of U.S. Forest Service data and represent more conservative estimates than the Forest Service's own projections. Figures for other regions and

countries are taken from Bob Flynn, "Latin America: The Future of Fiber Exports" and Mike Edwards, "The South African Forestry and Forest Products Industry: A Synopsis," both in *Proceedings of the International Woodfiber Conference, Atlanta, GA, May 13–14* (San Francisco: Miller Freeman, 1996).

35. This argument draws on an interpretation advanced by Gavin Wright. In his view, the institutions of tenancy and sharecropping combined with pervasive racial discrimination worked to create and sustain an isolated regional labor market that served as the primary constraint on economic development in the region. Wright, *Old South, New South*, 64–70. Such an interpretation resonates with Robert Brenner's work on the fundamental role of agrarian class structures in shaping economic development in preindustrial Europe. See Robert Brenner, "Agrarian Class Structure and Economic Development in Pre-Industrial Europe," *Past and Present*, no. 70 (1976).

36. Much of this "constructivist" work has been in response to the path-breaking scholarship of Alfred J. Chandler Jr. on American industrial development. In this book, Chandler certainly looms large, particularly in understanding the economic imperatives facing firms in large, mass-production industries and in the development of what he calls "organizational capabilities." See, e.g., "Organizational Capabilities and the Economic History of the Industrial Enterprise," *Journal of Economic Perspectives* 6, no. 3 (1992); *The Visible Hand: The Managerial Revolution in American Business* (Cambridge, MA: Belknap Press of Harvard University Press, 1977); and *Scale and Scope: The Dynamics of Industrial Capitalism* (Cambridge: Belknap Press of Harvard University Press, 1990). This books seeks to develop a more grounded approach to American industrialization than Chandler's narrative allows. The analytical perspective here is best characterized as "constructivist political economy"—a term borrowed from Gary Herrigel—which views industrialization as a historically situated process that is heterogeneous and contingent rather than driven solely by some underlying logic. In his study of German industrialization, Herrigel introduces the concept of "industrial order" as an alternative to more traditional conceptions of industrial organization. In his view, industrial orders are made rather than discovered. Understanding how they get made is the critical task facing scholars of industrialization and regional development. See Gary Herrigel, *Industrial Constructions: The Sources of German Industrial Power* (Cambridge: Cambridge University Press, 1996). Herrigel's notion of industrial order is similar to the regional industrial system concept used by Annalee Saxenian in *Regional Advantage: Culture and Competition in Silicon Valley and Route 128* (Cambridge, MA: Harvard University Press, 1994). For similar perspectives, see Charles F. Sabel and Jonathan Zeitlin, "Stories, Strategies, and Structures: Rethinking Historical Alternatives to Mass Production," in *World of Possibilities: Flexibility and Mass Production in Western Industrialization*, ed. Charles F. Sabel and Jonathan Zeitlin (Cambridge: Cambridge University Press, 1997); Gerald Berk, *Alternative Tracks: The Constitution of American Industrial Order, 1865–1917* (Baltimore, MD: Johns Hopkins University Press, 1994); Philip Scranton, *Endless Novelty: Specialty Production and American Industrialization, 1865–1925* (Princeton, NJ: Princeton University Press, 1997).

37. See Chandler, "Organizational Capabilities and the Economic History of the Industrial Enterprise."

38. Under the combined impact of New Deal agricultural and labor policies, farmers shifted to wage labor and mechanization, forcing more than a million tenant farmers to leave the land as southern agriculture concentrated on larger farms in the more productive regions. Between 1950 and 1975, total farms declined from 2.1 million to 722,000 while average farm size increased from 93 to 216 acres (excluding Texas and Florida). In the process, over 40 million acres of cropland became available for other uses. See Wright, *Old South, New South*; Bruce Schulman, *From Cotton Belt to Sun Belt: Federal Policy, Economic Development, and the Transformation of the South, 1938–1980* (New York: Oxford University Press, 1991); and Gilbert C. Fite, *Cotton Fields No More: Southern Agriculture, 1865–1980* (Lexington: University Press of Kentucky, 1984). With this "rationalization" of southern agriculture, new opportunities for forest regeneration emerged.

39. Many scholars have explored the relationship between nature and technology, emphasizing the difficulties of making hard and fast distinctions. Environmental historians such as Donald Worster, William Cronon, and Richard White have interrogated some of the ways nature gets incorporated into technological and political-economic systems. See, e.g., Donald Worster, *Rivers of Empire: Water, Aridity, and the Growth of the American West* (New York: Oxford University Press, 1985); William Cronon, *Nature's Metropolis: Chicago and the Great West* (New York: W. W. Norton, 1991); and Richard White, *The Organic Machine* (New York: Oxford University Press, 1995). See also the review essay by Jeffrey K. Stine and Joel Tarr, "At the Intersection of Histories: Technology and the Environment," *Technology and Culture* 39 (1998). Several historians and social scientists have also investigated the role of science and technology in the industrialization of agricultural systems. Jack Ralph Kloppenburg Jr., *First the Seed: The Political Economy of Plant Biotechnology, 1492–2000* (New York: Cambridge University Press, 1988), and Deborah Fitzgerald, *The Business of Breeding: Hybrid Corn in Illinois, 1890–1940* (Ithaca: Cornell University Press, 1990), have both demonstrated how a particular biological organism (hybrid corn) gets refashioned as an agricultural commodity and as a vehicle for capital accumulation. See also William Boyd, "Making Meat: Science, Technology, and American Poultry Production," *Technology and Culture* 42 (2001) for a discussion of the industrialization of chicken biology as a basis for the American poultry industry.

40. The term is from Richard White's book, *The Organic Machine*.

41. According to state-reported data for 1996, the pulp and paper industry employed some 227,000 workers in the South—176,000 of whom were production workers. Total payroll for the region was more than $8.4 billion—almost $6 billion of which went to wages for production workers. Total value of shipments from the region was more than $64 billion. Value added by manufacture was almost $30 billion. See American Forest & Paper Association, *Paper, Paperboard, and Wood Pulp Statistics* (Washington, DC, 1999), 50.

42. Tindall, *Emergence of the New South*.

Chapter 1 · Industrializing the Southern Forest

Epigraph. From a speech Calder, the company president, gave at a dinner in Savannah celebrating the formal opening of Union Bag's Savannah mill. Quoted in "Bright Future for Industry," *Savannah Morning News*, October 2, 1936.

1. Formally known as the "resin barrier," this problem consisted of the conviction that highly resinous young pine trees were unsuitable for papermaking. During the 1910s and 1920s, the success of early Kraft (also known as sulfate) mills in the South, as well as research at the U.S. Forest Products Laboratory in Madison, Wisconsin, demonstrated that suitable paper could be made from young pine trees using the Kraft process. See Jack P. Oden, "Origins of the Southern Kraft Paper Industry, 1903–1930," *Mississippi Quarterly* 30, no. 4 (1977): 573–74. Savannah's own Charles Holmes Herty, though sometimes mistakenly identified as the man who debunked the sap myth, played an active role in the 1920s and 1930s in promoting the virtues of southern pine for pulp and paper manufacture. See James I. Pikl Jr., *A History of Georgia Forestry*, Research Monograph no. 2 (Athens: University of Georgia Bureau of Business and Economic Research, 1966).

2. Warnings of an impending timber famine had been voiced since just after the Civil War and were issued with increasing frequency during the late nineteenth and early twentieth centuries. See Henry Clepper, *Professional Forestry in the United States* (Baltimore, MD: Johns Hopkins University Press, 1971), 136; and William Robbins, *American Forestry: A History of National, State, and Private Cooperation* (Lincoln: University of Nebraska Press, 1985), 89–90. On the specific issue of pulpwood supplies, the U.S. Department of Agriculture, *How the United States Can Meet Its Present and Future Pulpwood Requirements*, Dept. Bulletin no. 1241 (Washington, DC: Government Printing Office, 1924), pointed to a national sense of urgency regarding domestic supplies in the face of a growing dependence on imports from Canada and elsewhere.

3. The city of Savannah, Georgia, for example, paid $325,000 in cash for plant site purchase and improvements when Union Bag decided to locate its new mill in the city. In addition, William Murphy of the C&S Bank provided $1 million from his own bank and $1.5 million from other sources as start-up money for the mill. See Pikl, *History of Georgia Forestry*, 31. For a more general treatment of industrial recruitment in the South during this period, see James C. Cobb, *The Selling of the South: The Southern Crusade for Industrial Development, 1936–1990*, 2nd ed. (Urbana: University of Illinois Press, 1993).

4. The Southern Pine Association cited figures for the third quarter of 1930 that showed a substantial cost advantage (one-third to one-half the cost) for the delivered mill cost of pulpwood in the South relative to other areas. See Southern Pine Association, *Economic Conditions in Southern Pine Industry* (New Orleans, 1931), 88. Another report noted that Union Bag expected to save on the order of twenty dollars per ton by producing sulfate pulp in the South instead of importing it. See American Institute for Economic Research, *A Report on the Future of the Paper Industry in the Southeastern United States and the Effects on Stumpage Values* (Cambridge, MA, 1938), 69.

5. Southern Kraft mills were able to capitalize on a series of innovations, including the introduction of synchronized electric drives in 1919, which permitted substantial increases in production. See Oden, "Southern Kraft Paper Industry," 583. For a discussion of the relationships between increasing scale and locational change in the industry see Helen Hunter, "Innovation, Competition, and Locational Changes in the Pulp and Paper Industry: 1880–1950," *Land Economics* 31, no. 4 (1955). See also chapter 3 of this text.

6. By the mid-1990s, a 2,000-ton per day capacity mill was using about 1 million cords of wood, 14 billion gallons of water, and approximately 6 billion kilowatt-hours of energy per year. Input numbers are based on information from *Lockwood-Post's Directory of the Pulp, Paper, and Allied Trades* (San Francisco: Miller Freeman, 1995).

7. In contrast to the lumber industry, the pulp and paper industry was able to utilize much smaller, and hence younger, timber (i.e., timber with a shorter "rotation age"). In the early years, the typical rotation age for pulpwood was between fifteen and twenty years (though it could be shorter or longer depending on management needs, etc.), whereas the optimal rotation age for sawtimber was at least thirty-five to forty years. Such a rotation age differential made it that much easier for the pulp and paper industry to institute a forest management regime that treated timber as a crop.

8. "New Industrial Epoch Begins," *Savannah Morning News*, October 1, 1936.

9. "Mayor Praises Savannah Spirit," *Savannah Morning News*, October 2, 1936.

10. Quoted in F. Basil Abrams, "Paper from Georgia Pines," *Atlanta Journal*, February 12, 1933.

11. John Popham, "King Cotton Is Dead in Savannah, the Most Famous of the Old South's Cotton Ports, and the Stately Pine Tree Is the New Ruler," *New York Times*, July 1, 1947.

12. Clepper, *Professional Forestry*, 202, cites a 1928 editorial from the *Journal of Forestry*, which defined industrial forestry as "timber growing as a business enterprise"—a concept that, at the time, represented quite a departure from the status quo. The Forest Service noted that by the mid-1980s, the value of roundwood timber products sold in the South (approximately $6.1 billion in 1984) was twice the value of soybean or cotton production and three times the value of tobacco, wheat, or corn crops (valued at local points of delivery). U.S. Forest Service, *The South's Fourth Forest: Alternatives for the Future*, Forest Resources Report no. 24 (Washington, DC: Government Printing Office 1988), 18–19.

13. The term "artificial regeneration," widely used in forestry practice, refers simply to reforestation by direct planting or seeding. In contrast, "natural regeneration" refers to the practice of leaving seed trees standing after harvest in order to allow lands to restock "naturally." Obviously, "natural" regeneration is very much a product of conscious design and directed action and thus could easily be considered "artificial." Without getting into the semantic difficulties associated with these two terms, this book will simply use them as employed in forestry practice.

14. James Scott uses the example of German scientific forestry, which strongly influenced scientific forestry in the United States and elsewhere, to illustrate his point about how the commercial and bureaucratic logics of certain state projects

necessarily depend on a radical simplification of the natural (and social) world—a process of imposing legibility and regimentation in order to refashion the forest into a monocropped "one commodity machine." Scott also notes that such projects of simplification and legibility are by no means exclusive to states. Large firms and capitalist markets can be powerful agents for simplification and homogenization as well. See James C. Scott, *Seeing Like a State: How Certain Schemes to Improve the Human Condition Have Failed* (New Haven, CT: Yale University Press, 1998), 11–22. For a similar perspective on the role of state projects in "simplifying" nature, see Donald Worster, *Rivers of Empire: Water, Aridity, and the Growth of the American West* (Oxford: Oxford University Press, 1985).

15. Alfred D. Chandler Jr., *Scale and Scope: The Dynamics of Industrial Capitalism* (Cambridge: Belknap Press, 1990); Alfred D. Chandler Jr., "Organizational Capabilities and the Economic History of the Industrial Enterprise," *Journal of Economic Perspectives* 6, no. 3 (1992). On technological change as a form of collective learning, see Gavin Wright, "Can a Nation Learn? American Technology as a Network Phenomenon," in *Learning by Doing in Markets, Firms, and Countries*, ed. Naomi Lamoreaux, Daniel Raff, and Peter Temin (Chicago: University of Chicago Press, 1999); and Gavin Wright, "Towards a More Historical Approach to Technological Change," *Economic Journal* 107 (1997): 1560–66.

16. Gavin Wright, *Old South, New South: Revolutions in the Southern Economy since the Civil War* (New York: Basic Books, 1986), 172–75. I am indebted to Professor Wright for helping me see the emergence of industrial forestry in the South as an example of regional collective learning.

17. R. D. Forbes, "The Passing of the Piney Woods," *American Forestry* 29, no. 351 (1923): 134.

18. Rupert B. Vance, *Human Geography of the South: A Study in Regional Resources and Human Adequacy*, 2nd ed. (Chapel Hill: University of North Carolina Press, 1935), 124–25. See also George B. Tindall, *The Emergence of the New South, 1913–1945* (Baton Rouge: Louisiana State University Press, 1967), 82. Howard Odum noted in 1936, "Of the 125 million acres in virgin pine, 100 million have been cutover, ten and a half million absorbed into agriculture, and 33 million remain without new growth." See Howard W. Odum, *Southern Regions of the United States* (Chapel Hill: University of North Carolina Press, 1936), 80, 576.

19. J. F. Duggar, "Areas of Cultivation in the South," in *The South in the Building of the Nation* (Richmond, VA: Southern Historical Publication Society, 1910), 5:17, cited in C. Vann Woodward, *Origins of the New South, 1877–1913* (Baton Rouge: Louisiana State University Press, 1951), 118.

20. Quoted in Joel Williamson, *William Faulkner and Southern History* (New York: Oxford University Press, 1993), 369, 399–404. I was fortunate enough to take Professor Williamson's course on Faulkner and Southern History when I was an undergraduate history major at the University of North Carolina and I have always been struck by the phrase "the slain wood." Although Faulkner wrote often about the destruction of the southern forest and what it meant for southern society, the phrase "the slain wood" has proven difficult to find in his many works. He does

use it in *The Sound and the Fury* (1929; reprint, New York: Vintage, 1987), 320, but not to signify the "doomed wilderness" that he writes about so powerfully in *The Bear* and other stories. In any event, most Faulkner scholars would probably agree that even if the phrase is not used precisely as I am using it here, drawing on Professor Williamson's book, it does have a certain Faulknerian ring to it! I thank Professor Williamson for his many insights regarding Faulkner's views on the fate of the southern forest and for helping me see southern history in new and powerful ways.

21. Charles S. Sargent, *Report on the Forests of North America* (Washington, DC: Government Printing Office, 1884).

22. For discussions of the labor situation in the southern lumber and naval stores industries, see Vernon H. Jensen, *Lumber and Labor* (New York: Farrar & Rinehart, 1945); Nollie Hickman, *Mississippi Harvest: Lumbering in the Longleaf Pine Belt* (Oxford: University Press of Mississippi, 1962); Thomas F. Armstrong, "Georgia Lumber Laborers, 1880–1917: The Social Implications of Work," *Georgia Historical Quarterly* 67, no. 4 (1983); Thomas F. Armstrong, "The Transformation of Work: Turpentine Workers in Coastal Georgia, 1865–1901," *Labor History* 25, no. 4 (1984); Robert B. Outland III, *Tapping the Pines: The Naval Stores Industry in the American South* (Baton Rouge: Louisiana State University Press, 2004).

23. See, e.g., Overton W. Price, "Saving the Southern Forests," *World's Work* 5, no. 5 (1903), and Gifford Pinchot, "Southern Forest Products, and Forest Destruction and Conservation since 1865," in *South in the Building of the Nation*, vol. 6.

24. The formal title was U.S. Forest Service, *Timber Depletion, Lumber Prices, Lumber Exports, and Concentration of Timber Ownership: Report of Senate Resolution 311* (Washington, DC: Government Printing Office, 1920).

25. Ibid., 66–69.

26. Ibid., 20–21.

27. George P. Ahern, *Deforested America: Statement of the Present Forest Situation in the United States*, S. Doc. 70-216 (1929), 3.

28. Pinchot, "Southern Forest Products," 154–55.

29. In the case of U.S. minerals development, the construction of a "public knowledge infrastructure" (referring to the early geological surveys) was a critical first step in resource discovery and exploitation. Paul David and Gavin Wright, "Increasing Returns and the Genesis of American Resource Abundance," *Industrial and Corporate Change* 6 (1997). The early forest surveys played a somewhat similar role in the transition to industrial forestry and the growth of the modern forest products industry in the South. Because most of the timberland in the South was (and is) privately owned, however, the "public knowledge infrastructure" established through the forest surveys was of primary benefit to private timberland owners.

30. U.S. Department of Agriculture, *Forest Resources of South Georgia*, USDA Misc. Pub. no. 390 (Washington, DC: Government Printing Office, 1941), ii; Inman F. Eldredge, "The Forest Survey in the South," Occasional Paper no. 31 (New Orleans: Southern Forest Experiment Station, 1934).

31. The original director of the southern forest survey was G. H. Lentz. Yet Inman F. "Cap" Eldredge, who succeeded Lentz, was primarily responsible for

the effort. H. R. Josephson, *A History of Forestry Research in the Southern United States*, USDA Misc. Pub. no. 1462 (Washington, DC: Government Printing Office, 1989), 40–42. Since the initial surveys of the 1930s, updates have been issued roughly every ten years.

32. E. L. Demmon, "Economics for Our Southern Forests," Occasional Paper no. 59 (New Orleans: Southern Forest Experiment Station, 1937).

33. See, e.g., A. R. Spillers and I. F. Eldredge, *Georgia Forest Resources and Industries*, USDA Misc. Pub. no. 501 (Washington, DC: Government Printing Office, 1943), 2, 17, 20, 19.

34. Ninety-eight percent of the fire damage occurred on non-protected lands. At the time, some two-thirds of south Georgia timberlands had no fire protection. Ibid., 8–9, 36.

35. Stephen J. Pyne, *Fire in America: A Cultural History of Wildland and Rural Fire* (Seattle: University of Washington Press, 1997), 143.

36. Regional Committee on Southern Forest Resources, *The Southern Forests* (Atlanta: National Resources Planning Board Field Office, 1940), 13.

37. W. W. Ashe, "Forest Conditions in the Southern States and Recommended Forest Policy," *South Atlantic Quarterly* 22, no. 4 (1925): 299.

38. Pyne, *Fire in America*, 152.

39. Ibid., 155; U.S. Forest Service, *South's Fourth Forest*, 38–41, 43, 287–88 table 2.10; Robbins, *American Forestry*, chaps. 4–5. See also Samuel P. Hays, *Conservation and the Gospel of Efficiency: The Progressive Conservation Movement, 1890–1920* (Cambridge: Harvard University Press, 1959); and Michael Williams, *Americans and Their Forests: A Historical Geography* (Cambridge: Cambridge University Press, 1989).

40. U.S. Department of Agriculture, *Woods Burning in the South*, Leaflet no. 40 (Washington, DC: Government Printing Office, 1940), 2–3.

41. Pikl notes that in 1935, the Georgia Forestry Association passed a resolution urging judges of the superior court to tighten up on woodsburning. In the fiscal year 1941–42, some forty-seven cases were brought against arsons in the state. In 1956, the Georgia legislature passed the Notification of Intention to Burn Law, requiring that citizens notify state officials before engaging in the practice of woodsburning. See Pikl, *History of Georgia Forestry*, 31, 51.

42. Southern Forestry Educational Project, *Second Annual Report* (Washington, DC: American Forestry Association, 1930); William F. Jacobs, "The Dixie Crusade," *American Forests*, December 1978, 18–21, 38–46; and James E. Mixon, "Progress of Protection from Forest Fires in the South," *Journal of Forestry* 54, no. 10 (1956): 650–51. Mixon was the state forester of Louisiana.

43. Compounding economic depression, fires ravaged much of the upcountry South during the drought years of 1930–31. The following year, Georgia and Florida both sustained heavy losses. Pyne, *Fire in America*, 156.

44. At the height of CCC activity, there were some 311 forestry camps spread throughout the South—25 on national forest land and 186 operating primarily on private lands. U.S. Forest Service, *South's Fourth Forest*, 46; Mixon, "Progress of Protection from Forest Fires in South," 650.

45. Pyne, *Fire in America*, 156.

46. H. H. Chapman, "Forest Fires and Forestry in the Southern States," *American Forestry* 18 (1912).

47. H. H. Chapman, "Factors Determining Natural Reproduction of Longleaf Pine on Cut-over Lands in Lasalle Parish, Louisiana," *Yale University School of Forestry Bulletin*, no. 16 (1926). For another early statement on the need for controlled burning in the southern woods, see W. G. Wahlenberg, "Fire in Longleaf Pine Forests," Occasional Papers no. 40 (New Orleans: Southern Forest Experiment Station, 1935). For brief historical overviews of the changing views of fire in the southern forest, see E. V. Komarek Sr., "The Use of Fire: An Historical Background," and Roland M. Harper, "Historical Notes on the Relation of Fire to Forests," both in *Proceedings of the First Annual Tall Timbers Fire Ecology Conference, March 1–2* (Tallahassee: Florida State University, 1962).

48. A. B. Crow, *Fire Ecology and Fire Use in the Pine Forest of the South* (Baton Rouge: Louisiana State University School of Forestry and Wildlife Management, 1982).

49. In a national context, the southern forestry community became the first to adopt prescribed burning as a positive tool of forest management—more than a decade before the rest of the country. Williams, *Americans and Their Forests*, 485–87.

50. Pyne, *Fire in America*, 158; U.S. Forest Service, *South's Fourth Forest*, 47.

51. For background on Tall Timbers and the involvement of Stoddard, Neel, and Komarek, told from the perspective of one of the participants, see Leon Neel (with Paul Sutter and Albert G. Way), *The Art of Managing Longleaf: A Personal History of the Stoddard-Neel Approach* (Athens: University of Georgia Press, 2010), 103–47. Discussion of the first forest fire ecology conference in 1962 is at 128–29.

52. Crow, *Fire Ecology and Fire Use*, 14.

53. Part of this was due to the long and important history of fire in the South and, because of the specific ecology of the southern pine forest, the fact that fire was necessary for industrial forestry to proceed. In Stephen Pyne's assessment: "Wildland fire . . . [was] . . . progressively removed from the hands of folk practitioners and placed in the grasp of professional foresters . . . Forestry in the South would have been impossible if promiscuous and malicious woodsburning had continued. But equally, it would have been impossible if conducted on a dogmatic policy of fire exclusion" (*Fire in America*, 144–45).

54. Virginia, North Carolina, South Carolina, Georgia, Florida, Alabama, Mississippi, Tennessee, Arkansas, Louisiana, Oklahoma, and Texas.

55. U.S. Forest Service, *South's Fourth Forest*, 287, table 2.10.

56. Pikl, *History of Georgia Forestry*, 50–51; Robbins, *American Forestry*, 208.

57. The success of fire protection efforts in Georgia, for example, was indicated by the fact that in 1957, for the first time in the United States, the privately organized Forest Insurance Company announced that it would accept applications for timber insurance, with coverage up to fifty dollars per acre, on timberlands in counties protected through the Georgia Forestry Commission. Pikl, *History of Georgia Forestry*, 51.

58. Regional Committee on Southern Forest Resources, *Southern Forests*, 13. In July 1998, however, more than half a million acres of timberland burned in Florida

alone, serving as a reminder of just how powerful fire can be in the South, despite extensive fire protection and control. See National Oceanic and Atmospheric Administration, National Climatic Data Center, "Florida Wild Fires and Extreme Conditions" (1998), www.ncdc.noaa.gov/oa/climate/research/1998/fla/florida.html. Impacts from climate change, in particular changing precipitation patterns, could increase the intensity and frequency of wildfires in the future. See Craig Hanson et al., *Southern Forests for the Future* (Washington, DC: World Resources Institute, 2010), 46–48.

59. L. W. Orr and R. J. Kowal, "Progress in Forest Entomology in the South," *Journal of Forestry* 54, no. 10 (1956); George H. Hepting, "Forest Disease Research in the South," *Journal of Forestry* 54, no. 10 (1956); Robbins, *American Forestry*, 206, 215, 220–21.

60. Spillers and Eldredge, *Georgia Forest Resources*, 37.

61. Regional Committee on Southern Forest Resources, *Southern Forests*, 17; Resources for the Future, *Forest Credit in the United States: A Survey of Needs and Facilities* (Washington, DC: Resources for the Future, 1958). The distinction between industrial owners and small nonindustrial owners is important here. Clearly, access to credit was not a problem for large firms that could borrow in national capital markets.

62. Up until this time, forest land had been considered "unimproved property" and hence not acceptable as collateral by national banks. The amendment to section 24 of the Federal Reserve Act was sponsored primarily by Pacific Northwest banking and timber interests. It authorized national banks to make mortgage loans secured by first liens on forest tracts that were "properly managed in all respects." Resources for the Future, *Forest Credit*, 53.

63. Resources for the Future, *Forest Credit*, chap. 6. In the early 1950s, the Travelers Insurance Company made its first timberland loans in the South. By 1966, G. A. Fletcher, senior vice president of Travelers, estimated that the insurance industry as a whole had well over $100 million invested in loans on timberlands. G. A. Fletcher, "Timberlands as Long-Term Loan Investments for Insurance Companies," *Forest Farmer* 26 (1967): 15. As for southern banks, the C&S Bank of Georgia was one of the first and most active banks to get involved in timberland loans, often teaming up with life insurance companies such as Travelers. C. M. Chapman, "A Private Banker Looks at Timber Loans," *Forest Farmer* 26 (1967): 16. For smaller timberland owners, the Federal Land Bank and the Farmer's Home Administration were the primary sources of long term credit. See "Trends in Land Bank Credit for Forest Owners," *Forest Farmer* 26 (1967): 17, 35; "Farmers Home Administration Loans for Forestry Purposes," *Forest Farmer* 26 (1967): 18; and Albert Ernest, "Financing the South's Forest Development," *Forest Farmer* 41 (1982): 6–7, 32.

64. Vance, *Human Geography*, 132.

65. "Forest Taxes and Conservation," *Harvard Law Review* 53 (1940): 1018–24; Samuel Trask Dana and Sally K. Fairfax, *Forest and Range Policy: Its Development in the United States*, 2nd ed. (New York: McGraw-Hill, 1980), 286–87.

66. Ronald B. Craig, an economist with the Southern Forest Experiment Station, researched the issue of tax delinquency extensively during the 1930s. In a 1934 study, he found that some 31 million acres of rural land (approximately 14% of the gross land area) in eight southern states had been in tax default for more than three years. Some two-thirds of all tax delinquent land that reverted back to public ownership was forest land. See Ronald B. Craig, "Reversion of Forest Land for Taxes Increasing in the South," Occasional Paper no. 32 (New Orleans: Southern Forest Experiment Station, 1934), and "The Extent of Tax Default in the Gulf States," Occasional Paper no. 49 (New Orleans: Southern Forest Experiment Station, 1935). In 1939, Craig estimated that 22 million acres of rural land (roughly 10% of gross land area) in these eight states had been in default for two or more years, and that between 13 and 15 million acres of this was forest land. Ronald B. Craig, "The Forest Tax Delinquency Problem in the South," *Southern Economic Journal* 6, no. 2 (1939). Clearly the depression had exacerbated the problem of tax delinquency in the region. For Craig, however, the primary reason why such a large percentage of forest lands was in default had to do with the nature of the tax system: "There can be no doubt of the fact of overassessment of forest land in the South . . . This is one, if not the chief, reason why forest land forms such a high percentage of the total tax-forfeited area of the South." Craig, "Forest Tax Delinquency," 152.

67. The issue had actually been on the public policy agenda in various guises since the early twentieth century.

68. U.S. Forest Service, *Timber Depletion*, 70–71.

69. Clarke-McNary Act of 1924, ch. 348, § 3, 43 Stat. 653 (1924).

70. Fred R. Fairchild, *Forest Taxation in the United States*, USDA Misc. Pub. no. 218 (Washington, DC: Government Printing Office, 1935), 5–6. Fairchild had previously reported on forest taxation issues in 1909.

71. Ibid., 636, 8.

72. Ibid., 525, 533, 537–39. "It is this uncertainty that more than anything else makes the property tax a menace to forestry" (537).

73. Ibid., 636. Still, the report was quite clear in its advocacy for tax reform, though it did not provide a ringing endorsement of the much-heralded severance or yield tax. For not only did the yield tax present considerable complications for local public finance, it was also very difficult to administer. Fairchild pointed out, moreover, that there was little evidence to support such a tax scheme: "The fact that after twenty years of experiment no State has yet succeeded in setting up a satisfactory yield tax of broad application is evidence of the difficulties involved." Instead, the report argued for reform of the operation of the property tax system through better assessment, improved collection and enforcement, and more efficient administration. Fairchild also outlined three proposals for correcting some of the inherent defects of the property tax system as applied to timberlands: (1) an adjusted property tax; (2) a deferred timber tax; and (3) a differential timber tax. Ibid., 637–40.

74. R. Clifford Hall, "The Rise of Realism in Forest Taxation," *Journal of Forestry* 36, no. 9 (1938): 903.

75. Four basic kinds of special timber tax laws have been developed in southern states: exemptions, modified assessments, yield taxes, and severance taxes. U.S. Forest Service, *South's Fourth Forest*, 65–66; Margaret P. Hamel, ed., *Forest Taxation: Adapting in an Era of Change* (Madison, WI: Forest Products Research Society, 1988).

76. While property tax reform waned as an issue in the 1950s, it has resurfaced frequently and still remains one of the more contentious local issues associated with the forest products industry. During the 1940s, forest taxation under the federal income tax emerged as a prominent source of concern for timberland owners, particularly for corporations. As in the debates over the property tax, the basic argument concerning the federal income tax focused on "discriminatory" taxation and perverse incentives. For a discussion, see U.S. Forest Service, *South's Fourth Forest*, 65; Charles W. Briggs and William K. Condrell, *Tax Treatment of Timber* (Washington, DC: Forest Industries Committee on Timber Valuation and Taxation, 1978), 3–7; John Walter Myers, *Impact of Forestry Associations on Forest Productivity in the South*, USDA Misc. Pub. no. 1458 (Washington, DC: Government Printing Office, 1988), 13–14; and Hamel, *Forest Taxation*.

77. One historian remarked that by this time forest taxation was a "dead issue." Williams, *Americans and Their Forests*, 456.

78. Henry C. Wallace, "Forestry and Our Land Problem," *American Forestry* 29, no. 349 (1923): 15–16.

79. Price, "Saving the Southern Forests"; Pinchot, "Southern Forest Products"; Forbes, "Passing of the Piney Woods."

80. See, e.g., J. W. Toumey, "The Regeneration of Southern Forests," in *Proceedings of the Southern Forestry Congress, Asheville, NC, July 11–15* (Durham, NC: Seeman Printery, 1916), 144–53.

81. Southern Cut-Over Land Association, *The Dawn of the New Constructive Era* (New Orleans, 1917). See also Frank Heyward, "History of Industrial Forestry in the South," in *The Colonel William B. Greeley Lectures in Industrial Forestry* (Seattle: University of Washington College of Forestry, 1958), 16; and Thomas Clark, *The Greening of the South: The Recovery of Land and Forest* (Lexington: University Press of Kentucky, 1984), 30.

82. See note 13 above for discussion of the terms "natural" and "artificial" regeneration. For early statements by Hardtner, see Henry Hardtner, "A Practical Example of Forest Management in Southern Yellow Pine," in *Proceedings of the Southern Forestry Congress* (1916), 71–80; and Henry Hardtner, "President's Address," in *Proceedings of the Third Southern Forestry Congress, Atlanta, GA, July 20–22* (New Orleans: Weihing Printing, 1921), 11–15. Clepper refers to the Urania Lumber Company as "the true innovator in practical forestry in the South" (*Professional Forestry*, 236–37).

83. For an early discussion of the importance of artificial reforestation in the South, see Philip C. Wakeley, *Artificial Reforestation in the Southern Pine Region*, USDA Technical Bulletin no. 492 (Washington, DC: Government Printing Office, 1935). See also Philip C. Wakeley, "The South's First Big Plantation," *Forests and People* 23, no. 2 (1973); and Heyward, "History of Industrial Forestry."

84. U.S. Forest Service, *South's Fourth Forest*, 41, 48.

85. The whole debate was kicked off in 1919 by Pinchot's Society of American Foresters report, which predicted an imminent timber shortage and called for regulatory controls over private forest practices. As noted previously, this led to the so-called Capper report of 1920. Senator Capper, however, was not content with the rather weak proposals contained in the report that bore his name, and he introduced a series of bills calling for direct regulation of cutting on privately owned timberlands. Due to significant opposition, no hearings were ever held on any of Capper's bills. Meanwhile, Senator Bertrand Snell of New York introduced several of his own bills, which reflected the position of Forest Service chief William B. Greeley and others who favored the cooperative approach (federal encouragement of state legislation). The Snell legislation was also blocked, this time by Pinchot's allies, who feared that state legislation would be ineffective because of the industry influence in state governments. In 1924, Senator McNary and Rep. Clarke introduced bills in their respective chambers that contained many of the cooperative programs, particularly in fire control, of the earlier Capper report and the Snell bills. The result was the Clarke-McNary Act of June 1924, which laid the foundation for federal-state cooperation in forest protection and management. For analyses of the debate leading up to Clarke-McNary and the impact of the final legislation, see Dana and Fairfax, *Forest and Range Policy*, 126–27; Clepper, *Professional Forestry*, 164; and Robbins, *American Forestry*, 85–99.

86. This was particularly evident in the emerging discourse over the South's underdevelopment. Witness the famous 1938 *Report on Economic Conditions of the South*, which, at least in Roosevelt's view, identified the region as the nation's number one economic problem and pointed to the waste of natural resources, particularly soils, minerals, and forests, as both cause and effect of underdevelopment: "The paradox of the South is that while it is blessed by Nature with immense wealth, its people as a whole are the poorest in the country. Lacking industries of its own, the South has been forced to trade the richness of its soil, its minerals and its forests, and the labor of its people for goods manufactured elsewhere. If the South received such goods in sufficient quantity to meet its needs, it might consider itself adequately paid." National Emergency Council, *Report on Economic Conditions of the South* (Washington, DC: Government Printing Office, 1938), 8. The report has sometimes been seen as an example of the colonial economy thesis that shaped much of the thinking during the 1930s and 1940s on southern economic development, i.e., that absentee ownership and the South's dependent status were responsible for the region's underdevelopment. As Gavin Wright has argued, however, the colonial economy thesis confuses symptoms with causes and does not explain adequately the underlying reasons for the persistence of southern economic "backwardness" (*Old South New South*, 13–14, 156–97).

87. Dana and Fairfax, *Forest and Range Policy*, 169; Clepper, *Professional Forestry*, 146.

88. Heyward, *History of Industrial Forestry*, 32–33; Clepper, *Professional Forestry*, 147–49; Dana and Fairfax, *Forest and Range Policy*, 169–70.

89. Hamlin L. Williston, *A Statistical History of Tree Planting in the South, 1925–1985*, Misc. Report SA-MR 8 (Atlanta, GA: USDA Forest Service, 1980), 19.

90. F. A. Silcox, *Report of the Chief of the Forest Service, 1937* (Washington, DC: Government Printing Office, 1937), 13. A dedicated New Dealer, Silcox stepped up Forest Service proposals for public regulation of private forest practices and provoked considerable hostility among industry leaders, particularly in the South. Though Silcox concurred with the view that the primary objective of forest policy was to keep forest land continuously productive, he believed that regulation, in addition to cooperation, was necessary to achieve such a goal. Clepper, *Professional Forestry*, 149; Robbins, *American Forestry*, 131–33.

91. Franklin D. Roosevelt, *Message from the President of the United States Transmitting a Recommendation for the Immediate Study of the National Forest Problem*, H.R. Doc. No. 75-539 at 1–4 (1938).

92. The final report of the committee, *Forest Lands of the United States*, was submitted to Congress in March 1941 and constituted a major disappointment for the proponents of regulation. Of the fifteen recommendations, only one dealt explicitly with regulation. Though the issue persisted in one form or another until the early years of the Truman administration, the joint committee's report effectively marked the end of the push for federal regulation. Dana and Fairfax, *Forest and Range Policy*, 171–72; Robbins, *American Forestry*, 131; Clepper, *Professional Forestry*, 152–53, 162–63.

93. E. H. Wilson, *Savannah Woodlands Operations and Contract System of Wood Procurement as It Affects the Farmers and Landowners* (Savannah: Union Bag and Paper Corporation, 1938), 12. The policy also called for selective cutting on noncompany lands in order to promote natural regeneration.

94. T. W. Earle, "Southern Pulpwood Conservation Association at Work for Fifteen Years," in *Address to the Southern Pulpwood Conservation Association, January 19, 1955* (Atlanta, GA, 1955), 8–9, 12; Heyward, "History of Industrial Forestry," 39; Clepper, *Professional Forestry*, 251; Myers, *Impact of Forestry Associations*, 10–11.

95. Clepper, *Professional Forestry*, 14. By 1990, according to one source, 38,000 industry-supported tree farmers actively managed 54 million acres of southern timberlands—60% of the U.S. total. D. L. Knight, "Southern Forest Industry: Shaping a Better Life for Southerners," *Forest Farmer* 50 (1990): 7.

96. In Georgia, for example, the first state nursery was established at Athens in 1929. See J. T. May, "Reflections on Southern Forest Tree Nurseries," in *Proceedings of the Annual Meeting of the Southern Forest Nursery Association* (Charleston, SC: Southern Forest Nursery Association, 1988). In 1950, Georgia nurseries produced a record 46 million seedlings—putting the state at the forefront of southern regeneration efforts. Seedling production by Georgia state nurseries skyrocketed during the 1950s—more than doubling to 95 million seedlings in 1954 and then climbing to an astonishing 305 million in 1959, the peak year for forest regeneration under the Soil Bank Program. Pikl, *History of Georgia Forestry*, 49, 52, 70.

97. For southern farmers, such payments typically ranged between eight and ten dollars per acre per year. See W. S. Swingler, "Forestry in the Soil Bank," *Journal of*

Forestry 54, no. 11 (1956): 747–749; Larry Lee, "Soil Bank Lands: A Survey of ASCS Directors," *Forest Farmer* 26 (1967): 10–11, 36; L. F. Kalmar, "Soil Bank Lands as Industry Views the Situation," *Forest Farmer* 26 (1967): 12, 39–40; D. A. Craig, "Soil Bank Lands: The Federal Government's Point of View," *Forest Farmer* 27 (1967): 13, 38–39; Jack Cantrell, "Industry's Stake in Soil Bank Plantations," *Forest Farmer* 27 (1968): 10–12; and R. J. Alig, T. J. Mills, and R. L. Shackelford, "Survey of Soil Bank Plantations Shows Very High Retention Rate," *Forest Farmer* 39 (1980): 8–9, 13–14.

98. Georgia led the South and the nation in forest regeneration during these years. Between 1956 and 1961, almost 1.3 million acres were planted or seeded in the state—two-thirds of which were on nonindustrial private land. Williston, *Statistical History.*

99. Nursery production peaked in the South during the CRP program of the late 1980s at more than 2 billion seedlings per year—approximately 82% of total seedling production in the entire United States. See Clark W. Lantz, "Role of State Nurseries in Southern Reforestation—An Historical Perspective," in *National Proceedings: Forest and Conservation Nursery Associations 1996*, USDA General Technical Report PNW-GTR-389 (Portland: Pacific Northwest Research Station, 1997), 50.

100. Alig, Mills, and Shackelford, "Survey of Soil Bank Plantations."

101. For a critical discussion of the effect of the Soil Bank Program on small southern farmers, see Pete Daniel, "The Legal Basis of Agrarian Capitalism: The South Since 1933," in *Race and Class in the American South Since 1890*, ed. Melvyn Stokes and Rick Halpern (Oxford: Berg, 1994).

102. Cantrell, "Industry's Stake."

103. This de facto "securitization" of timberland received a major boost in the 1980s as institutional investors began buying huge acreages of timberland in the South. See C. S. Binkley, C. F. Raper and C. L. Washburn, "Institutional Ownership of US Timberland: History, Rationale, and Implications for Forest Management," *Journal of Forestry* 94, no. 9 (1996): 21–28.

104. According to Lantz, planting on company land peaked in 1986 at 1.2 million acres per year. During this year, industry nurseries in the South were producing over 1 billion seedlings per year. Clark W. Lantz, "Overview of Southern Regeneration," in *Proceedings of the Annual Meeting of the Southern Forest Nursery Association* (1988), 5; Lantz, "Role of State Nurseries," 49.

105. Lloyd Austin, "A New Enterprise in Forest Tree Breeding," *Journal of Forestry* 25, no. 8 (1927): 928.

106. Scott S. Pauley, "The Scope of Forest Genetics," *Journal of Forestry* 52, no. 9 (1954): 644.

107. Bruce J. Zobel, "Forest Tree Improvement—Past and Present," in *Advances in Forest Genetics*, ed. P. K. Khosla (New Delhi: Ambika Publications, 1981), 13.

108. Aldo Leopold, "Some Thoughts on Forest Genetics," *Journal of Forestry* 27, no. 6 (1929): 710.

109. Austin, "New Enterprise," 928.

110. According to Bruce Zobel, tree improvement involves the combination of genetic selection and improved silvicultural practices to increase the productivity of

forest trees. Major objectives include: (1) adaptability; (2) resistance to pests and diseases; (3) growth rate; (4) tree form and quality; and (5) wood qualities. Bruce J. Zobel, *Increasing Productivity of Forest Lands through Better Trees*, S. J. Hall Lectureship in Industrial Forestry (Berkeley: University of California, Berkeley School of Forestry and Conservation, 1974), 4–5.

111. G. Muller-Starck and H. R. Gregorius, "Analysis of Mating Systems in Forest Trees," in *Proceedings of the Second International Conference on Quantitative Genetics*, ed. B. S. Weir, E. J. Eisen, M. M. Goodman, and G. Namkoong (Sunderland, MA: Sinauer Associates, 1988), 573–74; J. P. van Buijtenen, "Quantitative Genetics in Forestry," in ibid., 549–54; C. Gaston, S. Globerman and I. Vertinsky, "Biotechnology in Forestry: Technological and Economic Perspectives," in *Technological Forecasting and Social Change* 50 (1995): 80; Bruce J. Zobel and Jerry R. Sprague, *A Forestry Revolution: The History of Tree Improvement in the Southern United States* (Durham, NC: Carolina Academic Press, 1993), 24–29.

112. Tree breeders often have to wait as long as twenty to thirty years (depending on the species) before mature traits can be evaluated. See William L. Olsen, "Molecular Biology in Forestry Research: A Review," in *Forest and Crop Biotechnology: Progress and Prospects*, ed. Frederick A. Valentine (New York: Springer-Verlag, 1988), 315–16.

113. S. E. McKeand and R. J. Weir, "Economic Benefits of an Aggressive Breeding Program," in *Proceedings of the 17th Southern Forest Tree Improvement Conference, June 6–19* (Athens, GA, 1983), 100.

114. The Eddy station was established in 1925, and was reorganized as the Institute for Forest Genetics in 1932. In 1935, the institute was donated to the government, to be operated by the U.S. Forest Service. See Austin, "New Enterprise," and Ryookiti Toda, "An Outline of the History of Forest Genetics," in Khosla, *Advances in Forest Genetics*.

115. Swedish researchers working in the 1930s and 1940s had developed a program for selecting superior trees and establishing seed orchards with the grafted ramets of these selected trees. It was not until 1950, however, that Sweden initiated a nationwide program of tree improvement based on a system of elite clonal seed orchards. Similar tree improvement programs were established in other countries in the 1950s, particularly in Germany, Great Britain, Australia, Japan, Hungary, and the United States. Toda, "History of Forest Genetics," 8.

116. For more details on the committee's early efforts, see U.S. Forest Service, *Report of the First Southern Forest Tree Improvement Conference, January 9–10* (Atlanta, GA, 1951); U.S. Forest Service, *Report of the Second Southern Forest Tree Improvement Conference, January 6–7* (Atlanta, 1953); and U.S. Forest Service, *Report of the Third Southern Forest Tree Improvement Conference, January 5–6* (New Orleans, 1955). The 24th Southern Forest Tree Improvement Conference was held in 1997.

117. Zobel and Sprague, *Forestry Revolution*, 59–61. See also Jonathan W. Wright, "The Role of Provenance Testing in Tree Improvement," in Khosla, *Advances in Forest Genetics*; and Philip C. Wakeley, "The Relation of Geographic Race to Forest Tree Improvement," *Journal of Forestry* 52, no. 9 (1954): 653.

118. Zobel and Sprague, *Forestry Revolution*, 4. In 1981, Zobel estimated "that over 30 percent of all tree improvement programs have been failures or have been only marginally successful because geographic variability within the species was ignored" ("Forest Tree Improvement," 16). See also Bruce J. Zobel and John Talbert, *Applied Forest Tree Improvement* (New York: John Wiley & Sons, 1984), 75–93.

119. Provenance or seed source studies focused on determining the genetic variation of trees due to their geographic origins. Provenance tests thus involved planting trees of the same species but from different geographic regions side by side to determine the existence and extent of "racial" differences. Early provenance tests were conducted in Europe on Scots pine in the late nineteenth century. In the United States, the earliest provenance tests were conducted during the 1910s on Ponderosa pine and Douglas fir in the West and on loblolly pine in Louisiana. See Wright, "Provenance Testing," 103; Philip C. Wakeley, "Importance of Geographic Strains," in *Report of the First Southern Forest Tree Improvement Conference*, 1–9; and Henry I. Baldwin, "Seed Certification and Forest Genetics," *Journal of Forestry* 52, no. 9 (1954): 654–55. Wakeley's study involved collecting seeds from each of the four southern species of pine in various areas of their natural range. A dozen or so plantations were then established in these areas to test for "racial" differences in survival, growth rate, and disease resistance within each species. See E. L. Demmon and P. A. Briegleb, "Progress in Forest and Related Research in the South," *Journal of Forestry* 54, no. 10 (1956): 674–82, 687–92.

120. Baldwin, "Seed Certification," 655; J. W. Duffield, "The Importance of Species Hybridization and Polyploidy in Forest Tree Improvement," *Journal of Forestry* 52, no. 9 (1954): 645.

121. Wakeley, "Geographic Race," 653.

122. The West Virginia Pulp and Paper Company developed the first industry seed selection program in 1949. By 1956, the company was planting all seedlings from selected stock. L. T. Easley, "Loblolly Pine Seed Production Areas," *Journal of Forestry* 52, no. 9 (1954): 672–73; Zobel and Sprague, *Forestry Revolution*, 84.

123. Clemens M. Kaufman, "Extensive Application of Genetics by the Silviculturist," *Journal of Forestry* 52, no. 9 (1954): 647–48.

124. Zobel and Sprague, *Forestry Revolution*, 17–21.

125. Thomas O. Perry, "The Cooperative Genetics Program at the University of Florida," in *Proceedings of the Third Southern Tree Improvement Conference*. Originally, the NC State cooperative had twelve charter members. In 1966 it was divided into two programs, one of which focused on hardwood and one on loblolly pine, with the pine program receiving the bulk of the attention and financial support. Zobel, "Increasing Productivity," 17. In addition, a host of other cooperatives focusing on various aspects of industrial forestry have been established in the South and, more recently, in other parts of the country.

126. Zobel, "Increasing Productivity," 4.

127. The university served as the headquarters and provided facilities, some operating funds, and salaries for staff members, including university faculty,

technicians, and support staff. Cooperative members covered major operational expenses. Zobel, "Increasing Productivity," 17–18.

128. Ibid., 17; Zobel and Sprague, *Forestry Revolution*, 50–55.

129. One hundred percent of the seedlings produced in Georgia, Alabama, and Louisiana in 1986 were genetically improved. Lantz, "Overview," 5; Lantz, "Role of State Nurseries."

130. This was roughly double the output of the other two tree improvement cooperatives. A. E. Squillace, "Tree Improvement Accomplishments in the South," in *Proceedings of the 20th Southern Forest Tree Improvement Conference* (Charleston, SC, 1989), 10.

131. Gains from the second and third generations are expected to exceed those of the first. See J. B. Jett, "Thirty-five Years Later: An Overview of Tree Improvement in the Southeastern United States," in *Proceedings of the Annual Meeting of the Southern Forest Nursery Association* (1988); and Steve McKeand and Jan Svensson, "Loblolly Pine: Sustainable Management of Genetic Resources," *Journal of Forestry* 95, no. 3 (1997): 5, 8–9.

132. The flip side of this transition to intensive forest breeding and management, of course, is the increased risk and vulnerability that accompanies any crop monoculture. Though loblolly pine has been growing in parts of the South in even-aged stands for centuries, the program of intensive selection and breeding undertaken since the 1950s, combined with the extensive practice of artificial regeneration, has rendered southern timberlands more vulnerable to insect and disease problems. As with high-input monocrop agriculture, maintaining high productivity on intensively managed timberlands requires increased applications of chemicals to combat pathogens, control competing vegetation, and make up for nutritional deficiencies. For discussions of some of the risks and vulnerabilities associated with forest tree improvement and the spread of pine monocultures in the South, see the collection of articles in Duke University School of Forestry, "Topic 1: Possible Consequences of Southern Pine Monocultures," *Proceedings of the Fourth Conference on Southern Industrial Forest Management* (Durham, NC, 1960). For a more general discussion, see John W. Duffield, "Forest Tree Improvement: Old Techniques and the New Science of Genetics," in *H. R. MacMillan Lectureship Address* (Vancouver: University of British Columbia, 1960). See also Scott, *Seeing Like a State*, 11–22.

133. See L. W. Orr and R. J. Kowal, "Progress in Forest Entomology in the South," *Journal of Forestry* 54, no. 10 (1956): 653–56; George H. Hepting, "Forest Disease Research in the South," *Journal of Forestry* 54, no. 10 (1956): 656–60; and A. J. Riker, "Opportunities in Disease and Insect Control Through Genetics," *Journal of Forestry* 52, no. 9 (1954): 651–52.

134. Demmon and Briegleb, "Progress in Forest Research."

135. Josephson, *History of Forestry Research.*

136. Interview with author, November 17, 1997.

137. In the words of Bruce Zobel: "The genetic improvement of forest trees is a long-term, expensive undertaking. It can be done best through Cooperative efforts because most organizations cannot afford a team of highly trained specialists. In a

Cooperative, one trained man can oversee a great deal of research. The need to keep the genetic base sufficiently broad is almost impossible for a single organization but is easily achieved in a cooperative effort. The funds and manpower required for a tree-improvement program for each member is minimal in a Cooperative, yet enables maximum genetic gains. Plant materials, methods, equipment and even manpower are exchanged amongst members to the benefit of all. In my view, naturally biased, it would seem wasteful and even foolish for each organization to strike out on its own with an expensive, inadequate, and inefficient program when faster and greater gains are assured through joint action. The Florida, Texas, and North Carolina Cooperatives are convincing demonstrations of how well combined action has succeeded" ("Increasing Productivity," 4).

138. John D. Rue, "The Development of Pulp and Paper-Making in the South," *Southern Lumberman*, December 1924, 143 (emphasis in original).

139. Jonathan Daniels, *The Forest is the Future* (New York: International Paper Company, 1957), 8.

140. The notion of regional collective learning is adapted from Wright, "Can a Nation Learn."

141. This notion of biological time-space compression is adapted from David Harvey, *The Condition of Postmodernity* (Oxford: Blackwell, 1990). See also William Boyd, Scott Prudham, and Rachel Schurman, "Industrial Dynamics and the Problem of Nature," *Society & Natural Resources* 14, no. 7 (2001).

Chapter 2 · Logging the Mills

Epigraph. James Willard Hurst, *Law and Economic Growth: The Legal History of the Lumber Industry in Wisconsin, 1836–1915* (Cambridge, MA: Belknap Press of Harvard University Press, 1964), 1.

1. Figures are for the states of Alabama, Arkansas, Florida, Georgia, Kentucky, Louisiana, Mississippi, North Carolina, South Carolina, Tennessee, and Virginia. Texas and Oklahoma were not included. For these eleven states, total land area was approximately 323 million acres while total forest land area was approximately 185 million acres. Douglas S. Powell et al., *Forest Resources of the United States, 1992*, USDA Technical Report RM-234 (Washington, DC: Government Printing Office, 1994), 22, table 1, 108, table 35.

2. See Warren A. Flick, "The Wood Dealer System in Mississippi: An Essay on Regional Economics and Culture," *Journal of Forest History*, July 1985. Flick is one of the very few who has written on the historical development of the procurement system in the South. See also Warren A. Flick, "From the Woods to the Mill" (unpublished manuscript, School of Forestry, Auburn University, 1987); and John C. Bliss and Warren A. Flick, "With a Saw and a Truck: Alabama Pulpwood Producers," *Forest and Conservation History* 38 (April 1994).

3. E. H. Wilson, a Union Bag manager, noted, "Southern pine is one of the most perishable raw materials used by a major industry producing a non-perishable commodity." The major cause of deterioration was a fungus known as blue stain, as well

as damage caused by insects and other diseases. E. H. Wilson, "Savannah Woodlands Operations and Contract System of Wood Procurement as It Affects the Farmers and Landowners, Prepared for the Union Bag and Paper Corporation" (1938), 6 (on file with author). See also I. James Pikl, "The Southern Woods-Labor 'Shortage' of 1955," *Southern Economic Journal* 27, no. 1 (1960): 43. Pikl observed that freshly cut pine produced a higher yield of pulp and pulp byproducts, such as turpentine and tall oil, further adding to the constraints on inventories. Although rapid deterioration of southern pine could be halted by "pickling" the wood in water, the costs of underwater storage were often prohibitive.

4. Not until the advent of widespread mechanization in the 1970s, did logging during wet weather become feasible in the South.

5. The Fourdrinier papermaking machines that came into widespread use during the nineteenth century could shift between different grades, providing an early example of flexibility in a mass production industry. For a discussion, see chap. 3.

6. Wilson, "Savannah Woodlands Operations," 7.

7. Flick, "Wood Dealer System," 135.

8. Used primarily to push mechanization, company crews accounted for only a small fraction of wood supply. By the 1980s, all but a few had been abandoned.

9. Warren Flick has argued that this resulted in a de facto subsidy to the mills: "Producers were generally ignorant of the relevant regulations, incapable of completing the forms, or unwilling to pay the taxes. Federal and state governments could not monitor the thousands of small producers who left no paper trails. Therefore, much of the wood delivered to the mills was produced illegally, and, by not bearing the taxes and associated accounting costs, cheaply" ("Wood Dealer System," 136). For more extended discussion of the role of law in shaping the procurement system, see Flick "From the Woods to the Mills."

10. This perspective is taken from Robert Gordon, who has argued for examining the "constitutive" role of law in social relationships: "legal relations . . . don't simply condition how . . . people relate to each other but to an important extent define the constitutive terms of the relationship, relations such as lord and peasant, master and slave, employer and employee, ratepayer and utility, and taxpayer and municipality." Gordon is very much opposed to the view that law is somehow peripheral or epiphenomenal to "real" social relations. In his view, "Power is a function of one's ability to form and coordinate stable alliances with others that will survive setbacks and the temptations of defection to satisfy opportunistic interests. Such organization and coordination are bound to involve something legal." Robert W. Gordon, "Critical Legal Histories," *Stanford Law Review* 36 (1984): 57, 102, 104. See also E. P. Thompson, whom Gordon quotes, "The Poverty of Theory or an Orrery of Errors," in *The Poverty of Theory and Other Essays* (London: Merlin, 1978). Arguing against the structural Marxism of Louis Althusser, Thompson invoked his own historical research on eighteenth-century England to expressly reject the idea that law operated only at a superstructural level: "For I found that law did not keep politely to a 'level' but was at every bloody level; it was imbricated within the mode of production and productive relations themselves (as property-rights, definitions of

agrarian practice) and it was simultaneously present in the philosophy of Locke; it intruded brusquely within alien categories, reappearing bewigged and gowned in the guise of ideology; it danced a cotillion with religion, moralising over the theatre of Tyburn; it was an arm of politics and politics was one of its arms; it was an academic discipline, subjected to the rigor of its own autonomous logic; it contributed to the self-identity both of rulers and of ruled; above all, it afforded an arena for class struggle, within which alternative notions of law were fought out" (96).

11. U.S. House of Representatives, Committee on Agriculture, *Pulpwood Investigation*, Report No. 145 to accompany H. J. Res. 15, 77th Cong., 1st Sess. (1941) at 5 (letter of Annie Mae Strickland, Atlanta, Georgia, March 2, 1940, submitted to the committee).

12. Ibid.; see also *Pulpwood Investigation: Hearings on H.J.R. 15 Before the House Comm. on Agric.*, 77th Cong., 1st Sess. (1941).

13. In addition to using timber that was too small for lumber manufacturing, pulp mills also used worked-out turpentine trees, tops from sawtimber, thinnings, poorly formed trees, and lower-grade trees such as loblolly and black pine. In effect, the mills created a new market for these types of wood.

14. Relying on the open market essentially meant that they would depend on farmers and farm laborers to harvest the wood and transport it to the mills—an approach subject to dramatic seasonal fluctuations.

15. Myron S. Kahler, "Logging Methods, Costs, and Woodlands Procedure of the Pulp and Paper Industry in the South" (unpublished report, Southern Forest Experiment Station, New Orleans, LA, 1936), 4–6 (on file with author). Kahler noted, "The pulp and paper industry in the South, has with few exceptions, adopted the contract system of logging their mills. Operating for the most part on timber purchased in the open market, the mills are conserving their own woodlands against a time when they will be more urgently needed; and as a matter of fact, it has generally been found cheaper to buy wood in the open market than to operate owned woodlands, lands which were acquired for their timber growing capacity and not for present logging purposes."

16. This particular hedging strategy proved important as pulp and paper firms were not in a position to hedge through futures contracts as other processors of raw materials do.

17. The first firms to establish operations in the South established the precedent for landownership. As time passed, the standard rule of thumb used by industry analysts and securities underwriters on Wall Street was 50% wood security; that is, each firm was expected to own or control enough acreage to supply at least half of its mills' fiber needs on a continuous basis. There appeared to be little economic rationale behind this particular number. International Paper, which was the biggest firm in the industry and had mills in nine southern states, is usually credited with creating the standard. Each firm, moreover, likely developed its own approach to landownership. Some might prefer to own more or less than the 50% rule of thumb implied. But most felt that a substantial land base was important for long-term security.

18. These are nominal prices. Kahler, "Logging Methods," 25, 24, 28.

19. Wilson, "Savannah Woodlands Operations," 12, 6.

20. In the eyes of some industry boosters, these leases would promote scientific forestry: "Forestry conservation and development to forestall any depletion of the already abundant supply are assured under the provisions of the new instrument of conveyance in forestry, known as the ninety-nine-year lease, whereby the lessor and the lessee mutually agree to see that scientific forestry methods are practiced, the lessor being guaranteed an income that is practically as sure as the interest on government bonds." "New Industrial Epoch Begins," *Savannah Morning News*, October 1, 1936.

21. For a discussion, see Henry W. Falk Jr., "Acquiring Long-Term Control of Timber," in *Proceedings of the Twelfth Industrial Forestry Seminar* (New Haven, CT: Yale School of Forestry, 1959), 3–5.

22. As noted in the previous chapter, the late 1930s and early 1940s were the high point of the push for government regulation of private forestry practices.

23. Contracting, of course, has also had a long history in southern agriculture.

24. Specifically, the document mentioned legislation related to taxes, forestry regulations, and wage and hour laws. Wilson, "Savannah Woodlands Operations," 5.

25. As Wilson continued, "Although it would be more economical to deal with a few well-financed contractors of long experience and with adequate financial resources and managerial capacity, the possibilities of collusion and increased cost of wood are so great that the present policy is deemed to prove least costly." Ibid., 6.

26. Ibid., 13–14.

27. One of Union Bag's largest contractors, for example, carried between eighty-five and one hundred subcontractors during the 1930s. Union Bag actually had three classes of contractors, depending on their scale of operations and their use of subcontractors. Payments per cord were adjusted slightly to reward the larger and more dependable contractors. Other less dependable contractors were paid the same rate that larger contractors paid their own subcontractors. Ibid., 14, 27.

28. During this time, prices paid for pulpwood stumpage varied from $0.40 to $1.50 per cord. Ibid., 19, 27–28, 34.

29. "By the same token, the allowable margin of profit to the contractor must be such that he will *average* a reasonable return over a period of time. The determination of what this margin should be can only be determined from a period of experience embracing the vagaries of climatic and economic conditions." Ibid., 7, 17.

30. Ibid., 13 (emphasis added), 17.

31. Interview with author, June 23, 1997.

32. Interview with author, June 4, 1997.

33. Wilson, "Savannah Woodlands Operations," 13, 31.

34. Ibid., 13, 24. Warren Flick notes that the "political" duties of wood dealers involved everything from getting roads built to bailing out producers from jail to clarifying land titles. Flick, "Wood Dealer System," 136–37.

35. See, e.g., F. A. Silcox, *Report of the Chief of the Forest Service* (Washington, DC: Government Printing Office, 1937), 13. See also the discussion in chapter 1.

36. Wilson, "Savannah Woodlands Operations," 31.

37. National Resources Committee, *Tomorrow's Timber: Forest Resources in the Development of the South* (Atlanta, GA, 1939), 46.

38. H.R.J. 15, 77th Cong. (1941).

39. *Pulpwood Investigation: Hearings.* Although some of the testimony at the hearings pertained to other regions, the vast majority focused on the southern pulpwood procurement system.

40. Ibid., 79 (testimony of George S. Wheeler, Senior Economist, U.S. Dept. of Labor).

41. See, e.g., ibid., 3–6 (report from Mr. Hicks, an investigator with the special congressional joint committee in U.S. House of Representatives). In response to the question: "Did you find that the various mills get together and fix the price?" Hicks responded, "It seems to be, as the price is the same and the agent organization is almost identical in every State and pulpwood area" (5).

42. Jack P. Oden, "Development of the Southern Pulp and Paper Industry, 1900–1970" (PhD diss., Mississippi State University, 1973), 284–87.

43. Pikl, "Southern Woods Labor Shortage," 47.

44. Interview with author, June 4, 1997.

45. Seven to eight thousand POWs were used in pulpwood production throughout the South. American Pulpwood Association (APA), *Fifty Years of Service to the Pulpwood Industry: A History of the American Pulpwood Association Inc., 1934–1984* (Washington, DC, 1984), 13.

46. Stumpage price figures from table 2.29 in U.S. Forest Service, *The South's Fourth Forest: Alternatives for the Future*, Forest Resources Report no. 24 (Washington, DC: Government Printing Office, 1988), 331.

47. These concentration yards were operated by companies and dealers and were little more than rail yards equipped with mechanical loaders. See A. I. Jeffords, "Trends in Pine Pulpwood Marketing in the South," *Journal of Forestry*, July 1956, 464. By 1955, there were 300 such yards in operation throughout the region. Pikl, "Southern Woods Labor Shortage," 48.

48. Pikl, "Southern Woods Labor Shortage," 49–50.

49. See ibid. for an analysis of the 1955 shortage and its relation to the structure and practice of southern pulpwood procurement.

50. Ibid., 49–50.

51. Based on data collected by the Southern Pulpwood Conservation Association. For the year 1956, some 16.4 million cords of pulpwood out of a total of 20.3 million cords came from independent owners. Southern Pulpwood Conservation Association, *Economic Analysis of the Southern Pulp and Paper Industry for 1956* (Atlanta, GA, 1957).

52. Willard S. Bromley, "Small Forest Operations—One Tree Generation Away," speech to the Annual Meeting of the Georgia Forestry Association, Atlanta, GA, May 6, 1959.

53. *To Amend the Fair Labor Standards Act: Hearings on H.R. 3935 Before the Special Subcomm. on Labor of the H. Comm. on Educ. and Labor*, 87th Cong. 524 (1961) (A. F. Hartung, statement before U.S. House of Representatives).

54. See Gavin Wright, *Old South, New South: Revolutions in the Southern Economy since the Civil War* (New York: Basic Books, 1986), 214–19; and Bruce J. Schulman, *From Cotton Belt to Sunbelt: Federal Policy, Economic Development, and the Transformation of the South* (New York: Oxford University Press, 1991), 54–58, 66–71.

55. For a discussion of the centrality of the "labor question" in American politics during this period, see Steve Fraser, "The Labor Question," in *The Rise and Fall of the New Deal Order, 1930–1980*, ed. Steve Fraser and Gary Gerstle (Princeton, NJ: Princeton University Press, 1989).

56. Alex Lichtenstein, *Twice the Work of Free Labor: The Political Economy of Convict Labor in the New South* (London: Verso, 1996).

57. Pete Daniel, *The Shadow of Slavery: Peonage in the South, 1901–1969* (Urbana: University of Illinois Press, 1972); Jerrell H. Shofner, "Forced Labor in the Florida Forests, 1880–1950," *Journal of Forest History*, January 1981; Robert B. Outland III, *Tapping the Pines: The Naval Stores Industry in the American South* (Baton Rouge: Louisiana State University Press, 2004) 168–71.

58. Vernon H. Jensen, *Lumber and Labor* (New York: Farrar & Rinehart, 1945).

59. Kahler, "Logging Methods," 6–7.

60. In the early days of the industry, railroad logging could achieve superior economies in dense stands of virgin timber. Large semipermanent sawmills depended on large-scale, industrialized logging operations. Most loggers lived in company towns or logging camps and worked in large crews as employees of the lumber company. Clear-cutting was the standard practice. As the virgin timber was cut, however, these large centralized operations proved increasingly inefficient.

61. Although some companies tried to integrate truck logging into company operations, most switched to contracting. These companies typically owned their own timber and contracted out the logging and hauling operations to small independent operators. See R. R. Reynolds, "Truck Logging of Pine in Mississippi and Louisiana," Occasional Papers no. 28 (New Orleans: Southern Forest Experiment Station, 1933). The author noted, "Many of the companies now obtaining a portion or all of their total mill cut of logs by truck have at one time or another tried truck logging with company-owned trucks. With very few exceptions, however, all have switched to the contract method" (2). Of the companies surveyed for the report, practically all owned their own timber and paid a flat fee to have it delivered to the mill. The fee paid to the contractor included costs associated with logging, loading, hauling, unloading, and supervision. During this time, most trucks cost about $750 and lasted for about a year and a half.

62. Interview with author, November 17, 1997.

63. Interview with author, June 26, 1997. Another former procurement manager offered a similar perspective: "The industry wanted to keep as far an arm's-length

relationship as it possibly could with those contractors . . . We didn't want to mess with the loggers." Interview with author, July 16, 1997.

64. Interview with author, June 4, 1997.

65. Interview with author, June 23, 1997.

66. See, e.g., Crossett Lumber Co. v. McCain, 170 S.W.2d 64 (Ark. 1943) (rejecting claim for unemployment compensation by pulpwood cutter on grounds that he was independent contractor); Huiet v. Brunswick Pulp & Paper Co., 39 S.E.2d 545 (Ga. Ct. App. 1946) (same); Miles v. W. Va. Pulp & Paper Co., 48 S.E.2d 26 (S.C. 1948) (rejecting workmen's compensation claim for injury suffered by pulpwood cutter on grounds that he was an independent contractor).

67. See Brown v. L.A. Penn & Son, 227 So. 2d 470 (Miss. 1969).

68. Wilson, "Savannah Woodlands Operations," 36.

69. Jensen, *Lumber and Labor*, 76. See also Thomas F. Armstrong, "Georgia Lumber Laborers, 1880–1917: The Social Implications of Work," *Georgia Historical Quarterly* 67, no. 4 (1983), and "The Transformation of Work: Turpentine Workers in Coastal Georgia, 1865–1901," *Labor History* 25, no. 4 (1984).

70. Wilson, "Savannah Woodlands Operations," 22.

71. R. E. Worthington and Joseph Yensco, "An Investigation into Pulpwood Production from Roundwood and Turpentined Longleaf Pine," Occasional Paper no. 58 (New Orleans: Southern Forest Experiment Station, 1936), 12, 41. In 1934, the average hourly earnings for all manufacturing industries in the United States was about 55 cents per hour. Average hourly wages for lumber sawmills in the United States was about 44 cents per hour, and for paper and pulp manufacturing it was about 52 cents per hour. See U.S. Department of Commerce, *Statistical Abstract of the United States, 1937* (Washington: Government Printing Office, 1938), 313–14, table no. 355. The national minimum wage was established in 1938 with the Fair Labor Standards Act at 25 cents per hour. Fair Labor Standards Act of 1938, ch. 676, 52 Stat. 1060 (1938) (codified as amended at 29 U.S.C.A. §§ 201 et seq.).

72. A 1997 survey found that logging was the most dangerous occupation in the country with over 128 deaths per 100,000 workers. This compared to a national average of 4.7 deaths per 100,000 workers. Eric F. Sygnatur, "Logging Is Perilous Work," *Compensation and Working Conditions*, Winter 1998, 3–4.

73. According to the Department of Labor study, roughly 3% of the disabling injuries to pulpwood loggers resulted in death or permanent disability. The study, which was conducted in 1943 and 1944, provided evidence that directly contradicted those who argued pulpwood logging was less hazardous than sawtimber logging because of the smaller and lighter logs produced in pulpwood operations. For the nation as a whole, the study found no significant difference between pulpwood logging operations (75.5 disabling injuries per million employee hours worked) and sawtimber logging (76.6 injuries per million employee hours). The all-manufacturing average for the United States was 18.4 disabling injuries per million employee hours. The accident rate for pulpwood logging in the southern region was 76.8, slightly lower than the Lake States' rate of 83.1 and higher than the northeastern rate of 70.3. Figures for the Pacific Northwest were not collected as the pulp and paper industry

in that region subsisted almost entirely on sawmill residues. U.S. Department of Labor, *Injuries and Accident Causes in the Pulpwood-Logging Industry, 1943 and 1944*, Bulletin no. 294 (Washington, DC: Government Printing Office, 1948).

74. See Marshall Courtney, "Workmen's Compensation Activity in Florida" (paper presented to the APA Southeastern Technical Committee Meeting, Jacksonville, FL, November 20–21, 1957).

75. Occupational Safety and Health Act of 1970, Pub. L. No. 91-596, 84 Stat. 1590 (1970) (codified at 29 U.S.C.A. §§ 651 et seq.).

76. M. N. Goldberg, *Worker Safety in Logging Operations*, HEW Paper no. (NIOSH) 74-103 (Cincinnati: National Institute for Occupational Safety and Health, 1974).

77. Fair Labor Standards Act of 1938, ch. 676, 52 Stat. 1060 (1938) (codified as amended at 29 U.S.C.A. §§ 201 et seq.). In 1940, the APA and twelve of its member companies were indicted for conspiracy to violate the FLSA. Eleven other company members promptly resigned from the APA for fear that such indictments might spread to them. In the fall of 1941, the government widened its indictments and brought FLSA "hot goods" injunction cases against several APA member companies. Several companies, faced with having their wood supply tied up indefinitely, entered into consent decrees with the government in which they agreed to make spot checks of producer compliance with the FLSA wage and hour provisions. In late 1941, the government's conspiracy indictment was dismissed, but legal proceedings continued for several years. For a brief history from the APA's perspective, see APA, *Fifty Years of Service*, 11–12. For representative court cases, see Fleming v. S. Kraft Corp., 37 F. Supp. 232 (S.D.N.Y. 1940); Walling v. W. Va. Pulp and Paper Co., 2 F.R.D. 416 (E.D.S.C. 1942); Walling v. Gulf States Paper Corp., 53 F. Supp. 619 (N.D. Ala. 1942); S. Advance Bag & Paper Co., Inc. v. United States, 133 F.2d 449 (5th Cir. 1943); and Walling v. Gulf States Paper Corp., 143 F.2d 301 (5th Cir. 1944).

78. See, e.g., *Pulpwood Investigation: Hearings*, 74–75 (Wheeler testimony). Wheeler observed that on July 1, 1940, an indictment was issued against the APA and several pulp and paper firms alleging conspiracy to violate the FLSA. Rep. Hampton Fulmer of South Carolina, who convened the hearings, suggested that pulp and paper mills circumvented many of the requirements of New Deal labor legislation "by buying from the regional agent [wood dealer]." The result, in his view, was that the workers were "not getting enough to buy the real necessities of life" (60).

79. Government investigators charged that the "hot goods provision" of the law made it illegal to ship goods across state lines that had been produced in violation of the wage and hours law, regardless of the actual employment relationship. Companies responded that they had made time studies of wood cutters and had attempted to increase their prices so as to match the minimum wage. Ibid., 77 (Wheeler testimony).

80. Ibid., 84 (testimony of John J. Babe, Principal Attorney, Solicitor's Office, U.S. Dept. of Labor).

81. For an early statement on segmented labor markets, see David Gordon, Robert Edwards, and Michael Reich, *Segmented Work, Divided Workers: The Historical Transformation of Labor in the United States* (Cambridge: Cambridge University Press, 1982). For a critical review and elaboration, see Jamie Peck, *Work Place: The Social Regulation of Labor Markets* (New York: Guilford Press, 1996). For an explicit discussion in the context of the southern pulp and paper industry, see Connor Bailey, Peter Sinclair, John Bliss, and Karni Perez, "Segmented Labor Markets in Alabama's Pulp and Paper Industry," *Rural Sociology* 61, no. 3 (1996).

82. See, e.g., *Miles*, 48 S.E.2d 26, 29–30 (discussing the layers of contractual relationships between the mills, the wood dealers, and the pulpwood producers and the manner in which this insulated the mills from the costs and liabilities associated with regular employees).

83. *Pulpwood Investigation: Hearings*, 98 (Wheeler testimony).

84. Pulpwood prices did rise during the war, but much of the increase was related to increased prices for stumpage.

85. For a discussion, see Wright, *Old South, New South*, 216–25.

86. See *Fair Labor Standards Act of 1937: Joint Hearings on S. 2475 and H.R. 7200 Before the S. Comm. on Educ. and Labor and the H. Comm. on Labor*, 75th Cong. Part 1 (1937). In particular, see the testimony and questioning of John E. Edgerton, president of the Southern States Industrial Council, at 760–807. Southern lumber interests, under the leadership of the Southern Pine Industry Association, which was founded in 1937 for the explicit purpose of combating the bill, actively opposed the legislation, arguing that because the industry was far more labor intensive than the more mechanized industry on the West Coast, uniform wage rates applied across the nation would discriminate against the South. See also George B. Tindall, *The Emergence of the New South, 1913–1945* (Baton Rouge: Louisiana State University Press, 1967), 533–35.

87. See Fair Labor Standards Act of 1938 § 3(f), 29 U.S.C.A § 203(f), for the full definition of "agriculture," and § 13(a)(6), 29 U.S.C.A § 213(a)(6), for the agricultural exemption.

88. Charles H. Livengood Jr., *The Federal Wage and Hour Law Including the Fair Labor Standards Amendments of 1949* (Philadelphia: American Law Institute Committee on Continuing Legal Education, 1951), 71–96, esp. 74–76.

89. *Amendments to the Fair Labor Standards Act of 1938: Hearings on H.R. 2033 Before the H. Comm. on Educ. and Labor*, 81st Cong. 659–61 (1949) (testimony of R. M. Eagle).

90. See, e.g., *Proposed Amendments to the Fair Labor Standards Act, Hearings Before the H. Comm. on Labor*, 79th Cong. 198–200 (1946) (letter from E. W. Tinker, executive secretary of the American Pulp and Paper Association). See also the letter from H. E. Brinckerhoff, executive secretary, APA, arguing that all pulpwood production should be included under the agricultural exemption, in *Fair Labor Standards Act Amendments of 1949, Hearings Before the S. Comm. on Labor and Pub. Welfare*, 81st Cong. 1074–78 (1949). See also *Amendments to the Fair Labor*

Standards Act of 1938, 531–35 (testimony of Carl Gibson, a pulpwood dealer from North Carolina representing a group of dealers from the South).

91. Fair Labor Standards Amendments of 1949, ch. 736, 63 Stat. 910 (1949). See also Livengood, *Federal Wage and Hour Law*, 12–16; and Bureau of National Affairs, *The New Wage and Hour Law* (Washington, DC, 1949).

92. The new section was codified at 29 U.S.C. § 213(a)(15). See also Bureau of National Affairs, *New Wage and Hour Law.*

93. The 1954 *Census of Manufactures* reported that 90% of loggers worked in crews of nine men or less. See U.S. Bureau of the Census, *United States Census of Manufacturers, 1954* (Washington, DC: Government Printing Office, 1957–58).

94. *Amending the Fair Labor Standards Act of 1938: Hearings Before the Subcomm. on Labor of the S. Comm. on Labor and Pub. Welfare*, 84th Cong. 17 (1956).

95. *Proposals to Extend Coverage of Minimum Wage Protection: Hearings Before the Subcomm. on Labor of the S. Comm. on Labor and Pub. Welfare*, 85th Cong. 432, 585–86 (1957). See also the statements of eight pulpwood dealers and producers that were included in Bromley's and Canfield's testimony, 591–604.

96. *Fair Labor Standards Act: Hearing on H.R. 7458, Before the Subcomm. of the Comm. on Educ. and Labor*, 85th Cong. 1036, 1051–68 Part 1 (1957). These men, along with APA representatives, also urged the Congress to clarify the definition of an "employee" as used by the FLSA, arguing that it was in contradiction with the more precise definition used in common-law rules and under the Social Security Act. The problem with the FLSA definition, in their view, was its ambiguity, which made it easier for supposedly "independent contractors" to be declared employees by the government, thereby increasing the potential liabilities of the "employer." See, e.g., ibid., 1394–1419, esp. 1414–15 (testimony of Willard Bromley).

97. S. 1046, introduced April 21, 1959, was known as the Proposed Fair Labor Standards Amendments of 1959. The bill's counterpart in the House was H.R. 4488.

98. According to the APA's own official history, "While everything was APA-coordinated, great care was taken to use different typewriters and various colors and sizes of paper to signify independent and widespread support for the exemption from many groups throughout the country." APA, *Fifty Years of Service*, 17.

99. Bromley, "Small Forest Operations," 3.

100. *To Amend the Fair Labor Standards Act: Hearings Before the Subcomm. on Labor of the S. Comm. on Labor and Pub. Welfare*, 86th Cong. 1027, 1037 (1959) (statement of Harry Scott, United Paper Makers & Paper Workers).

101. See *APA Legislative Bulletin* of July 29, 1960, describing the substitution of an amended bill, H.R. 12853, which retained the twelve-man exemption, for H.R. 12677, which had been previously approved by the committee and sought to repeal the exemption.

102. The overtime exemption was retained. Regarding the 1966 amendments, which renumbered the exemption section 13(a)(13), see Labor Relations Reporter, *The New Wage and Hour Law*, rev. ed. (Washington, DC: Bureau of National Affairs, 1967), 65. Regarding the repeal of the twelve-man rule in 1974, see Fair Labor Standards Amendments of 1974, P.L. 93-259, 88 Stat. 55 (1974).

103. U.S. Department of Labor, *Small Logging Operations: Data Pertinent to an Evaluation of the 13(a)(15) Exemption of the Fair Labor Standards Act* (Washington, DC, 1964), 2, 4, 21, 25.

104. See W. N. Haynes, "A Time and Cost Study of Pulpwood Harvesting Operations in Southeast and Central Georgia, Report Prepared for Savannah Woodlands Division, Union Bag and Paper Corporation" (December 1949), 1 (on file with author).

105. Livengood, *Federal Wage and Hour Law*, 5.

106. U.S. Department of Labor, *Small Logging Operations*, 4, 10.

107. See Oden, "Development of the Southern Pulp and Paper Industry," 298. Based on its 1963 survey of southern loggers, the Department of Labor concluded that more than 80% of southern loggers worked 40 hours a week or less, and nearly half worked less than thirty-five hours a week. U.S. Department of Labor, *Small Logging Operations*, 19–20.

108. One study of southern pulpwood production in the early 1960s found that some 17% of pulpwood producers sampled employed crews in which no one was literate, and a full 30% had only a couple of workers who were semiliterate. Oden, "Development of the Southern Pulp and Paper Industry," 300–301.

109. Thomas A. Walbridge, "The Effect of Increased Timber Demand upon Wood Procurement Systems," in *Proceedings of the Seventh Conference on Southern Industrial Forest Management: Problems of Timber Supply as Related to Expanding Use in the South, October 5–6, 1966* (Durham, NC: Duke University School of Forestry, 1967), 54.

110. G. E. Knapp, "New Developments Relating to Improved Pulpwood Procurement in the South," in *Pulpwood Stands, Procurement, and Utilization: Papers Presented at the Detroit Meeting of the Technical Association of the Pulp and Paper Industry, September 26–28, 1946* (New York, 1947).

111. In a study of pulpwood harvesting in Arkansas, one government forester found that the cost-effectiveness of a pulpwood crew in 1950 was basically the same as it had been in 1940, primarily because of the increased costs associated with wages and equipment. R. R. Reynolds, "Pulpwood Production Costs in Southeast Arkansas 1950," Occasional Paper no. 121 (New Orleans: Southern Forest Experiment Station, 1951), 12.

112. One estimate put the total losses to the industry from the 1955 and 1959 wood shortages at roughly $50 million. Battelle Memorial Institute, *The Application of Scientific Management and Improved Technology to Woodlands Problems* (Columbus, OH, 1963), 2.

113. T. A. Walbridge, "What Price Further Mechanization?" *Pulp and Paper* 31, no. 7 (1957): 100. See also Oden, "Development of the Southern Pulp and Paper Industry," 304.

114. Total cost of the study was about $500,000.

115. Henry R. Hamilton, *A Summary of the Southern Pulpwood Production Research Project* (Columbus, OH: Battelle Memorial Institute, 1963), i, 10.

116. The Battelle researchers conducted county-level analysis in several southeastern states and showed that there was no correlation between rural population

decline and pulpwood production for the periods in question. The report also rejected the argument that pulpwood workers were shifting to other forms of employment. Farm employment was on the decline and tended to pay even less than pulpwood production. Most manufacturing jobs, according to the report's authors, required a level of skill that pulpwood loggers did not have. In their view, what appeared to be a labor shortage was really a lag in hiring workers due to the sluggish nature of the system in responding to large increases in demand. There was no chronic labor shortage. H. R. Hamilton, R. G. Bowen Jr., R. W. Gardner, and J. J. Grimm, *Phase Report on Factors Affecting Pulpwood Production Costs and Technology in the Southeastern United States* (Columbus OH: Battelle Memorial Institute, 1961), 34–42.

117. Hamilton, *Summary of the Southern Pulpwood Production Research Project*, 10.

118. The pulpwood harvesting system involved felling (cutting down trees), limbing (removing limbs), measuring, bucking (cutting logs to size and stacking them), skidding (dragging the logs to a loading point), loading, hauling, and unloading. The report focused on the cost and productive efficiency of each constituent activity. Loading, in particular, was seen as a problem of "overriding . . . importance," accounting for some 45% of the total man-hours and 33% of the total costs of the harvesting operation. See Hamilton et al., Phase Report, 6.

119. Battelle estimated that the overall productivity of the harvesting system was approximately 0.27 cords per man-hour, which translates to about 2.16 cords per man-day for an eight-hour day. A typical crew of five men could expect to cut and haul about ten cords of wood per day. Hamilton et al., *Phase Report*, 6–7.

120. H. R. Hamilton and J. J. Grimm, *An Evaluation of Alternative Methods of Pulpwood Production* (Columbus OH: Battelle Memorial Institute, 1963), 1.

121. Ibid., 45, 51.

122. See Robert H. Donnelly, R. William Gardner, and Henry R. Hamilton, *Integrating Woodlands Activities by Mathematical Programming* (Columbus, OH: Battelle Memorial Institute, 1963).

123. As one former procurement manager reflected: "I remember one of the first things I did when I got this job . . . They gave me a stack of Battelle reports, and I had to read those things. That opened my eyes. I wasn't really trained in that sort of thing. I had to embrace some concepts that I wasn't really into." Interview with author, May 29, 1997.

124. For a discussion of this issue in the more general context of forest credit, see Resources for the Future, *Forest Credit in the United States: A Survey of Needs and Facilities* (Washington, DC, 1958), 42–45.

125. Walbridge, "What Price Mechanization," 102, 100–101.

126. International Paper first initiated company harvesting operations in response to the labor shortages during World War II. After the war, the company began experimenting with mechanized operations, primarily by adapting farm and construction machinery to logging needs. See Tom N. Busch, "The Busch Combine Operations," in *Proceedings of the Timber Harvesting and Procurement Short Course*

(Baton Rouge: Louisiana State University School and Forestry and Wildlife Management and the American Pulpwood Association, 1967).

127. Ibid. See also Oden, "Development of the Southern Pulp and Paper Industry," 306.

128. See Jack L. Arthur, "The Beloit Harvester System," in *Proceedings of the Timber Harvesting and Procurement Short Course.*

129. Mechanized pulpwood harvesting, from this perspective, entailed challenges not found in the harvesting of agricultural crops. Indeed, although timber plantations certainly offered a more uniform operating environment for the new machines, not unlike that of an agricultural row crop, the inherent variability in natural stands of timber required that these machines be extremely versatile. For overviews of mechanization, see James A. Altman, "Changing Trends Due to Labor and Mechanization," *Forest Farmer* 26 (November 1967); T. A. Walbridge and W. B. Stuart, "Outlook for Harvesting Mechanization," *Forest Farmer* 34 (November–December 1975); and Joe McNeel, "Harvesting Systems and the Forest Farmer," *Forest Farmer* 48 (February 1988).

130. Union Camp, one of the last companies to phase out its harvesting operations, shut down its last company crew in the early 1990s.

131. The APA actually initiated a Harvesting Research Project in 1968, headed by Tom Walbridge. This project was discontinued in 1973, when the Virginia Tech Harvesting Cooperative was established. See APA, *Fifty Years of Service*, 39–40.

132. Oden, "Development of the Southern Pulp and Paper Industry," 298–99. See also McNeel, "Harvesting Systems and the Forest Farmer."

133. See Leland R. Gauron, "Direct Purchase and Satellite Chipping," in *Current Challenges to Traditional Wood Procurement Practices*, ed. Margaret P. Hamel (Madison, WI: Forest Products Research Society, 1988), 30–32. Gauron was the wood procurement operation manager for Westvaco's Charleston, South Carolina, mill.

134. Flick, "From the Woods to the Mill," 16.

135. Most equipment had (and has) an operating life of one to five years. Loggers found it very difficult to amortize the original equipment within its usable life. The value of used equipment was so low that trade-ins were not a real possibility.

136. Productivity for 1960 (2 cords per man-day) was only about 10% higher than it had been in 1950 (1.8 cords per man-day). APA, *Southern Pulpwood Producer Survey Report 1968* (New York, 1969), 21.

137. See American Pulpwood Association, *1987 Pulpwood Logging Contractor Survey*, Technical Bulletin no. 162 (Mississippi State, MS, 1989): 3; APA, *Southern Pulpwood Producer Survey*, Technical Release, no. 97-R-22 (Rockville, MD, 1997), 2.

138. Flick, "From the Woods to the Mill," 11. Longwood logging also allowed for the integration of sawtimber and pulpwood logging. A mechanized logger could separate timber at the stump and haul sawtimber to sawmills and pulpwood to pulp mills.

139. APA, *1987 Pulpwood Logging Contractor Survey*, 3; APA, *Southern Pulpwood Producer Survey*, 1.

140. Alice H. Ulrich, *U.S. Timber Production, Trade, Consumption, and Price Statistics 1960–1988*, USDA Misc. Publication no. 1486 (Washington, DC, 1990), 43, table 29; Edgar L. Davenport, *Pulpwood Prices in the Southeast 1989* and *1990* (Asheville, NC: USDA Southeast Forest Experiment Station, 1989, 1992); Michael Howell, *Pulpwood Prices in the Southeast 1991* and *1992* (Asheville, NC: USDA Southeast Forest Experiment Station, 1993, 1994). The reports were discontinued in 1992. Prices reflect a weighted average for all pricing points and includes Virginia, North Carolina, South Carolina, Georgia, and Florida.

141. Stumpage price data are from Timber-Mart South, adjusted using the BLS Consumer Price Index.

142. APA, *Southern Pulpwood Producer Survey Report*, 8–9.

143. According to a 1968 survey, the mean number of operating days for southern loggers was 220 days per year. Ibid., 30.

144. Interview with author, May 29, 1997. Another procurement manager from a different company explained the capacity problem in similar terms: "The truth is, mother nature dictates whether you have excess capacity or not. The normal situation is that you've got to have more logging capacity than you really need because when it's dry you've got way too much, but when it's wet you never have enough. And so you gear your production based on the worst scenario rather than the best scenario. And, of course, that aggravates the hell out of the loggers, and I can understand that. If you need ten loggers and you've got twenty because when it's real wet, all of them are going to cut half as much as they normally would or could, so when it gets dry you got twenty loggers and you give every one of them a smaller piece of the pie and you put them on quotas and so they can't capitalize on the opportunities . . . and it's a yo-yo, I mean it really truthfully is." Interview with author, August 20, 1997.

145. Wilson, "Savannah Woodlands Operations," 13.

146. In 1979, according to one survey, the average southern pulpwood contractor had less than nine years of formal education. By 1987, the average had increased to 12.5 years. APA, *1987 Pulpwood Survey*, 1. In a survey of logging contractors in Georgia in 1996, 100% of the contractors interviewed were high school graduates, yet only a third had graduated from college. Less than half of the logging employees had graduated from high school. See W. Dale Greene, Ben D. Jackson, and David C. Woodruff, "Characteristics of Logging Contractors and Their Employees in Georgia," *Forest Products Journal* 48, no. 1 (1998): 49.

147. In the words of one logger: "Do you think for one damn minute we're businessmen? Everybody was told: 'cash flow, cash flow, cash flow.' That's all I've heard. Mill people will tell you, 'Oh you just need to have more cash flow.' Cash flow will put you out of business quicker than anything I know of." Interview with author, July 16, 1997.

148. Interview with author, August 20, 1997.

149. As one logger put it, "Do you know anybody who would let you cut their lawn with a Franklin logger? I mean what can you do with it?" Interview with author, July 16, 1997.

150. Interview with author, July 16, 1997.

151. Interview with author, August 7, 1997. Some loggers attempted to diversify by seeking out alternative markets (i.e., other dealers or mills) for their wood. Even as late as 1979, however, three out of four loggers still supplied only one market. By 1987, the number had dropped to one in three, with the average producer serving three markets. As mechanization proceeded, finding alternative outlets became an economic imperative for loggers seeking to maximize the use of their equipment. APA, *1987 Pulpwood Logging Survey*, 3, 7.

152. According to one logger, "That new machinery—it's great, but the man who owns it, he's in debt and he can't quit." Interview with author, July 16, 1997. Another logger offered a similar perspective: "You're so much in debt, you got to continue or you'll lose it." Interview with author, May 22, 1997.

153. A few firms, such as Brunswick Pulp and Paper in Georgia, had been using such a system since the end of World War II.

154. Westvaco, for example, initiated a program of direct purchase at its Charleston mill in 1969. Gauron, "Direct Purchase." Several companies that went with a direct purchase approach began to use written contracts as a way of outlining the exact terms of the relationship between the mill and the logger and ensuring that it would not be construed as an employer-employee relationship. See also Edward C. Lee, "Procurement Systems and Markets," *Forest Farmer* 46 (September 1987): 8–9.

155. As one procurement manager using direct purchase noted, "We're very sensitive to the arm's-length transaction, as much as we can possibly protect that relationship." Interview with author, August 20, 1997.

156. For discussions of differing approaches to southern wood procurement, see Robert S. Manthey, "Pulpwood Procurement Practices Viewed by Forest Economist" (paper presented at a meeting of the APA Southeastern Technical Committee, Atlanta, GA, April 13–14, 1965); Lee, "Procurement Systems and Markets"; Margaret P. Hamel, ed., *Current Challenges to Traditional Wood Procurement Practices* (Madison, WI: Forest Products Research Society, 1988); Flick, "From the Woods to the Mill"; and Flick, "Wood Dealer System." By the 1980s, southern mills procured about 30% of their wood through the direct purchase system and more than half through the dealer system. The remainder of the wood came from company harvesting and from "gatewood" purchases—people who showed up at the mill with wood to sell but with whom the mills had no formal relationship. Flick, "From the Woods the Mill," 2.

157. St. Regis Paper Company built the first satellite chip mill at Fargo, GA, in 1956.

158. Chip mills also allowed for better quality control on chips than the older chipping operations at the mill site. See Richard Peccie, "Satellite Chip Mills—Yours or Theirs" (paper presented to the Southern Forest Industries Conference, Birmingham, AL, May 13–14, 1997).

159. Figures are from John R. Luoma, "Whittling Dixie," *Audubon*, November–December 1997, 42.

160. Interview with author, July 1, 1997.

161. See Chris Lyddan, "Southern Woodchip Export Trends" (paper presented at the Southern Forest Industries Conference, Birmingham, AL, May 13–14, 1997). See also Gary Faulkner and Louis Hyman, "U.S. South Increases Hardwood Chip Exports to Meet Far East Demand," *Pulp and Paper* 63, no. 7 (1989): 74–78.

162. By the mid-1990s, there were at least twenty chip mills dedicated exclusively to the export market. Chip mill exports grew at about 20% per year during the 1989–1998 period. Ross W. Gorte, *The Chip Mill Industry in the South*, CRS Report 98-540 (Washington, DC, 1998).

163. This was more than triple the next largest export site. See Lyddan, "Woodchip Export Trends." See also Ken L. Patrick, "Scott Paper's Chip Exports to Asia Become Major, Profitable Operation," *Pulp and Paper* 68, no. 10 (1994). Between 1989 and 1999, residue exports from southern ports increased 369% for hardwood and 372% for softwood. In 1999, almost 40% of hardwood chip production in the South was exported. Jeffrey P. Prestemon and Robert C. Abt, "Timber Products Supply and Demand," in *Southern Forest Resource Assessment*, Gen. Tech. Rep. SRS-53, ed. David N. Wear and John G. Greis (Asheville, NC: Southern Research Station, 2002), 310.

164. Quoted in Patrick, "Scott Paper's Chip Exports," 81.

165. The mixed mesophytic forests of southern Appalachia are some of the most biologically diverse temperate forests in the world. See C. R. Hinkle et al., "Mixed Mesophytic Forests," in *The Biodiversity of the Southeastern United States: Upland Terrestrial Communities*, ed. W. H. Martin et al. (New York: Wiley, 1993). For a discussion of impacts, see Luoma, "Whittling Dixie"; Danna Smith, *Chipping Forests and Jobs: A Report on the Economic and Environmental Impacts of Chip Mills in the Southeast* (Chattanooga, TN: Dogwood Alliance, 1997); Eric Bates, "Exporting Southern Forests," *DoubleTake* 3 (1996); Commonwealth of Va., *Report of the Joint Subcommittee Studying the Impact of Satellite Chip Mills on Virginia's Economy and Environment*, H. Doc. No. 2001-62 (2001); Rex H. Schaberg et al., "Economic and Ecological Impacts of Wood Chip Production in North Carolina: An Integrated Assessment and Subsequent Applications," *Forest Policy and Economics* 7 (2005): 157.

166. Not surprisingly, the environmental implications of chip mills have been the most visible and contentious aspects of this relatively new feature of the southern wood procurement system. In 1993, at the urging of a local coalition of environmentalists, sawmill operators, and furniture makers, the Tennessee Valley Authority conducted the first environmental impact study of chip mills and ended up denying permits for three proposed chip mills. In the late 1990s, the EPA started to look into the issue and, in early 1999, joined with the state of North Carolina to fund a two-year investigation of chip mills and their impact on southern forests to be conducted by researchers from Duke University and North Carolina State University. The results of the study are reported in Schaberg et al., "Economic and Ecological Impacts of Wood Chip Production in North Carolina." In general, the study confirmed that chip mill expansion in the state led to increased harvest levels, particularly of hardwoods in the western part of the state. The study also found that chip mills resulted in more complete removal of biomass from harvest sites and

noted the growing importance and prevalence of intensively managed pine plantations (170–72). See also David N. Wear and John G. Greis, *The Southern Forest Resource Assessment: Summary Report*, Gen. Tech. Rep. SRS-54 (Asheville, NC: Southern Research Station, 2002), 15.

167. Sam Hodges, "Chipping the Best with the Rest," *Mobile Register*, October 27, 1996 (Special Supplement: "Alabama Forest Cut Short"); Smith, *Chipping Forests and Jobs*.

168. The majority of the chips produced in the South by satellite chip mills are used by the paper and composite wood panel industry in the region. However, close to 40% of southern hardwood chip production was exported in 1999. This compares to only 1.3% for southern softwood chip production, which accounted for the majority of total chip production in the region. Prestemon and Abt, "Timber Products Supply and Demand," 308–10.

169. By the late 1990s, satellite chip mills were processing approximately 27% of the total pulpwood harvested in the South. Ibid., 309.

170. Scott Paper Co. v. Gulf Coast Pulpwood Assoc'n, 1973 WL 881, at *9–10 (S.D. Ala. Sept. 21, 1973).

171. Ibid., at 8–9. See also Wayne Greenhaw, "Echoes of Change in the South's Backwoods: Woodcutters Organize," *Southern Changes*, December 1980; "Big Profits and Little Pay in South's Backwoods: Woodcutters Organize, Part II," in *Southern Changes*, February–March 1981; and "The Fight to Survive in the South's Backwoods, Part III," *Southern Changes*, July 1981.

172. *Scott Paper Co. v. Gulf Coast Pulpwood*, at 9.

173. In their arguments that the Gulf Coast producers were independent contractors, the plaintiffs also cited an earlier federal decision from Minnesota that identified a group of pulpwood producers as independent contractors. The judge, however, rejected the applicability of the Minnesota court's conclusions for the Gulf Coast case, pointing out the substantial differences between the two groups of producers, namely that the Minnesota producers dealt directly with the company on an annual basis, furnished all of their own equipment, and desired their status as independent operators. Boise Cascade Int'l v. N. Minn. Pulpwood Producers Assoc'n, 294 F. Supp. 1015 (D. Minn. 1968).

174. *Scott Paper Co. v. Gulf Coast Pulpwood*. Pittman also noted an earlier workmen's compensation case from Mississippi, which also held the producers to be employees of the dealers. Brown v. L.A. Penn & Son, 227 So. 2d 470 (Miss. 1969), in which the Mississippi Supreme Court, on the basis of substantially identical facts, held that the "right of control rather than the actual exercise" of it is an important factor in determining the status of pulpwood producers.

175. Scott Paper Co. v. Gulf Coast Pulpwood Assoc'n, 491 F.2d 119 (5th Cir. 1974).

176. For a discussion, see Wayne Greenhaw, "Big Profits and Little Pay," 16.

177. John E. Granskog, "Collective Bargaining in the Forest Industries," in *Manpower: Forest Industry's Key Resource, Papers Presented at the 39th Industrial Forestry Seminar, May 20–24, 1974*, ed. Lloyd C. Irland (New Haven, CT: Yale University School of Forestry and Environmental Studies, 1975), 78.

178. Norris-LaGuardia Act, ch. 90, 47 Stat. 70 (1932) (codified at 29 U.S.C.A §§ 101 et seq.).

179. *Scott Paper Co. v. Gulf Coast Pulpwood Assoc'n*, 491 F.2d 119 at 11.

180. The phrase is adapted from Durkheim: "For in a contract, not everything is contractual." Emile Durkheim, *The Division of Labor in Society*, trans. W. D. Halls (New York: Free Press, 1984), 158.

181. *Scott Paper Co. v. Gulf Coast Pulpwood Assoc'n*, 491 F.2d 119 at 7–8, 14, 16–17.

182. The Southern Woodcutters Assistance Project (SWAP) was formed in Thomastown, MS, in 1979 with backing from the United Church of Christ. Initially, SWAP operated as a purchasing cooperative, providing equipment and supplies to pulpwood producers at lower cost than they could get from their dealers or on the open market. During the same year, the United Woodcutters Association was also formed as a full-fledged producer association, like the Gulf Coast Pulpwood Association. Together with SWAP and Gulf Coast, this organization worked successfully to charter the first federal credit union for woodcutters in 1980 and to lobby for state legislation, most notably the Mississippi Fair Pulpwood Scaling and Practices Act, which went into effect in July 1982 and which established licensing requirements for wood dealers, uniform measurement procedures, complaint procedures, and third-party arbitration for dispute resolution. Wayne Greenhaw, "Woodcutters Organize," 16–18; Tom Israel and Randall Williams, "Ending the Short Stick in Mississippi's Woods," *Southern Changes*, June–July 1982, 16–18.

183. Cited in Greenhaw, "Big Profits and Little Pay," 16.

184. This represents a striking contrast, at least historically, with the situation facing loggers in the Pacific Northwest, where larger logging crews, company-run operations, and a history of labor militancy have all facilitated unionization and collective bargaining among woods workers. And, as chapter 3 elaborates, pulp and paper mills inside the South have historically been highly organized, further highlighting the segmented labor market within which southern woods workers operated. For a discussion, see Granskog, "Collective Bargaining in the Forest Industries."

185. Since the 1980s, logging associations have emerged in various parts of the South. Organized most often along state lines, some are "pure" loggers associations, in that their membership is restricted to loggers, while others are more open associations organized under state forestry associations. In general, these are "professional" associations intended to provide for cooperative purchasing, insurance pooling, and lobbying efforts among state legislatures.

186. See, e.g., United States v. Silk, 331 U.S. 704, 713 (1947); Rutherford Food Corp. v. McComb, 331 U.S. 722, 729 (1947); and Goldberg v. Whitaker House Coop., Inc., 366 U.S. 28, 33 (1961). In basing its enforcement decisions on this doctrine, the IRS used the so-called twenty common law factors test to determine whether an individual was an employee or an independent contractor.

187. The study, known as the IRS Employee / Independent Contractor Compliance Study, was released in 1979. Based on a random sample of some 7,000 workers from specific industries and occupations, the study found that 47% of workers sur-

veyed failed to report income for tax purposes and 62% failed to pay social security taxes. See *Independent Contractors: Hearings on H.R. 3245 Before the Subcomm. on Select Revenue Measures of the H. Comm. on Ways and Means*, 96th Cong. 14–16, 21–40 (1979) (statement submitted by Donald C. Lubick, assistant secretary of the Treasury for tax policy).

188. *Independent Contractor Tax Proposals: Hearing on S. 2369 Before the Subcomm. on Oversight of the IRS of the S. Comm. on Fin.*, 97th Cong. 257–58 (1982) (quoted in the statement submitted by the APA).

189. See, e.g., *Independent Contractors*, 333–52 (testimony of Lynn Abraham, president, Oregon Log Truckers Association; Robert Beausoleil, president, Vermont Timber Truckers and Producers Association; Bobby Seid, pulpwood dealer, on behalf of the Virginia Forestry Association; Thomas Fenley, pulpwood dealer, on behalf of the Texas Forestry Association; and others representing the timber industry).

190. *Classification Issues Relating to Independent Contractors: Hearing Before the Subcomm. on Selective Revenue Measures of the H. Comm. on Ways and Means*, 97th Cong. 146 (1982) (statement of Kenneth R. Rolston Jr., president of the APA).

191. Cited in ibid., 147.

192. Specifically, as outlined in I.R.C. § 530(D), this legislation established "safe harbor" conditions, which, if met, protected independent contractors from IRS efforts to reclassify them as employees. For a review, see the APA's 1987 guide, "Independent Contractor," an abbreviated version of its publication *How to Stay at Peace with Your Government* (Washington, DC, 1981).

193. See Marvin W. Pearson, "The Challenge of a New Safety Priority," in Irland, *Manpower*, and *Mechanization, Safety, and Manpower in Southern Forestry*, ed. Robert W. McDermid (Baton Rouge: Louisiana State University School of Forestry and Wildlife Management, 1972).

194. See J. Edwin Carothers, "Workmen's Compensation—New Developments," in Irland, *Manpower*.

195. Interview with author, May 22, 1997.

196. Endangered Species Act of 1973, Pub. L. No. 93-205, 87 Stat. 884 (1973) (codified at 16 U.S.C. §§ 1531 et seq.)

197. Declared endangered on October 13, 1970, 35 Fed. Reg. 16047.

198. Section 9(a)(1)(B) of the Endangered Species Act of 1973 makes it illegal for any person to "take" endangered or threatened species. The term "take" is defined as an action that would "harass, harm, pursue," "wound," or "kill," such a species § 3(19). The Interior Department further defined "harm" to include "significant habitat modification or degradation where it actually kills or injures wildlife" 50 CFR § 17.3 (1995). Claiming that application of the "harm" regulation to the red-cockaded woodpecker and the northern spotted owl had injured them economically, representatives for a group of small landowners, logging companies, and families dependent on the forest products industries in the Pacific Northwest and in the Southeast challenged this regulation in U.S. district court, claiming that Congress did not intend the word "take" to include habitat modification and that the secretary of the

interior had exceeded his authority. The district court granted summary judgment in favor of the secretary of the interior, but the court of appeals reversed the decision, concluding that "harm," like the other words in the definition of "take," should be read as applying only to the perpetrator's direct application of force against the animal taken. The Supreme Court took up the issue in 1995 and reversed the Court of Appeals. See Babbitt v. Sweet Home Chapter of Communities for a Great Oregon, 515 U.S. 687 (1995). For a discussion of the issue in the broader context of "takings," see Barton H. Thompson Jr., "The Endangered Species Act: A Case Study in Takings and Incentives," *Stanford Law Review* 49 (1997): 305.

199. Interview with author, May 22, 1997. He then went on to say, "We [the loggers] are an endangered species."

200. See Sustainable Forestry Initiative, www.sfiprogram.org/.

201. See W. Henson Moore, "The New Face of Forestry: Sustainable Forestry Initiative," *Pulp and Paper* 70, no. 7 (1996).

202. M. N. Goldberg, *Worker Safety in Logging Operations*, National Institute for Occupational Safety and Health, HEW Paper no. (NIOSH) 74-103 (Cincinnati, OH, 1974). See also David Ross Vendt, "An Evaluation of Injuries Incurred in Pulpwood Production in Louisiana and Mississippi" (master's thesis, School of Forestry and Wildlife Management, Louisiana State University, 1976).

203. Eric F. Sygnatur, "Logging Is Perilous Work," *Compensation and Working Conditions*, Winter 1998. In a survey of logging injuries in North Carolina, researchers also found that African American loggers had a higher fatality rate than their white counterparts: a relative risk of 1.31 compared with whites. Rosa L. Rodriguez-Acosta and Dana P. Loomis, "Fatal Occupational Injuries in the Forestry and Logging Industry in North Carolina, 1977–1991," *International Journal of Occupational and Environmental Health* 3, no. 4 (1997): 260.

204. Interview with author, May 29, 1997.

205. In a somewhat extreme case of paternalism, one wood dealer suggested that a logger would "always be an Indian, never be a chief . . . That's just the way it is. Most loggers were born to log." Interview with author, August 8, 1997.

206. A. V. Chayanov, *The Theory of Peasant Economy* (Madison: University of Wisconsin Press, 1986).

207. Interview with author, August 20, 1997.

208. Durkheim, *Division of Labor*, 149–75. Many of the noncontractual norms and practices that shaped the relationship between southern loggers and the dealers and mills with whom they contracted are discussed in court cases from across the region seeking to determine the nature of the relationship and the attendant allocation of various liabilities for taxes, workers compensation, negligence, etc. See, e.g., Brown v. L.A. Penn & Son, 227 So. 2d 470, 473–74 (Miss. 1969) (discussing the relations of debt and dependency that dominated the relationship between Napoleon Brown, an illiterate black pulpwood cutter, and his wood dealer, L. A. Penn); *Scott Paper Co. v. Gulf Coast Pulpwood*, 1973 WL 881 at *14 (discussing general economic dependence of pulpwood producers on wood dealers and paper mills); Slater v. Canal Wood Corp., 345 S.E.2d 71, 74–75 (Ga. Ct. App. 1986) (discussing

"paternalistic relationship whereby Canal Wood [a prominent wood dealer] provides for or assists Jackson [a pulpwood producer] in virtually all aspects of his business, from assisting in insurance claims and obtaining necessary credit, to paying advances and providing field communications").

209. In Weber's words:

> The development of legally regulated relationships toward contractual association and the law itself toward freedom of contract, especially toward a system of free disposition within stipulated forms of transaction, is usually regarded as signifying a decrease of constraint and an increase of individual freedom . . . The possibility of entering with others into contractual relations the content of which is entirely determined by individual agreement . . . has been immensely extended in modern law, at least in the spheres of exchange of goods and of personal work and services. However, the extent to which this trend has brought about an actual increase of the individual's freedom to shape the conditions of his own life or the extent to which, on the contrary, life has become more stereotyped in spite, or, perhaps, just because of this trend, cannot be determined simply by studying the development of formal legal institutions. The great variety of permitted contractual schemata and the formal empowerment to set the content of contracts in accordance with one's desires and independently of all official form patterns, in and of itself by no means makes sure that these formal possibilities will in fact be available to all and everyone. Such availability is prevented above all by the differences in the distribution of property as guaranteed by law . . . The result of contractual freedom, then, is in the first place the opening of the opportunity to use, by the clever utilization of property ownership in the market, these resources without legal restraints as a means of achieving power over others.
> Max Weber, "Economy and Law (Sociology of Law)," in *Economy and Society*, eds. Guenther Roth and Claus Wittich (Berkeley: University of California Press, 1978), 729–30.

210. He goes on to note, "Society, by proclaiming freedom of contract, guarantees that it will not interfere with the exercise of power by contract. Freedom of contract enables enterprisers to legislate by contract and, what is even more important, to legislate in a substantially authoritarian manner without using the appearance of authoritarian forms." Friedrich Kessler, "Contracts of Adhesion—Some Thoughts about Freedom of Contract," *Columbia Law Review* 43 (1943): 629, 640.

211. Robert L. Hale, "Bargaining, Duress, and Economic Liberty," *Columbia Law Review* 43 (1943): 603, 625–26.

212. This "relational" view of contracting has been developed by Ian Macneil as an alternative to the neoclassical or "transactional" view of contracting. In addition to precedents in classical social theory (Durkheim and Weber), this view of contract echoes earlier legal realist scholarship, such as that of Kessler and Hale. Ian Macneil, "Contracts: Adjustment of Long-Term Economic Relations under Classical,

Neoclassical, and Relational Contract Law," *Northwestern University Law Review* 72 (1978): 854. See also the comment on Macneil's work by Robert W. Gordon, "Macaulay, Macneil, and the Discovery of Solidarity and Power in Contract Law," *Wisconsin Law Review* (1985): 565. As Gordon notes,

> In the messy and open-ended world of continuing contract relations, where the contours of obligation are constantly shifting, the effects of power imbalances are not limited to the concessions that parties can extort in the original bargain. Such imbalances tend to generate hierarchies that can gradually extend to govern every aspect of the relation in performance . . . What starts out as mere inequity in market power can be deepened into persistent domination on one side and dependence on the other. This is not slavery, since the parties are legally free to exit; but the whole perspective of relational contract suggests that sunk costs can matter tremendously, that the trauma of abandoning a relationship around which a company has structured all its operations, hiring, investment, and planning decisions, can keep it tied into a dependence that its members experience as all the more corrupting because it is in some sense voluntary. (570)

213. James Willard Hurst, *Law and Economic Growth: The Legal History of the Lumber Industry in Wisconsin, 1836–1915* (Cambridge, MA: Belknap Press of Harvard University Press, 1964), 1.

214. Interview with author, July 16, 1997.

Chapter 3 · Making Paper

Epigraph. "Economics of Paper," *Fortune* 16, no. 4 (1937): 111.

1. "Union Bag & Paper Corp.," *Fortune* 16, no. 2 (1937); "Economics of Paper," *Fortune* 16, no. 4 (1937); "Fifteen Paper Companies," *Fortune* 16, no. 5 (1937); and "International Paper & Power," *Fortune* 16, no. 6 (1937).

2. "Economics of Paper," 111, 113.

3. As Philip Scranton points out in *Endless Novelty: Specialty Production and American Industrialization, 1865–1925* (Princeton, NJ: Princeton University Press, 1997), "American manufacturers did not define a uniform search for order, standardization, and corporate consolidation. Instead, order had different meanings from industry to industry, meanings that valorized certain sectoral technologies, social relations, and market tactics" (355). See also Charles F. Sabel and Jonathan Zeitlin, "Stories, Strategies, Structures: Rethinking Historical Alternatives to Mass Production," in *World of Possibilities: Flexibility and Mass Production in Western Industrialization*, ed. Charles F. Sabel and Jonathan Zeitlin (Cambridge: Cambridge University Press, 1996).

4. "Economics of Paper," 116.

5. Between 1887 and 1919, the annual rate of speed increase was 3.43%. Between 1919 and 1931, the annual rate was 4.17%, and between 1931 and 1940, it was 2.51%. Avi J. Cohen, "Technological Change as Historical Process: The Case of the U.S.

Pulp and Paper Industry, 1916–1940," *Journal of Economic History* 46, no. 3 (1984): 786–87.

6. "Economics of Paper,"184.

7. Cohen, "Technological Change as Historical Process," 791.

8. In 1939, for example, more than 80% of the pulp and some two-thirds of the paper and paperboard produced at U.S. mills was sold to customers by company sales forces. Alfred Chandler, *Scale and Scope: The Dynamics of Industrial Capitalism* (Cambridge, MA: Belknap Press of Harvard University Press, 1990), 112.

9. See Naomi R. Lamoreaux, *The Great Merger Movement in American Business, 1895–1904* (Cambridge: Cambridge University Press, 1985).

10. "Union Bag," 50; "International Paper & Power," 134–35.

11. Chandler suggests that the industry achieved a stable oligopolistic structure during the early twentieth century (*Scale and Scope*, 112–13). For a discussion of the problems with this interpretation, see Nancy Kane Ohanian, *The American Pulp and Paper Industry, 1900–1940: Mill Survival, Firm Structure, and Industry Relocation* (Westport, CT: Greenwood Press, 1993), 51–71. For a perspective on this issue during the 1990s, see Laurits Rolf Christensen and Richard E. Caves, "Cheap Talk and Investment Rivalry in the Pulp and Paper Industry," *Journal of Industrial Economics* 45, no. 1 (1997): 58–60.

12. "Economics of Paper," 184.

13. John Guthrie, "Price Regulation in the Paper Industry," *Quarterly Journal of Economics* 60, no. 2 (1946): 203–12.

14. Simon N. Whitney, *Antitrust Policies: American Experience in Twenty Industries* (New York: Twentieth Century Fund, 1958), 351.

15. Paul V. Ellefson and Robert N. Stone, *U.S. Wood-Based Industry: Industrial Organization and Performance* (New York: Praeger, 1984), 232–38.

16. In some cases, many of the best jobs went to outsiders who had the requisite skills. Conner Bailey, Peter Sinclair, John Bliss, and Karni Perez, "Segmented Labor Markets in Alabama's Pulp and Paper Industry," *Rural Sociology* 61, no. 3 (1996): 479–80.

17. Herbert R. Northup, "The Negro in the Paper Industry," in *Negro Employment in Southern Industry: A Study of Racial Policies in Five Industries*, by Herbert R. Northup, Richard L. Rowan, Darold T. Barnum, and John C. Howard, Industrial Research Unit, Wharton School of Finance and Commerce (Philadelphia: University of Pennsylvania, 1970), 47–48.

18. The key case, which is discussed below, was Local 189 v. United States 416 F. 2d 980 (5th Cir. 1969).

19. For an interesting discussion of the use of distributed control systems in the pulp and paper industry and its implications for the labor process, see Steven P. Vallas and John P. Beck, "The Transformation of Work Revisited: The Limits of Flexibility in American Manufacturing," *Social Problems* 43, no. 3 (1996): 346–47.

20. *Pulp & Paper: North American Fact Book 1996* (San Francisco: Miller Freeman, 1997), 52. Total employment at Union Camp's Savannah mill, for example, declined from 5,600 to 2,200 between the mid-1960s and the mid-1990s.

21. Mark X. Diverio, “Poor Financial Returns Are a Call to Action for the Paper Industry,” *Pulp & Paper* 73, no. 4 (1999): 71–79; James K. Flicker, “Structural Impediments to Profitability,” *Pulp & Paper* 72, no. 1 (1998): 142.

22. John Guthrie, “Price Regulation in the Paper Industry,” *Quarterly Journal of Economics* 60, no. 2 (1946): 195–96.

23. Developed in China almost 2,000 years ago, when Tsai L’un stamped the bark of the mulberry tree into crude sheets, papermaking spread westward reaching Baghdad and Damascus by the eighth century, Morocco by the end of the eleventh century, and Europe by the early twelfth. In 1691, William Rittenhouse, a German immigrant, first introduced papermaking to the American colonies, with his small mill near Germantown, Pennsylvania. The first southern paper mill was built in 1744 at Williamsburg, Virginia. For the early history of papermaking, see Nicholas A. Basbanes, *On Paper: The Everything of its Two Thousand Year History* (New York: Vintage, 2013); Dard Hunter, *Papermaking: The History and Technique of an Ancient Craft* (New York: Knopf, 1947); Louis Tillotson Stevenson, *The Background and Economics of American Papermaking* (New York: Harper & Brothers, 1940); David C. Smith, *History of Papermaking in the United States, 1691–1969* (New York: Lockwood, 1970), 1–12; and Jack P. Oden, “Development of the Southern Pulp and Paper Industry, 1900–1970” (PhD diss., Mississippi State University, 1973), 1–3.

24. The first such paper machine was patented in France in 1799. The technology was refined by the Fourdrinier brothers in England and patented there in 1806. The new Fourdrinier machine was first introduced into the United States in 1822. Hunter, *Papermaking*, 340–73. See also Judith A. McGaw, *Most Wonderful Machine: Mechanization and Social Change in Berkshire Paper Making, 1801–1885* (Princeton, NJ: Princeton University Press, 1987), 93–117; James D. Studley, *United States Pulp and Paper Industry*, U.S. Department of Commerce, Trade Promotion Series no. 182 (Washington, DC: Government Printing Office, 1938), 5–6.

25. See Kenneth W. Britt, “Sulfite Pulping,” and George E. Jackson, “Alkaline Pulping,” both in *Handbook of Pulp and Paper Technology*, ed. Kenneth W. Britt (New York: Reinhold Publishing, 1964). See also Stevenson, *American Papermaking*, 22–26; Ohanian, *American Pulp and Paper Industry*, 185–88; and Studley, *United States Pulp and Paper Industry*, 5–6, 27–28. By the early 1990s, almost 70% of North American pulping capacity utilized the sulfate or Kraft process. Kraft pulping proved advantageous because a variety of wood species could be used as furnish, and the process produced a strong, easily bleached pulp. Also, the spent cooking chemicals or liquor (known as black liquor) could be washed from the pulp, concentrated, and burned to produce both energy and chemicals. That said, the Kraft process also produces a distinctive odor (due to the presence of total reduced sulfur compounds) and the bleaching generates dioxins and other chlorinated organics. Because Kraft mills are generally expensive in terms of cost-per-ton of product produced, they are usually built on a very large scale as part of an integrated complex.

26. Cohen, “Technological Change as Historical Process.”

27. Ibid., 780–82.

28. Jackson, "Alkaline Pulping," 176–77.

29. See Alfred Chandler, *The Visible Hand: The Managerial Revolution in American Business* (Cambridge, MA: Belknap Press of Harvard University Press, 1977), 281–83; and Chandler, *Scale and Scope*, 24.

30. See Oliver E. Williamson, *Markets and Hierarchies* (New York: Free Press, 1975).

31. See Ronald H. Coase, "The Nature of the Firm," *Economica* 4 (November 1937).

32. Scranton, *Endless Novelty.*

33. Chandler, *Visible Hand*, 281–83. See also Lamoreaux, *Great Merger Movement*, 27–32.

34. "Economics of Paper," 184.

35. Cohen, "Technological Change as Historical Process," 792. This problem has hardly disappeared. One article from the industry's leading trade journal analyzing the "structural impediments" to profitability in the paper industry noted that "high-cost, inefficient machines built sixty or seventy years ago are still running today. Approximately 40 percent of the machines in the industry are over twenty years old . . . Incredibly, 20 percent of the machines are at least thirty to thirty-five years old." Flicker, "Structural Impediments," 142. Perhaps this was the revenge of the Fourdrinier brothers.

36. Guthrie, "Price Regulation," 195.

37. Figures from Cohen, "Technological Change as Historical Process," 796, table 6.

38. Jack P. Oden, "Origins of the Southern Kraft Paper Industry, 1903–1930," *Mississippi Quarterly* 30, no. 4 (1977): 574.

39. "International Paper & Power," 229.

40. "Union Bag & Paper Corp.," 129.

41. International Paper Co., *International Paper Company after Fifty Years* (New York, 1948), 82–83, 85.

42. Myron R. Watkins, *Industrial Combinations and Public Policy* (Boston: Houghton Mifflin, 1927), 176.

43. For a brief discussion of the late nineteenth century paper industry, see Lamoreaux, *Great Merger Movement*, 37–44. See also Smith, *History of Papermaking*, 153–218.

44. Lamoreaux, *Great Merger Movement*, 44.

45. Smith, *History of Papermaking*, 189–218.

46. Ibid., 166–76 (quotes at 170); "International Paper."

47. Another consolidation, the General Paper Company, was formed in June 1900 to act as a selling agent for twenty-six Midwestern mills (Wisconsin and Minnesota). It was dissolved by the government in 1906. Information on the various consolidations was taken from *Pulp and Paper Investigation Hearings: Hearings Before the H. Select Comm. on Pulp and Paper Investigation*, 60th Cong. 436–39 (1909). See also "Union Bag & Paper Corp.," 50; "International Paper"; and Smith, *History of Papermaking in the United States*, 166–76, 189–212.

48. These consolidations, however, also benefited from a set of external circumstances that came together during this period. In particular, the Supreme Court's interpretation of the antitrust laws seemed to favor consolidations among competitors rather than pooling agreements and other looser forms of combination, thus providing greater incentives for the types of mergers that occurred. Moreover, the depth attained by capital markets by this time also proved instrumental in financing the new consolidations. Finally, the very fact that so many industries were experiencing mergers during such a short period combined with the idea that consolidation was somehow inevitable suggested the possibility of a contagious, "follow-the-leader" phenomenon. Lamoreaux, *Great Merger Movement*, 107–17. See also Watkins, *Industrial Combinations and Public Policy*, and Ralph L. Nelson, *Merger Movements in American Industry, 1895–1956* (Princeton, NJ: Princeton University Press, 1959), 89–105.

49. For a discussion, see Lamoreaux, *Great Merger Movement*, 120–26. See also Richard E. Caves, Michael Fortunato, and Pankaj Ghemawat, "The Decline of Dominant Firms, 1905–1929," *Quarterly Journal of Economics* 99, no. 3 (1984).

50. Roscoe R. Hess, "The Paper Industry in Its Relation to Conservation and the Tariff," *Quarterly Journal of Economics* 25, no. 4 (1911): 659–60; Watkins, *Industrial Combinations and Public Policy*, 193–200; Stevenson, *American Papermaking*, 169; Simon N. Whitney, *Antitrust Policies*, vol. 1 (New York: Twentieth Century Fund, 1958), 330–84.

51. On the importance of trade associations, witness the words of one economist writing in 1911, "Whether or not there is actual restraint of trade, it must be said that, tho [*sic*] competition is often visible, there likewise comes to light evidence of combination or 'cooperation.' Social-business meetings have been remarkably frequent, and there is extreme friendliness among the manufacturers. Not only do they report their monthly production to the American Pulp and Paper Association, but there is in several instances a curious interlocking of interests." Hess, "Paper Industry," 659. Based on testimony of several newsprint firms contained in the FTC newsprint investigation reports from 1930, however, the economist Arthur Burns maintained that International Paper continued to set the price of newsprint in the United States (except for the West Coast) during the 1920s. Arthur R. Burns, *The Decline of Competition: A Study of the Evolution of American Industry* (New York: McGraw-Hill, 1936), 132.

52. Stevenson, *American Papermaking*, 218.

53. For a discussion of the NRA and various other New Deal efforts (some of which operated at cross purposes) to rationalize American industry, see Ellis W. Hawley, *The New Deal and the Problem of Monopoly: A Study in Economic Ambivalence* (Princeton: Princeton University Press, 1966). For a discussion of the paper industry during the depression, see Smith, *History of Papermaking*, 443–63, and Oden, "Development of the Southern Pulp and Paper Industry," 66–75. See also Morton Keller, "The Pluralist State: American Economic Regulation in Comparative Perspective, 1900–1930," and Ellis Hawley, "Three Facets of Hooverian Associationalism: Lumber, Aviation, and Movies, 1921–1930," both in *Regulation in Per-*

spective: Historical Essays, ed. Thomas K. McCraw (Cambridge, MA: Harvard University Press, 1981).

54. The words are from a letter Wilson wrote in late 1933. Cited in Smith, *History of Papermaking*, 448.

55. National Recovery Administration, *Code of Fair Competition for the Paper and Pulp Industry* (Washington, DC, 1933). See also Stevenson, *American Papermaking*, 206–11; Smith, *History of Papermaking*, 446–47; and Oden, "Development of the Southern Pulp and Paper Industry," 73–74.

56. Guthrie, "Price Regulation," 203–6; Stevenson, *American Papermaking*, 206–11.

57. Stevenson, *American Papermaking*, 220–21.

58. Guthrie, "Price Regulation," 203–6; Stevenson, *American Papermaking*, 206–11.

59. Schecter Poultry Corp. v. United States, 295 U.S. 495 (1935).

60. Guthrie, "Price Regulation," 206–7.

61. Ibid., 207–10; Stevenson, *American Papermaking*, 220–21, 223–24; Whitney, *Antitrust Policies*, 340–43. For a discussion of the rising antitrust sentiment within the second Roosevelt administration, see Hawley, *New Deal and the Problem of Monopoly*.

62. Guthrie, "Price Regulation," 210–11.

63. Oden, "Development of the Southern Pulp and Paper Industry," 143–60.

64. George J. Stigler, *Capital and Rates of Return in Manufacturing Industries* (Princeton, NJ: Princeton University Press, 1963), 4–5 table 1.

65. Bureau of Census data, 1917–1959, compiled by the American Paper and Pulp Association, *Statistics of Paper 1964* (New York, 1964); 1960–1994 data compiled in *Pulp & Paper: North American Fact Book 1996*.

66. Oden, "Development of the Southern Pulp and Paper Industry," 222–25 (Zellerbach quote at 223–24).

67. Northup, "Negro in the Paper Industry," 8.

68. Lloyd G. Reynolds and Cynthia H. Taft, *The Evolution of Wage Structure* (New Haven, CT: Yale University Press, 1956), 104–5; Smith, *History of Papermaking*, 449; Northup, "Negro in the Paper Industry," 27–28.

69. For historical background on trade unionism in the pulp and paper industry, see Reynolds and Taft, *Evolution of Wage Structure*, 103–4, and Smith, *History of Papermaking*, 595–96. In 1972, the two unions merged to form the United Paperworkers International Union (UPIU). By the early 1990s, the UPIU (based in Nashville, TN) represented some 280,000 workers throughout the United States—approximately 75,000 of which were primary pulp and paper workers. *Pulp & Paper: North American Fact Book 1996*, 49–51.

70. Reynolds and Taft, *Evolution of Wage Structure*, 106; James Youtsler, "Collective Bargaining Accomplishments in the Paper Industry," *Southern Economic Journal* 21 (April 1955): 441.

71. Interregional wage differentials also narrowed considerably during the 1940s and 1950s. While southern mills paid marginally lower wages, this typically reflected

the relative cheapness of "semi-skilled" and "unskilled" workers. Youtsler, "Collective Bargaining," 446–47; Reynolds and Taft, *Evolution of Wage Structure*, 107–8, 150.

72. Guthrie, *Economics of Pulp and Paper*, 221–28; Reynolds and Taft, *Evolution of Wage Structure*, 106–7.

73. Ray Marshall, *The Negro and Organized Labor* (New York: John Wiley & Sons, 1965), 183. As Herbert Northup put it: "When the industry entered the South in force, it had never had any real experience with Negro labor and, like industry generally, typically followed local custom" ("Negro in the Paper Industry," 75).

74. Northup, "Negro in the Paper Industry," 31–33.

75. Ibid., 33, 35.

76. Ibid., 45–46, 74–75.

77. Black workers ended up in segregated locals of the International Brotherhood of Pulp, Sulfite, and Paper Mill Workers (IBPSPW). Because of the jurisdictional agreement between the IBPSPW and the more powerful United Papermakers and Paperworkers, confining black workers to the IBPSPW reinforced their exclusion from the more lucrative paper machine jobs and further undermined their ability to exercise any power in union-management negotiations. For a broader discussion of these practices, see Herbert Hill, "Racism within Organized Labor: A Report of Five Years of the AFL-CIO, 1955–1960," *Journal of Negro Education* 30, no. 2 (1961): 109–18; and Marshall, *Negro and Organized Labor*, 89–105. For a discussion in the context of the pulp and paper industry, see Northup, "Negro in the Paper Industry," 29, 37.

78. Reynolds and Taft, *Evolution of Wage Structure*, 124; Northup, "Negro in the Paper Industry," 41–42, 76.

79. Northup, "Negro in the Paper Industry," 40–41, 76–78. See also Hill, "Racism within Organized Labor," 115.

80. Ray Marshall notes that as early as 1949, many southern blacks had come to suspect the CIO's professed egalitarianism in the area of race relations as "window dressing." Black leaders within the CIO were also attacked as "Uncle Toms" for allowing themselves to be used by the Union for public relations purposes. The merger of the AFL and the CIO simply made things worse, adding to the distance between organized labor and the black community. Marshall, "The Negro and Organized Labor," *Journal of Negro Education* 32, no. 4 (1963): 382–84. For a broader discussion the CIO's Operation Dixie, initiated in 1946, see Michael Goldfield, "The Failure of Operation Dixie: A Critical Turning Point in American Political Development," in *Race, Class, and Community in Southern Labor History*, ed. Gary M. Fink and Merl E. Reed (Tuscaloosa: University of Alabama Press, 1994).

81. Continuous digesters had been in use to produce lower grades for years, but they did not become available for Kraft pulping until the 1950s. Whereas batch digesters typically required several workers to maintain them, a large continuous digester could be operated by a single man. See U.S. Department of Labor, *Impact of Technological Change and Automation in the Pulp and Paper Industry*, Bulletin no. 1347 (Washington, DC: Government Printing Office, 1962), and Jackson, "Alkaline Pulping," 176–77.

82. U.S. Department of Labor, *Impact of Technological Change*; *Pulp & Paper: North American Factbook 1996*, 52.

83. Northup, "Negro in the Paper Industry," 47, 77.

84. Ibid., 47–48.

85. These figures, which were also used by Northup in his study of the paper industry, were taken from an annual time-series data set collected by the South Carolina Department of Labor documenting employment by race (and gender) in the state's manufacturing sector. For a discussion of the data set and its use to document how federal civil rights laws affected black workers, see James J. Heckman and Brook S. Payner, "Determining the Impact of Federal Antidiscrimination Policy on the Economic Status of Blacks: A Study of South Carolina," *American Economic Review* 79, no. 1 (1989), and John J. Donohue III and James Heckman, "Continuous versus Episodic Change: The Impact of Civil Rights Policy on the Economic Status of Blacks," *Journal of Economic Literature* 29, no. 4 (1991).

86. Gavin Wright, "The Civil Rights Revolution as Economic History," *Journal of Economic History* 59, no. 2 (1999); Gavin Wright, *Sharing the Prize: The Economics of the Civil Rights Revolution in the American South* (Cambridge, MA: Harvard University Press, 2013). For a more detailed discussion of this issue in the context of the pulp and paper industry, see Timothy J. Minchin, "Federal Policy and the Racial Integration of Southern Industry, 1961–1980," in *Journal of Policy History* 11, no. 2 (1999); and Timothy J. Minchin, *The Color of Work: The Struggle for Civil Rights in the Southern Paper Industry, 1945–1980* (Chapel Hill: University of North Carolina Press, 2001). Minchin's research on the struggle for civil rights in the pulp and paper industry supplements Northup's work with detailed archival research and interviews with key actors. It demonstrates the positive impact of civil rights legislation and litigation on the relative position of black workers in the southern pulp and paper industry.

87. Hill, "Racism within Organized Labor," 110–11. See also Marshall, *Negro and Organized Labor*, 183–85. For a discussion of the role of pulp and paper union locals in supporting George Wallace in Alabama, see Robert J. Norrell, "Labor Trouble: George Wallace and Union Politics in Alabama," in *Organized Labor in the Twentieth Century South*, ed. Robert H. Zieger (Knoxville: University of Tennessee Press, 1991), 250–72, esp. 264–65.

88. Northup, "Negro in the Paper Industry," 55–56.

89. Marshall, *Negro and Organized Labor*, 183–85; Northup, "Negro in the Paper Industry," 56, 79.

90. Quoted in Marshall, *Negro and Organized Labor*, 184.

91. Marshall, *Negro and Organized Labor*, 184.

92. Hill, "Racism within Organized Labor," 109. For a contrasting perspective on the relationship between black workers and the unions in the Jim Crow South, see Michael Honey, "Black Workers Remember: Industrial Unionism in the Era of Jim Crow," in Fink and Reed, *Race, Class, and Community in Southern Labor History*, and Rick Halpern, "Organized Labor, Black Workers, and the Twentieth-Century South: The Emerging Revision," in *Race and Class in the American South since 1890*, ed. Melvyn Stokes and Rick Halpern (Oxford: Berg, 1994).

93. Hill, "Racism within Organized Labor," 109; Marshall, "Negro and Organized Labor," 381.

94. Northup, "Negro in the Paper Industry," 57. For a discussion of the "organic society" in the antebellum and postbellum South and, more generally, an analysis of the psychology of race relations in the South, see Joel Williamson, *The Crucible of Race: Black-White Relations in the American South since Emancipation* (New York: Oxford University Press, 1984). See also C. Vann Woodward, *The Strange Career of Jim Crow*, 2nd rev. ed. (New York: Oxford University Press, 1966). Woodward, in particular, calls attention to the relatively late arrival of Jim Crowism in the South and thus to what he calls the "forgotten alternatives."

95. Northup, "Negro in the Paper Industry," 58.

96. Civil Rights Act of 1964, P.L. 88-352, 78 Stat. 241 (1964), codified as amended at 42 U.S.C. §§ 2000(e) et seq.

97. Minchin, "Federal Policy and Racial Integration," 162.

98. Northup, "Negro in the Paper Industry," 79.

99. See United States v. Local 189, United Papermakers & Paperworkers, 282 F. Supp. 39 (E.D. La., 1968) (enjoining discriminatory promotional policy at paper mill and ordering new seniority system that did not perpetuate past discriminatory practices), affirmed by Local 189, United Papermakers & Paperworkers v. United States, 416 F.2d 980 (5th Cir. 1969) cert. denied 397 U.S. 919.

100. For background on Bogalusa and a detailed discussion of the town's civil rights conflicts, see Adam Fairclough, *Race and Democracy: The Civil Rights Struggle in Louisiana, 1915–1972* (Athens: University of Georgia Press, 1995), 344–80.

101. Ibid., 346–47.

102. Ibid., 346–48; Northup, "Negro in the Paper Industry," 95–96.

103. Fairclough, *Race and Democracy*, 349.

104. Northup, "Negro in the Paper Industry," 97; Fairclough, *Race and Democracy*, 349.

105. Northup, "Negro in the Paper Industry," 97–98; Fairclough, *Race and Democracy*, 349.

106. Fairclough, *Race and Democracy*, 349–50.

107. Ibid., 350.

108. Ibid., 345.

109. Quoted in ibid., 355.

110. Ibid., 354–60.

111. Ibid., 370–76.

112. Ibid., 377.

113. Civil Rights Act of 1964, codified as amended at 42 U.S.C. §§ 2000(e) et seq.

114. *United States v. Local 189*, 282 F. Supp. 39; *Local 189 v. United States*, 416 F. 2d 980.

115. *United States v. Local 189*, 282 F. Supp. 39. See also Minchin, "Federal Policy and Racial Integration," 154.

116. *Local 189 v. United States*, 416 F.2d 980.

117. Ibid.

118. Northup, "Negro in the Paper Industry," 95.

119. Minchin, "Federal Policy and Racial Integration," 154–55. For a discussion of the Jackson Memorandum and its application to the International Paper mill in Mobile, see Stevenson v. Int'l Paper Co., 352 F. Supp. 230 (S.D. Ala. 1972); Fluker v. Locals 265 and 940, United Papermakers, 1972 WL 271 (S.D. Ala. 1972) and Stevenson v. Int'l Paper Co., 516 F.2d 103, 107–08 (5th Cir. 1975).

120. Minchin, "Federal Policy and Racial Integration," 154–55. See also Watkins v. Scott Paper Co., 530 F.2d 1159 (5th Cir. 1976).

121. See, for example, Myers v. Gilman Paper Corp., 556 F.2d 758 (5th Cir. 1977); *Watkins v. Scott Paper*, 530 F.2d 1159; Rogers v. Int'l Paper Co., 510 F.2d 1340 (8th Cir. 1975), remanded 423 U.S. 809 (1975) for further consideration in light of Albemarle Paper Co. v. Moody, 422 U.S. 405 (1975); Stevenson v. Int'l Paper Co., 516 F.2d 103 (5th Cir. 1975); Long v. Ga. Kraft Co., 450 F.2d 557 (5th Cir. 1971); United States v. Cont'l Can Co., 319 F. Supp. 161 (E.D. Va. 1970); and Miller v. Cont'l Can Co., 544 F. Supp. 210 (S.D. Ga. 1981). According to Kent Spriggs, an NAACP Legal Defense Fund lawyer who worked on paper industry cases during the 1960s and 1970s, pulp and paper was "the most litigated industry in the South." Quoted in Minchin, "Federal Policy and Racial Integration," 162.

122. See, e.g., *Watkins v. Scott Paper*, 530 F.2d at 1167–68: "The rule, adopted by this Court in Local 189 . . . and recently endorsed by the Supreme Court . . . is that blacks previously discriminated against must be given such remedial relief as to enable them to achieve their 'rightful place' in an employer's employment hierarchy."

123. For interviews with key actors supporting the critical role of civil rights law in changing the industry's employment practices, see Minchin, "Federal Policy and Racial Integration" and *Color of Work*.

124. Fairclough, *Race and Democracy*, 377.

125. For a discussion, see Vallas and Beck, "Transformation of Work Revisited," 339–61.

126. Ibid.

127. Bruce E. Kaufman, "The Emergence and Growth of a Nonunion Sector in the Southern Paper Industry," in *Southern Labor in Transition, 1940–1995*, ed. Robert Zieger (Knoxville: University of Tennessee Press, 1997), 319.

128. Ibid.

129. According to Bruce Kaufman, the push for new principles of work design and the corresponding de-emphasis on union representation began in 1975 with the Mead Corporation's new mill in Stevenson, Alabama. Ibid., 300–301, 309–18.

130. Ibid., 299–300, 320–21. For a discussion of International Paper's more antagonistic approach to labor relations in the context of federal labor law and industrial relations policy, see Julius G. Getman and F. Ray Marshall, "Industrial Relations in Transition: The Paper Industry Example," *Yale Law Journal* 102, no. 1803 (1993).

Chapter 4 · Appropriating the Environment

Epigraph. Eugene P. Odum, "The Strategy for Ecosystem Development," *Science* 164 (1969): 266–67.

1. Eugene P. Odum, "The Strategy for Ecosystem Development," *Science* 164 (1969).

2. See Eugene P. Odum, "The New Ecology," *Bioscience* 14 (1964) (discussing the new systems approach to ecology); Paul Charles Milazzo, *Unlikely Environmentalists: Congress and Clean Water, 1945–1972* (Lawrence: University Press of Kansas, 2006), 89–111 (discussing the rise of systems thinking in ecology and its influence on the environmental movement); Sharon E. Kingsland, *The Evolution of American Ecology, 1890–2000* (Baltimore, MD: Johns Hopkins University Press, 2005), 189–205 (discussing the systems approach to ecology developed by Odum and its influence on the early environmental movement).

3. See Eugene P. Odum, "The Emergence of Ecology as a New Integrative Discipline," *Science* 195 (1977): 1289, 1290 (discussing research in Georgia salt marshes).

4. Some even discarded the ecosystem concept itself as the relic of an older, more static, view of ecological processes. Where Odum once saw stability, order, and balance, the new generation saw dynamics, disturbance, and disequilibrium. In place of the self-maintaining ecosystem, one now had to reckon with "the lowly 'patch.'" Donald Worster, "The Ecology of Order and Chaos," *Environmental History Review* 14 (Spring–Summer 1990): 1–18 (quote at 10); Kingsland, *Evolution of American Ecology*, 244–45.

5. For a discussion of the impact of ecology on American environmental law, see Fred P. Bosselman and A. Dan Tarlock, "The Influence of Ecological Science on American Law: An Introduction," *Chicago-Kent Law Review* 69 (1994): 847; A. Dan Tarlock, "The Nonequilibrium Paradigm in Ecology and the Partial Unraveling of Environmental Law," *Loyola of Los Angeles Law Review* 27 (1994): 1121; Jonathan Baert Wiener, "Law and the New Ecology: Evolution, Categories, and Consequences," *Ecology Law Quarterly* 22 (1995): 325; Jonathan Baert Wiener, "Environmental Law Faces the New Ecology: Beyond the Balance of Nature," *Duke Environmental Law and Policy Forum* 7 (1996); and Milazzo, *Unlikely Environmentalists*.

6. Wallace was referring to the odor of a recently constructed paper mill some 20 miles from Montgomery. Cited in Marshall Frady, "The View from Hilton Head," *Harper's*, May 1970, 103.

7. "Pine Pulp Industry Proves Fastest Growing in the South," *Atlanta Constitution*, December 9, 1945.

8. James C. Cobb, *The Selling of the South: The Southern Crusade for Industrial Development*, 2nd ed. (Urbana: University of Illinois Press, 1993), 229–53, quotes at 229 and iii.

9. Reports of fish kills and contaminated shellfish beds downstream from pulp and paper mills, for example, had become common throughout the region. Some of the region's rivers had been transformed into little more than industrial sewers. According to a 1964 report from the Department of Health, Education, and Welfare,

for example, the lower Savannah River had become so "grossly polluted" because of the wastewater discharges from the Union Camp mill that it failed to support aquatic life over large stretches. Air pollution had become so bad at some mills that employees were given free daily car washes to reduce corrosion. In other cases, the resulting smog was so thick that nearby roads had to be closed. In Canton, North Carolina, the smog from the Champion paper mill was so dense in late November 1970 that it led to a multi-car collision in which more than twenty-five people were injured. Council on Economic Priorities, *Paper Profits: Pollution in the Pulp and Paper Industry* (Cambridge, MA: MIT Press, 1972), 3–9. See also U.S. Department of Health, Education, and Welfare (HEW), *Pollution of Interstate Waters of the Mouth of the Savannah River, GA* (Washington, DC, 1964).

10. Figures are from National Association of Manufacturers, *Water in Industry* (Washington, DC, 1965), 72–73.

11. Interview with author, August 13, 1997.

12. See Robert V. Percival, "Environmental Federalism: Historical Roots and Contemporary Models," *University of Maryland Law Review* 54 (1995): 1141; Robert V. Percival, "Regulatory Evolution and the Future of Environmental Policy," *University of Chicago Legal Forum* 159 (1997); and Richard N. L. Andrews, *Managing the Environment, Managing Ourselves: A History of American Environmental Policy* (New Haven, CT: Yale University Press, 1999), 227–54.

13. *Pulp & Paper: North American Factbook 1996* (San Francisco: Miller Freeman, 1997), 77.

14. "Bioaccumulation," sometimes called "biomagnification" or "biological amplification," refers to the process whereby the concentration of certain toxic compounds increases in living tissues as one moves up the food chain from one trophic level to another. This tendency is especially marked in the case of chlorinated hydrocarbons such as dioxins because of their high solubility in fatty tissues and their low solubility in water. Paul H. Ehrlich, Anne H. Ehrlich, and John P. Holdren, *Ecoscience: Population, Resources, Environment* (San Francisco: W. H. Freeman, 1977), 630.

15. This region, along with much of the rest of the southeastern coastal plain, sits atop one of the most prolific groundwater resources in the world and as a result has been especially attractive for pulp and paper mills, many of which depend on the regional aquifer to supply their massive water needs. By the 1950s, to take one example, Union Camp's Savannah mill was using close to 30 million gallons of groundwater a day. As early as 1938, the *Savannah Morning News* warned of the potential impacts on the groundwater resource from saline intrusion caused by overpumping at the Union Camp mill. *Savannah Morning News*, "Pollution Danger from Sea Water," August 15, 1938. In 1944, a report by the Georgia Geological Survey raised similar concerns and concluded that 25 million gallons per day was the maximum pumping level that could be sustained at Savannah in order to avoid saline intrusion. M. A. Warren, *Artesian Water in Southeastern Georgia with Special Reference to the Coastal Area*, Georgia Geological Survey Bulletin no. 49 (Atlanta, 1944). In 1964, scientists from the U.S. Geological Survey concluded on the basis of

new data and the then current pumpage rate of 62 million gallons per day that salt-water intrusion would begin to significantly affect outlying areas in less than a hundred years. M. J. McCollum and H. B. Counts, *Relation of Salt-Water Encroachment to the Major Aquifer Zones, Savannah Area, Georgia, and South Carolina*, U.S. Geological Survey Water-Supply Paper 1613-D (Washington, DC: Government Printing Office, 1964). The issue achieved national prominence in the early 1970s when James Fallows and Ralph Nader's "study group" came to Savannah to study the environmental crisis in the region, with particular attention to the groundwater issue. James Fallows, *The Water Lords* (New York: Grossman, 1971).

16. The state of South Carolina threatened to sue the state of Georgia in the early 1990s for allowing companies such as Union Camp to pump at unsustainable rates that threatened the long-term integrity of the groundwater resource. Carrie Teegardin, "The Southeastern Water Wars," *Atlanta Constitution*, June 19, 1991.

17. Robert U. Ayres and Allen V. Kneese, "Production, Consumption, and Externalities," *American Economic Review* 59, no. 3 (1969); Robert U. Ayres, "Industrial Metabolism," in *Technology and Environment*, ed. Jesse H. Ausubel, Robert A. Frosch, and Robert Herman (Washington, DC: National Academy Press, 1989). See also Mariana Fischer-Kowalski, "Society's Metabolism: The Intellectual History of Materials Flow Analysis, Part I: 1860–1970," *Journal of Industrial Ecology* 2, no. 1 (1998).

18. Ronald Coase, "The Problem of Social Cost," *Journal of Law and Economics* 3 (October 1960).

19. *Water Pollution Control and Abatement (Part 1A—National Survey): Hearings Before the Subcomm. on Natural Res. and Power of the H. Comm. on Gov't Operations*, 88th Cong. 682 (1964).

20. Formed over millions of years, the aquifer resides in a highly permeable, 400–500-foot-thick carbonate section known as the Ocala Limestone formation. In its "pre-development" state, groundwater flowed through the aquifer in an easterly direction before being discharged into the Atlantic Ocean. Recharge came from inland areas, where consistent rainfall and sandy, permeable soils provided for the natural replenishment of the resource. By the time the water reached Savannah and coastal Georgia, it was several centuries old. Mary A. Elfner, "Groundwater Use and Conservation in the Savannah Area," *Proceedings of the 1997 Georgia Water Resources Conference*, ed. Kathryn J. Hatcher (Athens: University of Georgia, 1997).

21. "The intake for the Champion mill diverts 46.4 million gallons a day and returns approximately 45 million gallons a day. During low flow conditions, the mill diverts virtually all of the flow of the Pigeon River." Champion Int'l v. EPA, 648 F. Supp. 1390, 1391 (W.D. N.C. 1986).

22. *Water Pollution Control and Abatement*, 680–81. Chisholm was later challenged on this point by Rep. John S. Monagan: "It [the pulp and paper industry] may not consume it, but it has rendered it unfit for future community use, so it might as well have been consumed" (689).

23. National Manufacturers Association, *Water in Industry: A Survey of Water Use in Industry* (New York, 1965), 73.

24. Hutchins & Hutchins v. Int'l Paper Co., 6 F.R.D. 510 (W.D. La. 1947).

25. *Champion Int'l v. EPA*, 648 F. Supp. 1390, 1391.

26. David C. Smith, *History of Papermaking in the United States, 1691–1969* (New York: Lockwood, 1970), 630; Jack P. Oden, "Development of the Southern Pulp and Paper Industry, 1900–1970" (PhD diss., Mississippi State University, 1973), 455.

27. This is known formally as biochemical oxygen demand (BOD), which refers to the quantity of oxygen required by bacteria to oxidize organic waste to form carbon dioxide and water. A standard measure of BOD is the number of milligrams of oxygen consumed per liter of water over a period of five days at a temperature of 20°C. This is called BOD_5. A BOD of 0.17 pound, or 77 grams, is sometimes referred to as a "population equivalent," being roughly equal to the daily BOD requirement for the domestic wastes produced by a single person. Ehrlich, Ehrlich, and Holdren, *Ecoscience*, 160–61, 556.

28. See, e.g., *Hutchins & Hutchins v. Int'l Paper*, 6 F.R.D. 510; Maddox v. Int'l Paper Co., 105 F. Supp. 89 (W.D. La. 1951).

29. *Hutchins & Hutchins v. Int'l Paper*, 6 F.R.D. 510. According to the judge, this was one of some thirty cases in which plaintiffs were suing International Paper for damages caused by its pollution of the Bodeau Bayou.

30. See, e.g., Spyker et al. v. Int'l Paper Co., 173 La. 580 (1931) (judgment in favor of International Paper); Young v. Int'l Paper Co., 179 La. 803 (1934) (judgment in favor of International Paper); and Maddox v. Int'l Paper Co., 47 F. Supp. 829 (W.D. La. 1942) (judgment in favor of plaintiffs). International Paper was also indicted by a grand jury in Louisiana's Bossier Parish for "polluting the waters and killing the fish" of the Bodeau Bayou. The charge was brought under the provisions of a 1932 state law that made it a criminal offense to discharge into state waters any substance that killed fish or rendered the water unfit for the maintenance of fish life. For a description of the grand jury proceedings, see State v. Int'l Paper Co., 201 La. 870 (1942).

31. See Craig E. Colten and Peter N. Skinner, *The Road to Love Canal: Managing Industrial Wastes before EPA* (Austin: University of Texas Press, 1996), 70–75. See also Christine Rosen, "Differing Perceptions of the Value of Pollution Abatement across Time and Place," *Law and History Review* 11 (1993).

32. In Weston Paper Co. v. Pope, 155 Ind. 394, 400–402 (1900), for example, the Supreme Court of Indiana upheld damages and injunctive relief against a paper company on the grounds that the manufacturer had no "right . . . to establish his plant upon the banks of a non-navigable stream and pollute its waters by a business wholly brought to the place, entirely disconnected with any use of the land itself, and which he may just as well conduct elsewhere." In reaching its conclusion, the court rejected any notion of balancing: "The fact that appellant has expended a large sum of money in the construction of its plant and that it conducts its business in a careful manner and without malice can make no difference in its rights to the stream" (at 401). Following this line of reasoning, the New York Court of Appeals adopted a similar rule in Whalen v. Union Bag and Paper Co., 208 N.Y. 1 (1913). The plaintiff in *Whalen* owned a farm a few miles downstream from the Union Bag mill.

According to the court, the mill, which "represent[ed] an investment of more than a million dollars" and employed some "four hundred to five hundred" workers, "discharg[ed] into the waters of the creek a large quantity of a liquid effluent containing sulphurous acid, lime, sulphur, and waste material consisting of pulpwood, sawdust, silvers, knots, gums, resins, and fibre, . . . [which] greatly diminished the purity of the water" (at 3). Finding "a clear case of wrongful pollution of the stream," the court emphasized that balancing the "great loss to the defendant [mill] by the granting of the injunction as compared to the small injury done to plaintiff's land cannot be justified by the circumstances of this case" (at 3–4). As the court reasoned, "Although the damage to the plaintiff may be slight as compared with the defendant's expense of abating the condition, that is not a good reason for refusing an injunction. Neither courts of equity nor law can be guided by such a rule, for if followed to its logical conclusion it would deprive the poor litigant of his little property by giving it to those already rich" (at 5). And so the Union Bag mill was forced to shut down, despite the fact that it represented an investment of more than a million dollars and provided employment for hundreds of workers. See Driscoll v. Am. Hide and Leather, 170 N.Y.S. 121, 122 (1918) (noting that the Union Bag mill had been shut down as a result of the 1913 decision).

33. See, e.g., Spyker v. Int'l Paper Co., 138 So. 109 (La. 1931) (denying injunctive relief on grounds that plaintiff failed to prove that wastes from International Paper mill had caused trees on plaintiffs land to die); and Rhodes v. Int'l Paper Co., 139 So. 755 (La. 1932) (setting aside damages award to plaintiffs for destruction of timber on grounds that plaintiff failed to prove causation).

34. *Rhodes v. Int'l Paper*, 139 So. 755, 757.

35. See Christine Meisner Rosen, "'Knowing' Industrial Pollution: Nuisance Law and the Power of Tradition in a Time of Rapid Economic Change, 1840–1864," *Environmental History* 8 (2003): 565 (discussing the difficulties courts had in coming to terms with the increased scale and new forms of industrial pollution during the mid-nineteenth century).

36. *Young v. Int'l Paper*, 179 La. 803, 810 (1934). The decision affirmed the trial court's denial of injunctive relief on grounds that shutting down the mill "would subject the defendant to grossly disproportionate hardship."

37. *Maddox v. Int'l Paper*, 47 F. Supp. 829, 832, 830.

38. W. Page Keeton et al., *Prosser and Keeton on Torts*, 5th ed. (St Paul: West, 1984), 631–32 (discussing the balancing doctrine in nuisance law). See also Christine Rosen, "Differing Perceptions of the Value of Pollution Abatement across Time: Balancing Doctrine in Pollution Nuisance Law, 1840–1906," *Law & History Review* 11 (1993): 303 (discussing use of balancing doctrine by American judges in pollution nuisance cases during the late nineteenth century); Morton J. Horwitz, *The Transformation of American Law, 1780–1860* (Cambridge, MA: Harvard University Press, 1977), 74–99 (discussing changes in the application of nuisance law to accommodate industrial development in early nineteenth century).

39. Hampton v. N.C. Pulp Co., 27 S.E. 2d 538, 541–43 (N.C. 1943).

40. Ibid., at 549. As the court elaborated, "The defendant would be in a better position upon such an argument [to weigh the economic consequences]—if indeed it could have anything to do with the law of the case—if it had not admitted to an agreement with a State Department that it would not permit the discharge of waste matter containing any noxious chemical or matter deleterious to fish life into the river, and if it had not repudiated this agreement as unenforceable, or outlawed by expiration of time."

41. Ibid., at 550.

42. Nat'l Container Corp. v. Stockton, 189 So. 4, 5 (Fla. 1939).

43. Ibid., at 11–12.

44. 7 Ala. Code § 1088 (1940).

45. Stone Container Corp. v. Stapler, 263 Ala. 524 (1955).

46. Ricou v. Int'l Paper Co., 117 F. Supp. 128, 131 (W.D. La. 1953). See also Connell v. Int'l Paper Co., 99 F. Supp. 699, 701 (W.D. La. 1951): "If industry is to be permitted a reasonable chance to develop for the benefit of the whole community, some inconvenience must be endured by its inhabitants with the right always to compensation for damages of any special or unreasonable type." See also Busby v. Int'l Paper Co., 95 F. Supp. 596, 597–98 (W.D. La. 1951): "Defendant has constructed and operates a large industrial plant which cost many millions of dollars, gives employment to several hundreds of people, and furnishes a market for large quantities of pulpwood to farmers and other owners of timber used for making paper. It is, therefore, a substantial asset to the community . . . [W]here, as here, the facts reveal that, at most, the plaintiff has been deprived of the ability to cultivate a small portion of the lands so inundated, such injuries are readily compensable in money, and a balancing of benefits and conveniences, would induce a court of equity to deny the harsh remedy of injunction."

47. *Maddox v. Int'l Paper*, 105 F. Supp. 89, 91, 93.

48. Smith, *History of Papermaking*, 630.

49. Oden, "Development of the Southern Pulp and Paper Industry," 455–57.

50. Ibid., 467. North Carolina, Mississippi, Tennessee, and Virginia all passed new laws.

51. See Alabama Session Law, Act of Aug. 25, 1949, No. 460, p. 667 (creating a Water Improvement Advisory Commission).

52. Ibid. at 670 (exempting existing industrial discharges from regulation except in cases of health hazard). See also *Water Pollution Control and Abatement (Part 7—Alabama, Georgia, Mississippi, and Tennessee): Hearings Before the Subcomm. on Natural Res. and Power of the H. Comm. on Gov't Operations*, 88th Cong. 3645 (1965) (statement of Ira L. Meyers, chairman of the Alabama Water Improvement Commission).

53. Alabama Session Law, Act of Aug. 25, 1949, No. 460 at 670 (emphasis added).

54. Other southern states pursued similar strategies. In 1947, the Florida legislature designated the Fenholloway River as an "industrial river" as part of a campaign to persuade Proctor & Gamble to build a pulp and paper mill in Perry, Florida. See 24952 Fla. Stat. § 1338 (1947).

55. See *Water Pollution Control and Abatement (Part 7)*, 3682–83 (statement of Governor George C. Wallace). Wallace pledged, "I have promised the people of Alabama that I would do everything in my power as their Governor to seek new industries. My record speaks for itself. I shall continue to seek industry because it is needed for the development of this great State and the wise utilization of its vast supply of natural resources . . . I have also given my support to the prevention and elimination of water pollution in Alabama . . . We often hear the question asked, 'Do you want jobs or clean waters?' We want both jobs and clean waters and we can have both . . . When industries locate in Alabama we want them to stay in Alabama. We do not want them moving away because of a lack of adequate clean waters for their operations. Industry must have a satisfactory supply of water for its survival."

56. In September 1971, Governor Wallace signed into law a new water pollution control statute designed to reduce pollution of the state's waterways. The new act included within its scope industries covered by the "grandfather clause" in the 1949 legislation. See Alabama Water Pollution Control Act, Ala. Acts No. 1260 (1971). The Act was codified in 1975 and amended in 1979 to bring the state into compliance with the federal water pollution control act of 1972.

57. The Refuse Act, which was enacted as Section 13 of the 1899 Rivers and Harbors Act, made it illegal to dump "refuse" into navigable waters without a permit. 30 Stat. 1121, 1151 (1899).

58. Water Pollution Control Act, 62 Stat. 1155 (1948).

59. A. Myrick Freeman III, "Water Pollution Policy," in *Public Policies for Environmental Protection*, ed. Paul R. Portney (Washington, DC: Resources for the Future, 1990), 98–101. See also Andrews, *Managing the Environment*, 205.

60. Water Pollution Control Act Amendments of 1956, 70 Stat. 498 (1956). The enforcement conference provision was contained in Section 8 of the 1956 law (504).

61. For a discussion of the 1956 amendments, see Freeman, "Water Pollution Policy," 99.

62. *Water Pollution Control and Abatement (Part 7)*, 3648–53 (statement of Ira L. Meyers). Meyers noted that the Mobile River within the harbor area and for a distance of some five miles upstream from its mouth had been defined by the commission "as an area in which a minimum dissolved oxygen concentration of two parts per million will be accepted. This decision was reached by the Commission after a thorough evaluation of the problems encountered and a determination that the public interest would not be compromised by such a decision. This concentration of dissolved oxygen will permit fish to travel up the Mobile River and will not interfere with other users of the waters" (3649–50). Meyers failed to note, however, that warm-water fish normally require at least four parts per million of dissolved oxygen in order to survive and propagate.

63. See *Water Pollution Control and Abatement (Part 7)*, 3687 (statement of J. W. Howell, president, Mississippi Wildlife Federation).

64. Ibid., 3718 (statement of Howard D. Zeller, Georgia Game and Fish Division).

65. HEW, *Pollution of Interstate Waters of the Mouth of the Savannah River.*

66. *Water Pollution Control: Hearings on S. 649, S. 737, S. 118, and S. 1183, Before a Special Subcomm. on Air and Water Pollution of the S. Comm. on Pub. Works,* 88th Cong. 1 (1963). Muskie also noted in his opening statement that although considerable progress had been made with regard to municipal waste treatment, "Industry ha[d] been laggard in its responsibilities."

67. Specifically, S. 649 "A Bill to Amend the Water Pollution Control Act," which would create a new Federal Water Pollution Control Administration within the Department of Health, Education, and Welfare and provided federal authority to establish water quality standards for interstate or navigable waters. *Water Pollution Control,* 3–5.

68. Muskie's legislative proposal, which was eventually passed in modified form as the Water Quality Act of 1965, led to a series of hearings held throughout the country during the tumultuous years of 1963, 1964, and 1965. For a discussion of Muskie's activities in this respect, see Robert F. Blomquist, "'To Stir Up Public Interest:' Edmund S. Muskie and the U.S. Senate Special Subcommittee's Water Pollution Investigations and Legislative Activities, 1963–66—A Case Study in Early Congressional Environmental Policy Development," *Columbia Journal of Environmental Law* 22 (1997).

69. *Water Pollution Control and Abatement (Part 1A—National Survey): Hearings Before the Subcomm. on Natural Res. and Power of the H. Comm. on Gov't Operations,* 88th Cong. 212 (1964).

70. *Water Pollution: Hearings Before a Special Subcomm. on Air and Water Pollution of the S. Comm. on Pub. Works,* 89th Cong. (1965).

71. *Water Pollution Control and Abatement,* 684. See the testimony of George Olmstead Jr., also representing the American Paper and Pulp Association, during Muskie's 1963 Senate hearings, *Water Pollution Control,* 528–44.

72. *Water Pollution Control and Abatement,* 707, 726, 737.

73. *Water Pollution Control and Abatement (Part 7),* 3683 (statement of George C. Wallace).

74. Ibid., 3700–701 (statement of P. A. Bachelder).

75. Ibid., 3724 (statement of Mrs. Atherton Hastings).

76. Ibid., 3844, 3846 (statement of Ralph Richards).

77. Georgia Water Quality Control Act of 1964, Ga. Laws. p. 416 §§ 1 et seq.

78. HEW, *Pollution of Interstate Waters of the Mouth of the Savannah River.*

79. U.S. Public Health Service, *Conference on the Pollution of the Interstate Waters of the Lower Savannah River and Its Tributaries, South Carolina–Georgia, February 2, 1965* (Washington, DC: U.S. Department of Health, Education, and Welfare, 1965), 12. The population figures were based on a population equivalent of 0.17 pounds of BOD per person.

80. Ibid., 129–35 (quote at 134) (statement of James R. Lientz).

81. Ibid., 131.

82. Ibid., 137 (statement of Murray Stein).

83. Ibid.

84. In signing the legislation, Johnson stated that "we are going to reopen the Potomac for swimming by 1975 and within the next 25 years we are going to repeat this effort in lakes and streams and other rivers across the country." Cited in *Congressional Quarterly Almanac 1965* (Washington, DC: CQ–Roll Call Group, 1966), 750. The act required that states establish water quality standards for interstate waters and develop implementation plans for reducing discharges in order to meet these standards. Freeman, "Water Pollution Policy," 99–102.

85. For a discussion, see Freeman, "Water Pollution Policy," 102–3.

86. Milazzo, *Unlikely Environmentalists*, 145–46 (discussing the impact of the Santa Barbara oil spill on the sense of environmental crisis and momentum behind new water pollution legislation), 145 (discussing the Cuyahoga incident and its impact on the environmental movement).

87. Federal Water Pollution Control Administration, *Second Session of the Conference in the Matter of Pollution of the Interstate Waters of the Lower Savannah River and Its Estuaries, Tributaries and Connecting Waters in the States of Georgia and South Carolina, October 1969* (Washington, DC, 1970); Georgia Water Quality Control Board, *Water Quality Data Lower Savannah River* (Atlanta, 1969), 10–12.

88. In a report presented at the enforcement conference, the Georgia Water Quality Control Board concluded that portions of the Lower Savannah River below Savannah were "grossly polluted by large volumes of untreated or inadequately treated domestic and industrial wastes" and that for the stretch of river "classified for Industrial and Navigation Uses, the River consistently fails to meet the minimum dissolved oxygen criteria for this classification from June through September each year." Georgia Water Quality Control Board, *Water Quality Data Lower Savannah River*, 73.

89. In Odum's words: "Many parts of the estuarine system are normally operating at close to full capacity with regard to oxygen. There isn't much leeway here. Since most types of pollution, both domestic and industrial, have a high oxygen demand, a relatively small added input can cause the system to go anaerobic, which means that it becomes, biologically speaking, useless." Quoted in Fallows, *Water Lords*, 225. These sorts of environmental insults, moreover, rendered some species even more vulnerable to other external shocks, such as toxics or bacterial contamination.

90. See ibid., 236–43, esp. table 12.3, which documents reductions in harvests of commercial fish in Georgia. See also G. Robert Lunz, director, South Carolina Wildlife Resources Department, "Comments on the Pollution of South Carolina Marine Waters in the Savannah River Basin" (January 28, 1965), on file with author; and J. David Clem, Georgia Public Health Service, "National Shellfish Sanitation Program Statement of Interest," in U.S. Public Health Service, *Conference on the Pollution of the Interstate Waters of the Lower Savannah.*

91. Cited in Fallows, *Water Lords*, 236.

92. Federal Water Pollution Control Administration, *Second Session of the Conference on the Lower Savannah.*

93. "Union Camp's Savannah Mill Invests $60 Million in Environmental Program," *Southern Pulp and Paper Manufacturer*, October 1975, 13–20. The $60 million total included $21 million for water pollution control ($17 million for the aerated lagoon and $4 million for a clarifier installed in 1968) and $39 million for air pollution control. At the time, this was the largest financial commitment for pollution abatement at a single pulp and paper complex.

94. Peter C. Yeager, *The Limits of Law: The Public Regulation of Private Pollution* (Cambridge: Cambridge University Press, 1991), 130.

95. Federal Water Pollution Control Act Amendments of 1972, Pub. L. No. 92-500, 86 Stat. 816 (1972) codified at 33 U.S.C.S. §§ 1251 et seq. In 1977, when Congress passed yet another set of amendments and renamed the existing law the Clean Water Act, it once again invoked the notion of balance in emphasizing "the protection and propagation of a balanced population of shellfish, fish, and wildlife in the establishment of effluent limitations." Clean Water Act of 1977, Pub. L. No 95-217, 91 Stat. 1566 (1977).

96. Freeman, "Water Pollution Policy," 103–8.

97. See Oliver Houck, "Of Bats, Birds, and B-A-T: The Convergent Evolution of Environmental Law," *Mississippi Law Journal* 63 (1994): 403 (discussing role of technology-based approaches to pollution control in early statutes).

98. The law stipulated that the EPA eventually transfer responsibility for issuing permits to the states once the states had met certain conditions. Freeman, "Water Pollution Policy," 103.

99. The 1972 amendments provided little guidance on toxic water pollution. The 1977 amendments, however, did distinguish between conventional and toxic water pollutants and focused the attention of the EPA on the toxics problem. Freeman, "Water Pollution Policy."

100. Later, in the 1980s, facing growing concern over toxics, the EPA promulgated another set of effluent guidelines for toxic pollutant discharges for the industry. U.S. EPA, *Profile of the Pulp and Paper Industry*, EPA Office of Compliance Sector Notebook Project (Washington, DC: Government Printing Office, 1995), 91–92.

101. See, e.g., Am. Paper Inst. v. Train, 543 F.2d 328 (D.C. Cir. 1976) (upholding effluent limitation regulations for the "unbleached" segment of the pulp and paper industry); and Weyerhaeuser v. Costle, 590 F.2d 1011 (D.C. Cir. 1978) (upholding effluent limitation regulations for the "bleached" segment of the pulp and paper industry).

102. *Pulp & Paper: North American Factbook 1996*, 77.

103. See, e.g., Richard A. Bartlett, *Troubled Waters: Champion International and the Pigeon River Controversy* (Knoxville: University of Tennessee Press, 1995).

104. For a defense of "command and control" regulatory programs to address water pollution, see Drew Caputo, "A Job Half Finished: The Clean Water Act after 25 Years," *Environmental Law Reporter* 27 (November 1997): 10574, 10578–80; Wendy E. Wagner, "The Triumph of Technology-Based Standards," *University of Illinois Law Review* 2000 (2000): 83; and Sidney A. Shapiro and Thomas O. McGarity, "Not So Paradoxical: The Rationale for Technology-Based Regulation," *Duke Law Journal* 40 (1991): 729.

105. See William L. Andreen, "The Evolution of Water Pollution Control in the United States—State, Local, and Federal Efforts, 1789–1972: Part I," *Stanford Environmental Law Review* 22 (2003): 145; Richard L. Revesz, "Rehabilitating Interstate Competition: Rethinking the 'Race-to-the-Bottom' Rationale for Federal Environmental Regulation," *New York University Law Review* 67 (1992): 1210; and Kirsten H. Engel, "State Environmental Standard-Setting: Is There a 'Race' and Is It 'To the Bottom'?" *Hastings Law Journal* 48 (1997): 271.

106. Henry Louis Gates Jr., *Colored People* (New York: Knopf, 1994), 6.

107. Certain types of particulate matter, such as asbestos fibers, can be very toxic in and of themselves. Other particulates, such as soot, can serve as vehicles for more toxic substances. Sulfur dioxide, for example, is far more damaging in the presence of particulate matter, indicating a synergism between these two types of pollutants. SO_2 molecules can "adsorb" onto small particulates, which penetrate further into the lung than SO_2 gas alone. For a discussion, see Ehrlich, Ehrlich, and Holdren, *Ecoscience*, 544–49.

108. For a discussion, see U.S. EPA, *Kraft Pulping: Control of TRS Emissions from Existing Mills* (Washington, DC, 1979).

109. This is not to suggest that the industry was unaware of the potential for health problems associated with air its emissions. See, e.g., H. Bergstrom, "Pollution of Water and Air by Sulfate Mills," *Pulp and Paper Magazine Canada* 54 (November 1953): 135–40 (discussing concentrations of pulp mill gases, such as hydrogen sulfide, sufficient to cause poisoning of mill personnel and even death in some instances).

110. *Air Pollution—1967 (Air Quality Act): Hearings Before the Subcomm. on Air and Water Pollution of the S. Comm. on Pub. Works*, 90th Cong. 2362 (1967).

111. For a brief discussion, see Paul Portney, "Air Pollution Policy," in Portney, *Public Policies for Environmental Protection*, 32–33.

112. Living downwind from the mill had a massive impact on Catherine Rotureau's health: "When the wind blows from the direction of the Union Camp mill, the odor is terrible and the air is so heavy that I can hardly breathe." Quoted in Charles Seabrook, "Pulp, Paper, and Pollution: Mills Pour 30 Million Pounds of Toxins in Air," *Atlanta Constitution*, October 30, 1990.

113. The average concentration of particulates in the neighborhood near the mill was around 160 micrograms per cubic meter, compared to 80 micrograms for downtown Savannah and 35 micrograms for an affluent, white, suburban neighborhood. Figures cited in Fallows, *Water Lords*, 136.

114. This is based on a search of reported cases in the state and federal court reporters. The Florida case discussed above, *Nat'l Container v. Stockton*, 189 So. 4, did include air emissions in addition to water pollution as the alleged source of the public nuisance suffered by plaintiffs.

115. On the inadequacies of common law approaches to air pollution, see William A. Campbell and Milton S. Heath Jr., "Air Pollution Legislation and Regulations," in *Air Pollution*, 3rd ed., vol. 5: *Air Quality Management*, ed. Arthur C. Stern (New York: Academic Press, 1977), 355–59. See also Andrews, *Managing the Environment*, 126–29. For a history of common law approaches to air pollution and the

implications of common law approaches for future regulation, see Noga Morag-Levine, *Chasing the Wind: Regulating Air Pollution in the Common Law State* (Princeton, NJ: Princeton University Press, 2003).

116. For a discussion of early statutory attempts to deal with the air pollution problem, see Jan G. Laitos, "Legal Institutions and Pollution: Some Intersections between Law and History," *Natural Resources Journal* 15 (July 1975). See also Arthur C. Stern, "History of Air Pollution Legislation in the United States," *Journal of the Air Pollution Control Association* 32, no. 1 (1982): 44–47.

117. California actually passed a law in 1947 authorizing counties to regulate air pollution. Oregon's law (1951) was the first such law to provide statewide authority to a particular state agency for air pollution control. By 1960, eight states had passed air pollution control legislation. Ten years later, all fifty states had enacted some form of legislation regarding air pollution, largely because of increased federal pressure. For a review of state efforts, see Stern, "Air Pollution Legislation," 47–48.

118. Congress passed the Air Pollution Control Act in 1955, authorizing funds and technical assistance to assist state and local air pollution control agencies in research and training efforts. Act of July 14, 1955, ch. 360, 69 Stat. 322 (1955). The law was amended in 1959 and 1962. Stern, "Air Pollution Legislation," 48–50. See also J. Clarence Davies III and Barbara S. Davies, *The Politics of Pollution*, 2nd ed. (Indianapolis, IN: Pegasus, 1975), 44. Part of the motivation behind the act stemmed from heightened public concern generated by several high-profile air pollution incidents from the late 1940s and early 1950s, including Donora, Pennsylvania, in 1948, when a heavy industrial smog enveloped the town for four days, leading to the deaths of twenty people and over six thousand illnesses; Los Angeles in 1951; London in 1952; and New York in 1953. Davies and Davies, *Politics of Pollution*, 19–20. Presidential pressure also pushed Congress to act. In January 1955, in both his State of the Union address and in a special health message, President Eisenhower remarked on the growing public concern with air pollution problems and asked Congress to address the issue. Stern, "Air Pollution Legislation," 49.

119. Davies and Davies, *Politics of Pollution*, 45.

120. Stern, "Air Pollution Legislation," 51–52.

121. *Air Pollution: Hearings Before a Subcomm. on Pub. Health and Safety of the H. Comm. on Interstate and Foreign Commerce*, 88th Cong. 1 (1963).

122. U.S. Public Health Service, *Proceedings of the Second National Conference on Air Pollution* (Washington, DC, 1963).

123. *Air Pollution*, 291. See also statement of Dr. E. R. Hendrickson, chairman of Florida Air Pollution Control Commission (123).

124. Ibid., 29, 34, 295.

125. Jean J. Schuenenman, "Air Pollution Problems and Control Programs in the United States" (Paper no. 62-84, U.S. Department of Health, Education, and Welfare, Public Health Service, Cincinnati, OH, April 1962), in *Air Pollution*, 48, 52–53.

126. Per capita expenditures ranged from less than 1 cent to 57 cents per year. Ibid., 53–54.

127. Clean Air Act of 1963, 77 Stat. 392 (1963).

128. Ibid., section 5, 396.

129. S. Rep. No. 88-638 (1963).

130. Field hearings were held in Los Angeles, Denver, Chicago, Boston, New York, and Tampa. *Clean Air: Hearings Before a Special Subcomm. on Air and Water Pollution of the S. Comm. on Pub. Works*, 88th Cong. (1964).

131. See, e.g., the statement submitted by Secretary Gardner of the Department of Health, Education and Welfare and the statement of Dr. William H. Stewart, Surgeon General of the United States, in *Air Pollution—1966: Hearings Before the Subcomm. on Air and Water Pollution of the S. Comm. on Pub. Works*, 89th Cong. 22, 113 (1966).

132. The proposal also called for federal assistance for state automobile inspection systems, a major increase in federal air pollution research, and federal registration of certain motor fuel additives. Davies and Davies, *Politics of Pollution*, 49–50.

133. *Air Pollution—1967 (Air Quality Act): Hearings Before the Subcomm. on Air and Water Pollution of the S. Comm. on Pub. Works*, 90th Cong. 1153 (1967).

134. Fallows, *Water Lords*, 146–47.

135. Ga. Code §§ 88-901 (1933), enacted by 1967 Ga. Laws p. 581, § 1.

136. Fallows, *Water Lords*, 147–48.

137. Cited in Cobb, *Selling of the South*, 230.

138. Kittle delivered these remarks to the 1966 National Conference on Air Pollution. His remarks are excerpted in *Air Pollution—1967*, 1836.

139. Middleton also elaborated on the important distinction between air quality criteria and standards, on the one hand, which prescribed pollutant levels for specific areas, and emission standards, on the other hand, which provided the mechanism for limiting discharges from specific sources in an effort to meet quality standards. *Air Pollution—1967*, 1155.

140. Ibid., 1827.

141. For a discussion of Randolph and his role in shaping the legislation that ultimately passed, see Davies and Davies, *Politics of Pollution*, 50–51.

142. *Air Pollution—1967*, 1829–32.

143. Remarks made to the 1966 National Conference on Air Pollution, in ibid., 1837.

144. American Paper Institute, "Financial Assistance for the Construction of Air Pollution Control Facilities, 1967," in ibid., 1840.

145. Among the proponents of such assistance, Florida senator George Smathers argued that through tax credits and other programs the federal government could "join forces with private enterprise in all-out war on pollution." In his view, the insatiable demand of the consumer was the driving force behind the pollution problem: "To satisfy the demand of our American consumer public, we are literally destroying the very substance we need for existence—air and water. It is my belief that total war against pollution and ultimate victory can be achieved, but it will only be realized when our entire industrial sector wholeheartedly joins the battle." *Air Pollution—1967*, 1349.

146. Davies and Davies, *Politics of Pollution*, 51–53. See also Stern, "Air Pollution Legislation," 52–54.

147. Davies and Davies, *Politics of Pollution*, 53.

148. See Clean Air Amendments of 1970, Pub. L. No 91-604, 84 Stat. 1676 (1970). By the summer of 1970, Senator Muskie, not wanting to be outdone by the president, had abandoned his previous opposition to national standards and embraced an even stronger set of amendments than those proposed by the administration. When the final amendments passed in December, Muskie's proposals were largely intact, despite intense lobbying from interest groups. Davies and Davies, *Politics of Pollution*, 52–56. See also Stern, "Air Pollution Legislation," 55.

149. These included new provisions for national ambient air quality standards 42 U.S.C. § 7409, new source performance standards 42 U.S.C. § 7411, hazardous air pollutants 42 U.S.C. § 7412, and mobile sources 42 U.S.C. §§ 7521 et seq (among others).

150. The NAAQS included primary standards for protecting human health and secondary standards to protect property and other non-health values. Primary standards were to be nationally uniform and were required to be set at a level to protect human health with "an adequate margin of safety." See 42 U.S.C § 7409. For a discussion of these issues and the problems they entail, see Portney, "Air Pollution Policy," 31–36.

151. The deadline was later extended to 1977, then 1982, 1987, and 1988 for some pollutants.

152. 42 U.S.C. § 7410.

153. 42 U.S.C. § 7412 (hazardous air pollutants); 42 U.S.C. §§ 7521–7553 (mobile sources).

154. 42 U.S.C. § 7411.

155. Portney, "Air Pollution Policy," 37–38; 42 U.S.C. § 7411(a)(1).

156. Standards of Performance for Kraft Pulp Mills, 43 Fed. Reg. 7572 (February 23, 1978). See also U.S. EPA, *Profile of the Pulp and Paper Industry*, 88–89. These NSPS standards for Kraft pulp mills were reviewed and revised in April 2014. See Kraft Pulp Mills NSPS Review, 79 Fed. Reg. 18952 (April 4, 2014).

157. Section 111 of the Clean Air Act defines a "performance standard" as a standard that "reflects the degree of emission limitation achievable through the application of the best system of emission reduction which (taking into account the cost of achieving such reduction and any non–air quality health and environmental impact and energy requirements) the Administrator determines has been adequately demonstrated." 42 U.S.C. § 7411(a)(1). This standard is sometimes referred to as the "best demonstrated technology" standard.

158. See 42 U.S.C. § 7411(d) (NSPS provisions for existing facilities); 42 U.S.C. §§ 7470–7479 (NSR provisions for prevention of significant deterioration areas); 42 U.S.C. §§ 7501–7514a (NSR provisions for nonattainment areas).

159. For a discussion of these issues, see Portney, "Air Pollution Policy," 38–39.

160. O.C.G.A. § 12-9-2 (1998).

161. Major sources were defined as those having the potential to emit 100 tons or more per year of any single air pollutant. Georgia Conservancy, *Air Pollution in Savannah: A Report on the State of the Air in Chatham County, Georgia and What Can Be Done to Improve It* (Savannah, GA, 1979), 5.

162. The report referred to the mill as "the major air pollution factor in the area." Ibid.

163. U.S. EPA, National Emission Data Systems, Region IV, January 1979, cited in Georgia Conservancy, *Air Pollution in Savannah*, 7.

164. The problem with using 24-hour intervals for ambient air measurements is that it assumes that conditions (such as rate of emissions, wind direction, etc.) do not change significantly during the period—a questionable assumption in a place such as Savannah where a given monitor may be upwind from a major emission source for part of the day and downwind at other times. Georgia Conservancy, *Air Pollution in Savannah*, 9,11, 14 (quote).

165. Pat Ramsey, "Clean Air Debate Intensifies," *Savannah Morning News*, May 31, 1987.

166. Kraft Pulp Mills; Final Guideline Document, 44 Fed. Reg. 28828 (May 22, 1979) (noticing the final guideline document U.S. EPA, *Kraft Pulping*) The EPA had designated TRS as a "welfare-related" pollutant rather than a "health-related" pollutant and sought to regulate it under the NSPS provisions contained in section 111. According to the agency, "states [would] have substantial flexibility to consider factors other than technology and costs in establishing plans for the control of welfare-related pollutants if they wish." U.S. EPA, *Kraft Pulping*, 1–3.

167. Ibid., 2–8.

168. See Ramsey, "Clean Air Debate."

169. Georgia submitted its plan to control TRS emissions from Kraft pulp mills in 1982. See Georgia—Plan for Control of Designated Pollutants from Existing Facilities [Section 111(d) Plan], 48 Fed. Reg. 31402 (July 8, 1983). But the state did not submit a compliance schedule for sources subject to the plan until 1988. See 40 C.F.R. §62.2600(b)(3).

170. Georgia Conservancy, *Air Pollution in Savannah*, 14.

171. Ibid., 57–62.

172. "An Air Affair," *Georgia Gazette and Journal Record*, December 3, 1979.

173. See "Union Camp Elects Vice President," *PR Newswire*, January 25, 1983 (noting that Governor Busbee joined the Union Camp board of directors on January 25, 1983, after finishing his second term as governor on January 11, 1983).

174. "Group Gets Set to Fight for Air Quality," *Savannah Morning News*, January 15, 1987; Ramsey, "Clean Air Debate." For an insider's history, see John Northup, "History of the Citizens for Clean Air" (unpublished ms., November 9, 1990), on file with author. Much of the information on the Citizens for Clean Air campaign in Savannah was also gathered from interviews conducted by the author with group members, local residents, industry representatives, and government officials during the spring and summer of 1997.

175. "Panel's Challenge: Clean the Air on Dirty Air," *Savannah Evening Press*, November 30, 1986.

176. Quoted in "Group Gets Set to Fight for Air Quality."

177. "Union Camp to Study Sulfur Odors," *Savannah Morning News*, January 30, 1987.

178. "Audubon Chapter Joins Clean Air Coalition," *Savannah Morning News*, February 6, 1987.

179. Quoted in Ramsey, "Clean Air Debate." See also Pat Ramsey, "TRS Emissions at Center of Debate," *Savannah Morning News*, May 31, 1987.

180. Facilities were given until 1992 to comply with the new rules.

181. Charles Craig, "Emissions Cutback Part of Huge Outlay," *Savannah Evening Press*, March 1, 1988. See also Charles E. Swann, "Union Camp Completes Rebuild, Starts Up New Machine at Savannah," *American Papermaker*, April 1991.

182. Craig, "Emissions Cutback."

183. This legislation was part of the Superfund Amendments and Reauthorization Act of 1986, Pub. L. No. 99-499, 100 Stat. 1613 (1986).

184. Chatham County also ranked as the sixteenth most polluted county in the country. Brad Swope, "Chatham Remains High on List of Air Polluters," *Savannah Evening Press*, August 1, 1989.

185. See, e.g., Brad Swope, "Cleaning the City's Air: Under Harsh Criticism as No. 1 Air Polluter, Union Camp Says Controls Well Under Way," *Savannah Evening Press*, August 22, 1989. See also Seabrook, "Pulp, Paper, and Pollution."

186. Initially, the EPA required reporting on over 300 different chemicals. Reporting requirements have changed from year to year, with new chemicals being added and others being deleted. In 1995, the EPA significantly expanded the TRI chemical list to include more than 600 different chemicals. Thus comparisons between years need to be done carefully.

187. The county's two pulp and paper mills, Union Camp and Stone Container, were the primary sources of these chemicals. In 1988, for example, 14.9 million pounds of methanol were released into Chatham County's air, along with 1.6 million pounds of acetone, 767,000 pounds of sulfuric and hydrochloric acid, and 460,000 pounds of toluene. Methanol and acetone are both respiratory irritants. Sulfuric and hydrochloric acids cause corrosion and respiratory damage. And toluene is a solvent capable of causing cell mutations and birth defects. Brad Swope, "Toxic Clouds Above," *Savannah News-Press*, May 6, 1990.

188. Quoted in ibid.

189. Quoted in Charles Seabrook, "Your Toxic Neighbors: Disclosures Spark Improvements," *Atlanta Constitution*, August 22, 1991.

190. Air emissions accounted for almost 90% of Union Camp's total releases in 1997. The 1988 figures are from Charles Seabrook, "Ga. Firms Reducing Toxic Discharges," *Atlanta Constitution*, March 15, 1991. The 1997 figures are from Georgia Environmental Protection Division, *Toxic Release Inventory Report, 1997* (Atlanta: Department of Natural Resources, 1998).

191. See, e.g., Brad Swope, "Air Pollution Levels Drop, Report Says," *Savannah Morning News*, February 6, 1995; Gail Krueger, "Chatham's Air Toxicity Declines, 1994 Report Says," *Savannah Morning News*, April 30, 1996; and Gail Kreuger, "Union Camp Drops to Third on Pollution List," *Savannah Morning News*, May 23, 1997.

192. Interview with author, July 30, 1997.

193. Interview with author, August 13, 1997.

194. Rachel Carson, *Silent Spring* (1962; reprint, New York: Houghton Mifflin, 1994), 6.

195. Al Gore, introduction to ibid., xv.

196. Muskie statement in *Water Pollution Control*, 1–2.

197. For overviews, see Michael Shapiro, "Toxic Substances Policy," in Portney, *Public Policies for Environmental Protection*; and Robert Gottlieb, Maureen Smith, and Julie Rocque, "By Air, Water and Land: The Media-Specific Approach to Toxics Policies," in *Reducing Toxics: A New Approach to Policy and Industrial Decisionmaking*, ed. Robert Gottlieb (Washington, DC: Island Press, 1995), 25–57.

198. Pulp and paper industry representatives had long maintained that the air and water pollution generated by their mills was not toxic and did not pose health hazards. As William H. Chisholm of the American Pulp and Paper Association put it during the 1963 hearings: "I want to point out here that pulp and paper mill wastes are organic in nature and completely nontoxic. They present no threat to human health." *Water Pollution Control and Abatement*, 682. In fact, the toxicity of pulp and paper mill effluent had been a subject of some concern in industry trade journals and in the scientific literature going back to the 1940s. See, e.g., W. A. Chipman, "Physiological Effects of Sulfate Pulp Mill Wastes on Shellfish," *Paper Trade Journal* 127 (September 16, 1948): 47–49 (reporting on experiments finding that pulp mill effluent concentrations of 50 parts per million were sufficient to depress the normal activity of oysters); and Willis M. Van Horn, "Stream Pollution Abatement Studies in the Pulp and Paper Industry," *Transactions of the Wisconsin Academy of Science, Arts and Letters* 39 (August 1949): 105–14 (citing research in the United States and Scandinavia from the late 1930s confirming the existence of "toxic materials in Kraft pulping wastes"). But it was only in the 1970s that considerable interest and attention was directed to the potential toxicity of pulp and paper mill effluent, a development that was instigated in large part by the 1977 amendments to the Clean Water Act, which contained new provisions dealing with toxic water pollutants. See, e.g., P. E. Wrist, "The Clean Water Act of 1977: Its Implications for the Paper Industry," *Paper Trade Journal* 162, no. 8 (April 1978): 34 (noting that the challenge facing industry was to make as much progress in the next five years with toxic pollutants as was made with conventional pollutants in the previous five years). See also C. C. Walden and T. E. Howard, "Toxicity of Pulp and Paper Mill Effluents," *TAPPI* 60, no. 1 (January 1977): 122 (summarizing research on effluent toxicity).

199. The most toxic and best known of the dioxin compounds is 2,3,7,8-tetrachl orodibenzo-p-dioxin or 2,3,7,8 TCDD. When most people speak of "dioxin" they are referring to 2,3,7,8 TCDD, a convention followed here.

200. Bioaccumulation refers to the process whereby the concentration of particular substances or compounds in living tissues increases as one moves up the food chain from one trophic level to another.

201. Mark R. Powell, *Science at EPA: Information in the Regulatory Process* (Washington, DC: Resources for the Future, 1999), 329; Joe Thornton, *Pandora's Poison: Chlorine, Health, and a New Environmental Strategy* (Cambridge, MA: MIT Press, 2000), 57–103; U.S. EPA, *Exposure and Human Health Reassessment of 2,3,7, 8-Tetrachlorodibenzo-p-dioxin (TCDD) and Related Compounds, Part III: Integrated Summary and Risk Characterization for 2,3,7,8-Tetrachlorodibenzo-p-diox in (TCDD) and Related Compounds* (Washington, DC, 2000), 99–158; U.S. EPA, "Health Assessment Document for 2,3,7,8-Tetrachlorodibenzo-p-dioxin (TCDD) and Related Compounds," external review draft, EPA/600/BP-92/001c (Washington, DC, 1994). See also John A. Moore, Renate D. Kimbrough, and Michael Gough, "The Dioxin TCDD: A Selective Study of Science and Policy Interaction," in *Keeping Pace with Science and Engineering*, ed. M. Uman (Washington, DC: National Academy Press, 1993), 221–42; and Resources for the Future, "The Clean Water Act and Lead in Soil at Superfund Mining Sites: Two Case Studies of EPA's Use of Science," Discussion Paper no. 97-08 (Washington, DC: Resources for the Future, 1997), 1–2.

202. In 1997 the International Agency for Research on Cancer classified 2,3,7,8-TCDD, the best studied member of the dioxin family, as a known human carcinogen. See U.S. EPA, *The Exposure and Human Health Reassessment of 2,3,7,8-Tetrach lorodibenzo-p-dioxin (TCDD) and Related Compounds* (Washington, DC, 2003); and National Research Council, *Health Risks from Dioxins and Related Compounds, Evaluation of the EPA Reassessment* (Washington, DC: National Academies Press, 2006).

203. U.S. EPA, "Health Assessment Document for 2,3,7,8-Tetrachlorodibenzo-p -dioxin (TCDD) and Related Compounds," 9.43-51. On dioxin's potential as an endocrine disruptor, see Theo Colburn, Dianne Dumanoski, and John Peterson Myers, *Our Stolen Future* (New York: Plume, 1997), 110–21; and Thornton, *Pandora's Poison*, 85–93.

204. The study was formally released in 1987. U.S. EPA, *National Dioxin Study—Report to Congress*, EPA/530-SW-87-025, Office of Solid Waste and Emergency Response (Washington, DC, 1987).

205. *Dioxin Contamination of Food and Water: Hearings Before the Subcomm. on Health and Environment of the H. Comm. on Energy and Commerce*, 100th Cong. 157 (1989) (statement of Charles L. Elkins, director of the EPA's Office of Toxic Substances). According to the EPA, measured background levels of dioxin in fish were 0-2 parts per trillion (ppt). Dioxin concentrations of more than 50 ppt were found in fish in a Wisconsin reservoir and concentrations of up to 85 ppt were found in samples from Maine and Minnesota. Powell, *Science at EPA*, 357 n3. According to the toxicological profile for chlorinated dibenzo-p-dioxins (CDDs) from the Agency for Toxic Substances and Diseases Registry, the FDA recommends against eating fish and shellfish with levels of 2,3,7,8,-TCDD greater than 50 ppt. See U.S. Department of

Health and Human Services, *Toxic Substances and Disease Registry, Toxicological Profile for Chlorinated Dibenzo-p-Dioxins* (Atlanta, GA, 1998), 16.

206. *Dioxin Contamination*, 159–61 (Elkins statement). Part of the reason for the "unexpected" nature of this discovery stemmed from the fact that dioxin formation was (and is) only partially understood. Moreover, the technology capable of measuring concentrations at the parts per trillion level was not widely available until this time.

207. U.S. EPA, *U.S. Environmental Protection Agency / Paper Industry Cooperative Dioxin Screening Study*, EPA-440/1-88-025 (Washington, DC, 1988).

208. Furans, or polychlorinated dibenzofurans (PCDFs), are a set of related compounds that have a similar structure to dioxin. Ibid., vi–ix, 135–39 (summarizing results of the five-mill study).

209. Powell, *Science at EPA*, 353.

210. U.S. EPA, *National Dioxin Study*.

211. Philip Shabecoff, "Traces of Dioxin Found in Range of Paper Goods," *New York Times*, September 24, 1987.

212. Ibid.

213. During the late 1980s and early 1990s, when the dioxin controversy was at its peak, there were 104 bleaching mills throughout the country. U.S. EPA, *Summary Report on USEPA / Paper Industry Cooperative Dioxin Study: The 104 Mill Study* (Washington, DC, 1990).

214. Michael Weiskopf, "Paper Industry Campaign Defused Reaction to Dioxin Contamination," *Washington Post*, October 25, 1987. See also *Dioxin Pollution in the Pigeon River, North Carolina and Tennessee: Hearing Before the Subcomm. on Water Res. of the H. Comm. on Pub. Works and Transp.*, 100th Cong. 137–40, 228–327 (1988) (statement of Greenpeace representative Carol Van Strum regarding the various EPA, API, and court documents, including depositions, that she included with her statement and that supported her claims of collusion between the EPA and API).

215. Quoted in Weiskopf, "Paper Industry Campaign."

216. Ibid.

217. Carol Van Strum v. EPA, 680 F. Supp. 349 (D.Or. 1987).

218. The study was released in July 1990. See U.S. EPA, *Summary Report*; *Dioxin Contamination*, 162–63 (Elkins statement).

219. The 1988 consent decree came out of a lawsuit filed by the EDF and NWF against the EPA in 1985 after the agency denied a citizens petition the organizations had filed under the Toxic Substances Control Act (TSCA) in 1984 requesting that the EPA regulate dioxins and furans from all known sources. At the time, bleaching pulp mills were not recognized as a source of these substances. See EDF v. Thomas, 657 F. Supp. 302 (D.C. 1987). By the time of the consent decree, of course, bleaching mills had been identified as a source of dioxins and furans, hence the requirement that the EPA conduct the 104-mill study. The consent decree also required that the agency propose regulations under TSCA to control the disposal of pulp mill residual sludges and under the Clean Water Act to deal with discharges of dioxins and furans into surface waters. This proposal became part of a larger set of rules regarding water effluent and emissions standards that has come to be known

as the "cluster rule." See Powell, *Science at EPA*, 330. The final "cluster rule" was released in April 1998. See National Emissions Standards for Hazardous Air Pollutants for Source Category: Pulp and Paper Production; Effluent Limitations Guidelines, Pretreatment Standards, and New Source Performance Standards: Pulp, Paper, and Paperboard Category [Cluster Rule], 63 Fed. Reg. 18504 (April 15, 1998).

220. For a detailed history, see Bartlett, *Troubled Waters.*

221. *Dioxin Pollution in the Pigeon River, North Carolina and Tennessee: Hearings Before the Subcomm. on Water Res. of the H. Comm. on Pub. Works and Transp.*, 100th Cong. 5, 3 (1988).

222. Ibid., 137–40, 228–327 (statement of Carol Van Strum).

223. Ibid., 9–77 (testimony, statements, and questioning of EPA officials Bruce Barrett and Martha Prothro).

224. Ibid., 81, 329.

225. Ibid., 5.

226. Ibid., 353. See also the statements from other local citizens (328–71).

227. Ibid., 371–98 (testimony and statements of Red Cavaney, president of API, and Richard Diforio, vice president at Champion).

228. *Dioxin Contamination of Food and Water: Hearings Before the Subcomm, on Health and Environment of the H. Comm. on Energy and Commerce*, 100th Cong. 157 (1989).

229. Ibid., 1–2.

230. Ibid., 3, 5, 14. Janet Heiber of Greenpeace offered testimony on alternatives to chlorine bleaching then being used in Scandinavia (142–44). See also U.S. Office of Technology Assessment, *Technologies for Reducing Dioxin in the Manufacture of Bleached Wood Pulp*, OTA-BP-0-54 (Washington, DC, 1989).

231. See "Interim Strategy for the Regulation of Pulp and Paper Mill Dioxin Discharges to Waters of the United States," memorandum from Rebecca W. Hammer, EPA Acting Administrator for Water, to Water Management Division Directors and NPDES States Directors, August 9, 1988, cited in Kimberly M. Thompson and John D. Graham, "Producing Paper without Dioxin Pollution," in *The Greening of Industry: A Risk Management Approach*, ed. John D. Graham and Jennifer Kassalow Hartwell (Cambridge, MA: Harvard University Press, 1997), 229.

232. Thompson and Graham, "Producing Paper without Dioxin," 229–33.

233. The guidance criteria of 0.013 ppq came from the EPA's *Ambient Water Quality Criteria for 2,3,7,8-Tetrachlorodibenzo-p-dioxin* (1984). In that document, which summarized the science on dioxin toxicity up to that point, the agency recommended the water quality standard of 0.013 ppq for waters used as a source of drinking water and edible fish. See NRDC v. EPA, 16 F.3d 1395, 1399 (4th Cir. 1993). See also "Board of Natural Resources Adopts Dioxin Standard," *PR Newswire*, March 29, 1990.

234. Charles Seabrook, "Georgia Board of Natural Resources Tightens Dioxin Standard," *Atlanta Journal & Constitution*, January 24, 1991.

235. Background information on the development of the Maryland and Virginia standards can be found in *NRDC v. EPA*, 16 F.3d at 1398–99, in which the Fourth

Circuit Court of Appeals rejected the NRDC's challenge to the EPA's approval of the Maryland and Virginia dioxin standards.

236. Ibid. at 1399.

237. Thompson and Graham, "Producing Paper without Dioxin," 230–33.

238. The final summary report was released in July 1990: U.S. EPA, *Summary Report.* The draft fish study was released in 1989: EPA, "Bioaccumlative Pollutants in Fish—A National Study," Draft Report, Office of Water Regulations and Standards (Washington, DC, 1989). The final report was released in 1992: EPA, *National Study of Chemical Residues in Fish*, EPA 823-R-92-008a (Washington, DC, 1992). The overall conclusion that emerged from the 104-mill study and the broader study of chemical residues in fish was that dioxins in pulp and paper mill wastewater effluent represented the greatest public health concern. Pulp mill sludge, which was sometimes used as fertilizer on crop lands, was identified as a secondary source of health concern but as a primary source of ecological concern. The EPA did not consider dioxin in pulp and paper products to be a major risk but rather "on the borderline of concern" and referred the issue to the FDA. Powell, *Science at EPA*, 355.

239. See "2 EPA Studies Confirm Threat to Fish of Dioxin From Paper Plants," *New York Times*, March 14, 1989.

240. "EPA Releases Risk Assessments for Eating Dioxin-Contaminated Fish," EPA Press Office Release R-158, Washington, DC, September 24, 1990.

241. Dioxin concentration in the effluent at the Georgetown mill was 0.64 parts per trillion. The median concentration for the mills where dioxin was measurable was 0.024 parts per trillion. "2 EPA Studies"; Peter Applebome, "Town Agonizes over Dioxin Levels," *New York Times*, September 19, 1989.

242. U.S. EPA, *Risk Assessments for Eating Dioxin-Contaminated Fish* (Washington, DC, 1990). See also Michael Weisskopf, "EPA Seeking to Reduce Dioxin in White Paper; Cancer Risk Said to Justify Mill Restrictions," *Washington Post*, May 1, 1990; and Philip Shabecoff, "Government Says Dioxin Poses No Major Danger," *New York Times*, May 1, 1990. Shabecoff noted that EPA and FDA officials had concluded that dioxin levels in bleached pulp and paper products were so low as to present a "negligible" health risk but that individuals who ate significant quantities of fish downstream from certain pulp and paper mills did have an elevated cancer risk.

243. Quoted in Applebome, "Town Agonizes."

244. Quoted in "International Paper Challenges EPA's Dioxin Risk Assessments," *Business Wire*, September 25, 1990. Phillips also expressed frustration at the fact that the EPA did not acknowledge the progress it had made at its Georgetown plant in reducing the dioxin levels in the mill's effluent. See also George Lobsenz, "Paper Companies Dispute EPA Dioxin Risk Study," United Press International, September 27, 1990.

245. Quoted in "International Paper Challenges EPA's Dioxin Risk Assessments."

246. Much of the initial controversy emerged from the 1990 Banbury conference, where a new "consensus" was reached on the biological action of dioxins and dioxin-like compounds at the molecular level. Drawing on research conducted since the mid-1980s, conference participants agreed that dioxin-like compounds gain en-

try into cells by binding to a particular protein on cells known as the "Ah receptor." Hypothesizing that there was a threshold level of dioxin exposure necessary to initiate the cascade of effects leading to cancer, some participants concluded that low-level exposure to dioxin (such as that existing at background levels in the environment) posed a negligible cancer risk. Others contested the notion that such biologically based models could be used to predict the cancer risks associated with dioxin. There was no agreement, in other words, on what the new understanding of the receptor-based mechanism of action meant in terms of human health risks. For a discussion of the Banbury conference, see Leslie Roberts, "Dioxin Risks Revisited," *Science* 251 (1991): 624–26.

247. The Banbury conference did lend credibility to those inside and outside of the EPA who had been calling for a reassessment of the dioxin cancer potency criteria used by the agency. In 1988, for example, a group of senior EPA scientists concluded that "reliance on the linearized model [of dose-response] may be less appropriate for TCDD than for many other chemicals, . . . and the model may overestimate the upper-bound on the risk by some unknown amount." John A. Moore, Renate D. Kimbrough, and Michael Gough, "The Dioxin TCDD: A Selective Study of Science and Policy Interaction," in *Keeping Pace with Science and Engineering*, ed. M. Uman (Washington, DC: National Academy Press, 1993), 229. In January 1991, arguments in favor of a reassessment received additional support from a study of more than 5,000 American chemical workers who had been exposed to dioxin on the job. This study suggested that dioxin, at low levels of exposure, might not be as potent a carcinogen among humans as previously thought. The study was published in the *New England Journal of Medicine* and was directed by Marilyn Fingerhut of the National Institute for Occupational Safety and Health. It included virtually all chemical workers in the United States who were exposed to dioxin between 1942 and 1984 and was one of the largest such studies ever undertaken. For those men with high levels of exposure for more than one year (and whose exposure occurred at least twenty years prior to 1987), the death rate from cancer was 46% higher than would have been expected. For those with lower levels of exposure, however, the report found a cancer mortality rate that was equivalent to the general population. M.A. Fingerhut et al., "Cancer Mortality in Workers Exposed to 2,3,7,8-Tetrachlorodibenzo-p-dioxin," *New England Journal of Medicine* 324 (1991). For press reaction, see Karen Klinger, "Dioxin Does Not Appear Potent Cancer Agent," United Press International, January 24, 1991. Here, it seemed, was the epidemiological data necessary to bolster the Banbury findings regarding the biological action of dioxin and the possibility that there might be a safe "threshold" dose.

248. Quoted in "U.S. to Review Dioxin Risk Given New Studies," *New York Times*, April 16, 1991.

249. Quoted in ibid.

250. Quoted in Keith Schneider, "U.S. Backing Away from Saying Dioxin Is a Deadly Peril," *New York Times*, August 15, 1991.

251. Ibid.

252. Ibid.

253. As defined at the EPA, risk assessment proceeds in four steps: (1) hazard assessment, which determines whether a particular agent is linked causally with a given health effect; (2) dose-response assessment, which seeks to determine the quantitative relation between exposure to a certain dose level of the agent in question and the incidence and/or severity of a response in test animals, which then provides a basis for making inferences for humans; (3) exposure assessment, which seeks to determine the population(s) exposed to the agent in question and the routes, magnitudes, frequencies, and durations of exposure; and (4) risk characterization, which attempts to characterize or summarize the nature of the risk in light of uncertainties involved and major assumptions employed. In principle, risk assessment is supposed to be separate from risk management, which is the process by which a regulatory agency such as the EPA decides what to do with the results of a risk assessment. In practice, however, such a separation is not always possible. See Milton Russell and Michael Gruber, "Risk Assessment in Environmental Policy-Making," *Science* 236 (1987); and Richard Andrews, "Risk-Based Decision Making," in *Environmental Policy in the 1990s*, 2nd ed., ed. Norman J. Vig and Michael E. Kraft (Washington, DC: Congressional Quarterly Press, 1994).

254. For important EPA statements on risk assessment and its relationship to the agency's mission, see U.S. EPA, *Unfinished Business: A Comparative Assessment of Environmental Problems* (Washington, DC, 1987); and EPA Scientific Advisory Board, *Reducing Risk: Setting Priorities and Strategies for Environmental Protection* (Washington, DC, 1990). See also Sheila Jasanoff, "Science, Politics, and the Renegotiation of Expertise at EPA," *OSIRIS* 7 ser. 2 (1992); and Andrews, *Managing the Environment*, 268–70. For a discussion of the rise of economic analyses at the EPA and their relation to risk assessment, see the collection edited by Richard D. Morgenstern, *Economic Analyses at EPA: Assessing Regulatory Impact* (Washington, DC: Resources for the Future, 1997). Of course, there were (and are) problems associated with the practice of risk assessment. To begin with, any attempt to assess the risks and calculate expected losses from some sort of environmental disturbance, such as dioxin contamination, entails assumptions about physiological, toxicological, and ecological processes (causal mechanisms) that are often not well understood. Significant uncertainties are also encountered in any effort to extrapolate between high-dose exposure and low-dose exposure and between responses in animals and responses in humans. Moreover, accommodating latent effects and latency periods, special sensitivities in exposed populations, synergistic effects with other substances, and temporal and spatial variability in exposure levels threatens to overwhelm any claims of being "scientific." Reducing risks to a common calculus (say excess cancer deaths) obviously leaves out or downplays other sorts of health risks (immune system impairment, neurotoxicity, etc.) that may not lend themselves as easily to quantification, not to mention ecological effects. The purely aggregate approach to risk, by focusing on averages, also tends to ignore the maximally exposed population. Finally, different agencies, and even different divisions

within the same agency, often employ different models, assumptions, and approaches to risk assessment. For criticisms of risk assessment, see Donald T. Hornstein, "Reclaiming Environmental Law: A Normative Critique of Comparative Risk Analysis, *Columbia Law Review* 92 (1992): 571–72; and Howard Latin, "Good Science, Bad Regulation, and Toxic Risk Assessment," *Yale Journal on Regulation* 5 (1988): 91–92. In his history of environmental policy, Richard Andrews has criticized the risk assessment approach at EPA as an attempt to use "scientific knowledge to mask fundamentally political decisions, and to allow policy to be controlled by an EPA subgovernment rather than by a broader political process" (*Managing the Environment*, 269). For a history of the origins and development of quantitative risk assessment in U.S. health, safety, and environmental law, see William Boyd, "Genealogies of Risk: Searching for Safety, 1930s-1970s," *Ecology Law Quarterly* 39 (2012): 895.

255. The quote came from a 1983 article by Ruckelshaus in *Science*. Quoted in Russell and Gruber, "Risk Assessment," 286.

256. For a discussion of some of the early lawsuits, particularly in Mississippi, see Nicholas Varchavar, "Muddy Waters," *American Lawyer*, July–August 1993, 52. See also Paul Kemezis, "Lawsuits Could Mount for Papermakers," *Chemical Week*, February 13, 1991, 24.

257. Procedural history of the initial lawsuit by Wesley Simmons is discussed in Leaf River Forest Products, Inc. v. Simmons, 697 So. 2d 1083 (Miss. 1996).

258. On discovering dioxin in the effluent and sludge produced by the Leaf River mill, the Mississippi Department of Wildlife and Fisheries closed the Leaf, Pascagoula, and Escatawpa Rivers (all of which received waste water from the Leaf River mill) to commercial fishing from October 1990 to January 1991 and issued consumption advisories for fish caught from the Leaf and Pascagoula Rivers. The state subsequently lifted the consumption advisory for the Pascagoula River, in December 1990, but not for the Leaf River. For background, see Leaf River Forest Products, Inc. v. Ferguson, 662 So. 2d 648, 650–51 (Miss. 1995).

259. Simmons's claims are discussed in *Simmons*, 697 So. 2d at 1083–86. See also Varchavar, "Muddy Waters," 57–59.

260. Varchavar, "Muddy Waters," 55, 57–58.

261. Figures are reported in *Simmons*, 697 So. 2d at 1083.

262. Kemezis, "Lawsuits Could Mount for Papermakers," 24.

263. Ibid.

264. Varchavar, "Muddy Waters," 59.

265. J. A. Shults & Joan Shults v. Champion Int'l Corp., 821 F. Supp. 517 (E.D. Tenn. 1992). For a discussion of the case, see Bartlett, *Troubled Waters*, 265–75.

266. Bartlett, *Troubled Waters*, 265–66.

267. Lead plaintiffs were Thomas and Bonnie Jane Ferguson. For background, see *Leaf River Forest Products, Inc. v. Ferguson*, 662 So. at 648, 650–51.

268. Varchavar, "Muddy Waters," 69.

269. Ibid., 61, 63–64.

270. "$100 Billion Suit Accuses Paper Firms of Dioxin Poisonings," *Los Angeles Times*, Mar. 6, 1992. See also "Class-Action Lawsuit Dropped," *Houston Chronicle*, April 3, 1992.

271. Bartlett, *Troubled Waters*, 265–75.

272. Ibid.

273. Ibid., 273–75.

274. "$6.5 Million Settlement Announced in Dioxin Case," *Mealey's Litigation Reports: Toxic Torts* 2, no. 7 (1993).

275. "$5 Million Settlement Approved in Dioxin Property Damage Action," *Mealey's Litigation Reports: Emerging Toxic Torts* 5, no. 4 (1996).

276. *Leaf River Forest Products, Inc. v. Ferguson*, 662 So. 2d at 648. In rejecting claims for both intentional and negligent infliction of emotional distress, the court stated emphatically that it had "never allowed or affirmed a claim of emotional distress based on a fear of contracting a disease or illness in the future, however reasonable" (at 658). Pointing to the failure of the plaintiffs to offer any evidence that they had actually been exposed to dioxin from the Leaf River mill, the court refused to create any such tort out of the facts of the Ferguson case (at 657–60). See also "Emotional Distress Claims Rejected in Mississippi Exposure Action," *Mealey's Litigation Reports: Toxic Torts* 4, no. 15 (1995).

277. "Mississippi Judge Dismisses Last of Dioxin Suits against International Paper Co.," *Toxic Chemicals Litigation Reporter*, August 6, 1996.

278. *Leaf River Forest Products, Inc. v. Simmons*, 697 So. 2d at 1083. Here the court relied on its holding in the Ferguson case to dismiss the plaintiff's emotional distress and nuisance claims. In rejecting Simmons's trespass claims, the court found that Simmons had failed to meet the burden of proving an actual physical invasion of his property by dioxin discharged from a mill located 40 miles upstream (at 1085–86). See also "Second Dioxin Decision by MS Supreme Court Favors Wood Pulp Mill," *Hazardous Waste Litigation Reporter*, January 10, 1997.

279. See, e.g., Prescott v. Leaf River Forest Products, Inc., 740 So. 2d 301 (Miss. 1999) (affirming summary judgment for defendants); Herrington v. Leaf River Forest Products, Inc., 733 So. 2d 744 (Miss. 1999) (same); Angaldop v. Leaf River Forest Products, Inc., 716 So. 2d 543 (Miss. 1998) (same).

280. For example, in *Ferguson*, Georgia-Pacific personnel testified that the company had made a concerted effort to reduce dioxin pollution from the Leaf River mill during the early 1990s largely through substitution of chlorine dioxide for chlorine as the principal bleaching agent used at the mill. *Leaf River Forest Products, Inc. v. Ferguson*, So. 2d at 654 (recounting testimony of Georgia-Pacific employees Warren Richardson and Acker Smith). For a discussion of the difficulties involved in using the regulatory compliance defense in the products area, see Robert L. Rabin, "Reassessing Regulatory Compliance," *Georgetown Law Journal* 88 (2000): 2049.

281. For a discussion of bleaching technologies and the benefits of substituting chlorine dioxide for elemental chlorine as a bleaching agent, see U.S. EPA, *Profile of the Pulp and Paper Industry* (Washington, DC, 1995), 30–35; and U.S. EPA, *Summary of Technologies for the Control and Reduction of Chlorinated Organics from the*

Bleached Chemical Pulping Subcategories of the Pulp and Paper Industry (Washington, DC, 1990).

282. Figures are for 2,3,7,8 TCDD only and are compiled in U.S. EPA, *Sources of Dioxin-like Compounds in the United States* (Washington, DC, 2000), table 8.3. According to the American Forest & Paper Association, the new industry trade association created through the merger of the American Paper Institute and the American Pulp and Paper Association, the industry spent some $1 billion between 1988 and 1992. Cindy Skrzycki, "Pulp Friction: The EPA's Tussle over Paper Pollutant Rules," *Washington Post*, April 11, 1997.

283. The proposed "cluster rule" was published in the *Federal Register* in December 1993. See Effluent Limitation Guidelines, Pretreatment Standards, and New Source Performance Standards: Pulp, Paper, and Paperboard Category; National Emission Standards for Hazardous Air Pollutants for Source Category: Pulp and Paper Production, 58 Fed. Reg. 66,078 (Dec. 17, 1993). The proposed TCF requirement for papergrade sulfite mills is at 58 Fed. Reg. 66,114 to 66,115. The bleaching requirements for other mills are at 66,110 to 66,114. EPA's estimates of compliance costs, mill closures, and job impacts for the entire rule are at 66,153 to 66,154. EPA's estimate of baseline dioxin and furan discharges in the industry's wastewater of 410 grams/year is at 66,101. EPA's estimates of the reductions in such discharges (354 grams/year) based on its proposed rules are at 66,130 to 66,131. See also Kelly H. Ferguson and Kirk J. Finchem, "The 'Cluster Rule' Continues: Industry Asks, What Now?" *Pulp & Paper* 71, no. 11 (November 1997): 39–41.

284. Bill Nichols, "EPA's Proposed Cluster Rules Shape U.S. Paper Industry's Near Future," *Pulp & Paper* 68, no. 9 (September 1994); Ferguson and Finchem, "'Cluster Rule' Continues," 39–41.

285. Quoted in "Paper Industry Braces for Costly Anti-Pollution Rules," *Atlanta Journal & Constitution*, October 29, 1993.

286. Ferguson and Finchem, "'Cluster Rule' Continues," 40.

287. U.S. EPA, "Health Assessment Document."

288. On the issue of dioxin's carcinogenicity, the draft report concluded, "In summary, publication of additional studies of human populations exposed to dioxin and related compounds since the last EPA assessment has strengthened the inference, based on all the evidence from mechanistic, animal, and epidemiologic studies, that these compounds are appropriately characterized as probable human carcinogens." Ibid., 9.42-51.

289. This was the equivalent of 40–60 picograms of dioxin per gram of lipid.

290. U.S. EPA, "Health Assessment Document," 9.76, 9.85.

291. Ibid., 9.18-22.

292. Environ Dioxin Risk Characterization Expert Panel, "EPA Assessment Not Justified," *Environmental Science & Technology* 29, no. 1 (1995): 31–32.

293. Quoted in Sharon Begley and Mary Hager, "Don't Drink the Dioxin," *Newsweek*, September 19, 1994.

294. Richard Clapp, Peter deFur, Ellen Silbergeld, and Peter Washburn, "EPA on the Right Track," *Environmental Science and Technology* 29, no. 1 (1995): 29–30.

295. U.S. EPA Science Advisory Board, *Dioxin Reassessment Review* (Washington, DC, 1995), 1–6.

296. See U.S. EPA, *Exposure and Human Health Reassessment, Part III*, 99–158.

297. U.S. EPA, "EPA's Science Plan for Activities Related to Dioxins in the Environment" (2013), http://cfpub.epa.gov/ncea/cfm/recordisplay.cfm?deid=209690. More information on the EPA's various efforts related to the dioxin risk assessment is available at www.epa.gov/dioxin/.

298. The first, supported by industry, called for complete substitution of chlorine dioxide for chlorine. The second, which was the EPA's original 1993 proposal, called for complete substitution combined with oxygen delignification. Based on data collected from various mills employing these two processes, the agency noted that once the new rule went into effect, dioxin levels in pulp mill effluent would be reduced by 95% and 99% respectively (depending on which option was chosen). In a departure from previous regulatory models, the EPA also suggested that its final rule would contain a voluntary incentives program aimed at rewarding those mills that exceeded regulatory requirements in reducing discharges. Powell, *Science at EPA*, 362. This was a welcome development for those mills that had already installed oxygen delignification processes and were thus worried that they might end up at a competitive disadvantage if that process was not part of the new rule.

299. For a discussion, see Maureen Smith, *The U.S. Paper Industry and Sustainable Production: An Argument for Restructuring* (Cambridge, MA: MIT Press, 1997), 120–32. See also Thornton, *Pandora's Poison*, 378–83.

300. Cindy Skrzycki, "Pulp Friction: The EPA's Tussle over Paper Pollutant Rules," *Washington Post*, April 11, 1997; John H. Cushman Jr., "EPA Seeks Cut in Paper-Mill Pollution, But Not Elimination," *New York Times*, May 21, 1997.

301. The final rule was published in April 1998: Cluster Rule, 63 Fed. Reg. 18504 (April 15, 1998).

302. U.S. EPA, "EPA Eliminates Dioxin, Reduces Air and Water Pollutants from Nation's Pulp and Paper Mills," press release, November 14, 1997. While the title of the press release suggested that the rule would eliminate dioxin, the text stated more accurately that it would "virtually eliminate dioxin discharges into waterways." See also U.S. EPA, *EPA's Final Pulp, Paper and Paperboard 'Cluster Rule'—Overview*, Office of Water (Washington, DC, 1997).

303. "EPA Orders $1.8 Billion Plan to Clean Up Mill's Discharges," *Washington Post*, November 15, 1997.

304. Quoted in Cathy Cooper, "For Pulp and Paper Mills, Reasonable Cluster Rules," *Chemical Engineering* 105, no. 1 (1998).

305. Quoted in "EPA Orders $1.8 Billion Plan."

306. Powell, *Science at EPA*, 366.

307. Quoted in John Holusha, "Pulp Mills Turn Over a New Leaf," *New York Times*, March 9, 1996.

308. Thompson and Graham, "Producing Paper without Dioxin Pollution," 254–59.

309. Robert D. Bullard, *Dumping in Dixie: Race, Class, and Environmental Quality* (Boulder, CO: Westview, 1990), 117.

310. Although the issue has been around in various forms for some time, environmental justice did not really emerge as a full-blown national issue until the 1980s. Building on an earlier government report on the siting of hazardous waste facilities in the southeastern United States, the United Church of Christ Commission on Racial Justice published a report in 1987 that put the issue squarely on the national agenda, positing race as the most significant factor in determining the location of commercial hazardous waste facilities. See United Church of Christ Commission on Racial Justice, *Toxic Wastes and Race in the United States: A National Report on the Racial and Socioeconomic Characteristics of Communities with Hazardous Waste Sites* (New York, 1987). The basic contention of the environmental justice movement is that poor and minority communities experience greater exposure to environmental hazards than wealthier and/or white communities at least in part because of discrimination on the basis of race and/or class. Much of the empirical work used to support the claims of the movement has focused on the siting and location of hazardous waste facilities and other locally undesirable land uses. There is considerable and ongoing controversy regarding the underlying causes of environmental inequities. Not surprisingly, the literature on environmental justice is large and growing. For a sample, see Bullard, *Dumping in Dixie*; Bunyan Bryant and Paul Mohai, eds., *Race and the Incidence of Environmental Hazards: A Time for Discourse* (Boulder, CO: Westview Press, 1992); Richard J. Lazarus, "Pursuing 'Environmental Justice': The Distributional Effects of Environmental Protection," *Northwestern University Law Review* 87 (1992): 787; and Vicki Been, "Locally Undesirable Land Uses in Minority Neighborhoods: Disproportionate Siting or Market Dynamics," *Yale Law Journal* 103 (1994): 1383. For statements from the federal government, see U.S. EPA, *Environmental Equity: Reducing Risk for All Communities*, Office of Policy, Planning, and Evaluation (Washington, DC, 1992); and U.S Council on Environmental Quality, "Environmental Justice," chap. 6 in *Environmental Quality, 1994–1995*, 25th Anniversary Report (Washington, DC, 1996).

311. George B. Tindall, *Emergence of the New South, 1913–1945* (Baton Rouge: Louisiana State University Press, 1967).

312. See, e.g., the differing views espoused in Bullard, *Dumping in Dixie*, and Been, "Locally Undesirable Land Uses."

313. Lazarus, "Pursuing 'Environmental Justice'"; Richard Lazarus, "Fairness in Environmental Law," *Environmental Law* 27 (1997): 705. See also Rob Nixon, *Slow Violence and the Environmentalism of the Poor* (Cambridge, MA: Harvard University Press, 2011).

314. This information was compiled from 1990 U.S. Census Data summary tape files 1a and 3a, available at www.census.gov.

Chapter 5 · New South, New Nature

Epigraph. Rupert Vance, *Human Geography of the South*, 2nd ed. (Chapel Hill: University of North Carolina Press, 1935), 511.

1. See, e.g., Rupert Vance, *All These People: The Nation's Human Resources in the South* (Chapel Hill: University of North Carolina Press, 1945), 196–97; and Rupert Vance, "The Old Cotton Belt," in *Regionalism and the South: Selected Papers of Rupert Vance*, ed. John Shelton Reed and Daniel Joseph Singal (Chapel Hill: University of North Carolina Press, 1982), 120–21.

2. Cf. Alfred D, Chandler Jr., *The Visible Hand: The Managerial Revolution in American Business* (Cambridge, MA: Belknap Press of Harvard University Press, 1977). For an important corrective to the notion that the "mass production" paradigm represented the dominant organizational form during the late nineteenth and early twentieth centuries, see Philip Scranton, *Endless Novelty: Specialty Production and American Industrialization, 1865–1925* (Princeton, NJ: Princeton University Press, 1997).

3. Karl Marx, "Economic and Philosophical Manuscripts," in *Early Writings* (Harmondsworth, UK: Penguin, 1992), 355.

4. This is not to suggest that a focus on markets and the commodification of nature is somehow unimportant. Indeed, such a perspective has been central to some of the most important works in environmental history. See, e.g., William Cronon, *Nature's Metropolis: Chicago and the Great West* (New York: W.W. Norton, 1991).

5. Obviously, all industries are ultimately "nature-based" and, as David Harvey reminds us, all processes of industrialization are ecological or socio-ecological projects. See David Harvey, *Justice, Nature, and the Geography of Difference* (Oxford: Blackwell, 1996). Yet, some industries are obviously more directly engaged in the transformation of biophysical properties and processes. See William Boyd, W. Scott Prudham, and Rachel A. Schurman, "Industrial Dynamics and the Problem of Nature," *Society and Natural Resources* 14 (2001): 555.

6. Recent work in economic geography, however, argues for a view of space and time as central *strategic* concerns for corporate managers rather than as external constraints. Erica Schoenberger makes the point explicit, suggesting that the "management of temporal processes is always caught up in the management of spatial processes. Both are invested with strategic value, and both are the subject of social conflict and shifting distributions of social power." Erica Schoenberger, "The Management of Time and Space," in *Oxford Handbook of Economic Geography*, ed. Gordon L. Clark, Meric S. Gertler, and Maryann P. Feldman (Oxford: Oxford University Press, 2000), 324. For a more extended treatment, see Erica Schoenberger, *The Cultural Crisis of the Firm* (Oxford: Basil Blackwell, 1997). Schoenberger's work has antecedents in economic geography. See, in particular, David Harvey, *The Limits to Capital* (Oxford: Basil Blackwell, 1982); David Harvey, *The Condition of Postmodernity* (Oxford: Basil Blackwell, 1990); and Michael Storper and Richard Walker, *The Capitalist Imperative: Territory, Technology, and Industrial Growth* (Oxford: Blackwell, 1989).

7. Marx used the term *Stoffweschel* to characterize the labor process as a metabolic interaction between society and nature. Karl Marx, *Capital*, vol. 1 (1867;

reprint, New York: International, 1967), 173–74. For a discussion, see John Bellamy Foster, "Marx's Theory of Metabolic Rift: Classical Foundations for Environmental Sociology," *American Journal of Sociology* 105, no. 2 (1999). This is somewhat different from the more popular view of industrial metabolism advocated by Robert Ayres, which emphasizes the material flows, mass balances, and residuals associated with industrial production. While Ayres advocates a view of pollution that goes beyond the traditional economic conception of externalities, he tends to leave out politics altogether. For a review of the history of the concept, see Mariana Fischer-Kowalski, "Society's Metabolism: The Intellectual History of Materials Flow Analysis, Part I: 1860–1970," *Journal of Industrial Ecology* 2, no. 1 (1998). See also Robert U. Ayres and Allen V. Kneese, "Production, Consumption, and Externalities," *American Economic Review* 59, no. 3 (1969); and Robert U. Ayres, "Industrial Metabolism," in *Technology and Environment*, ed. Jesse H. Ausubel, Robert A. Frosch, and Robert Herman (Washington, DC: National Academy Press, 1989). Christine Rosen has argued that the environmental impacts of industry are as much a part of industrial systems as are extraction and manufacturing. In her view, business-environment-society relationships should thus be an integral part of business history. Christine Rosen, "Industrial Ecology and the Greening of Business History," *Business and Economic History* 26 (1997). See also Jeffrey K. Stine and Joel A. Tarr, "At the Intersection of Histories: Technology and the Environment," *Technology and Culture* 39, no. 4 (1998).

8. Samuel P. Hays, *Conservation and the Gospel of Efficiency: The Progressive Conservation Movement, 1890–1920* (Cambridge, MA: Harvard University Press, 1959); Richard White, *Organic Machine: The Remaking of the Columbia River* (New York: Hill & Wang, 1995).

9. James Scott uses the example of scientific forestry in Germany and elsewhere to illustrate his arguments regarding state projects aimed at simplification and legibility, in *Seeing Like a State: How Certain Schemes to Improve the Human Condition Have Failed* (New Haven, CT: Yale University Press, 1998). David Demerit, "Scientific Forest Conservation and the Statistical Picturing of Nature's Limits in the Progressive-Era United States," *Environment and Planning D: Society and Space* 19 (2001), has argued that certain Progressive era practices of "statistical picturing" aimed at making American forests calculable can best be understood as projects intended to subject forests to disciplinary forms of state power. See also William Boyd, "Ways of Seeing in Environmental Law: How Deforestation Became an Object of Climate Governance," *Ecology Law Quarterly* 37 (2010): 843 (discussing role of calculability in making forests a viable part of climate policy).

10. In 2005, ArborGen, LLC, a joint venture between International Paper, MeadWestvaco, and a New Zealand company, Rubicon Limited, received permits from the U.S. Department of Agriculture's Animal and Plant Health Inspection Service (APHIS) to import a transgenic "freeze tolerant" eucalyptus hybrid from New Zealand and to conduct field tests in Baldwin County, Alabama. Subsequent permits for additional field testing, including permits authorizing the flowering of the transgenic eucalyptus, were issued in 2007. See *ArborGen, LLC, Availability of an Environmental Assessment and Finding of No Significant Impact for a Controlled*

Release of Genetically Engineered Eucalyptus Hybrids, 72 Fed. Reg. 35215 (Jun. 27, 2007). The company received additional permits in 2008, 2009, and 2010 to conduct more extensive testing at twenty-eight sites in seven southern states. *ArborGen, LLC, Availability of an Environmental Assessment and Finding of No Significant Impact for a Controlled Release of Genetically Engineered Eucalyptus Hybrids*, 75 Fed. Reg. 26708 (May 12, 2010). In October 2011, a federal district court granted summary judgment to the government in a case brought by several environmental groups challenging the permits. See Ctr. for Biological Diversity v. Animal and Plant Health Inspection Serv., 2011 WL 4737405 (S.D. Fla. 2011). ArborGen subsequently filed a petition with APHIS for a determination of "non-regulated status" for its transgenic Eucalyptus trees, which would allow it to commercialize the technology. See ArborGen Inc., "Petition for Determination of Non-regulated Status for Freeze Tolerant Hybrid Eucalyptus Lines," received January 19, 2011, www.aphis.usda.gov/brs/aphisdocs/11_01901p.pdf. In February 2013, APHIS issued a notice of intent to prepare an environmental impact statement as part of its effort to determine whether to grant ArborGen's petition for nonregulated status. See *ArborGen Inc.; Availability of Petition, Notice of Intent to Prepare an Environmental Impact Statement for Determination of Nonregulated Status of Freeze Tolerant Eucalyptus Lines, and Notice of Virtual Public Meetings*, 78 Fed. Reg. 13309 (Feb. 27, 2013). For a general discussion of biotechnology and tree improvement in U.S. forestry during the post–World War II period, see William Boyd and Scott Prudham, "Manufacturing Green Gold: Industrial Tree Improvement and the Power of Heredity in the Postwar United States," in *Industrializing Organisms: Introducing Evolutionary History*, ed. Susan R. Schrepfer and Philip Scranton (New York: Routledge, 2004). See also the essays in *The Bioengineered Forest: Challenges for Science and Society*, ed. Steven H. Strauss and H. D. Bradshaw (Washington, DC: Resources for the Future, 2004).

11. Charles F. Sabel and Jonathan Zeitlin, "Stories, Strategies, Structures: Rethinking Historical Alternatives to Mass Production," in *World of Possibilities: Flexibility and Mass Production in Western Industrialization* (Cambridge, MA: Cambridge University Press, 1996), 5.

12. As Sabel and Zeitlin note, "In human societies and in economies more specifically, adaptation occurs through the reconstruction of existing institutions to meet new demands rather than the wholesale replacement of one set of institutions by another." Ibid., 9.

13. Michael Storper and Robert Salais, *Worlds of Production: The Action Framework of the Economy* (Cambridge, MA: Harvard University Press, 1997). See also Michael Storper, "Conventions and the Genesis of Institutions," in *Institutions and the Role of the State*, ed. Burlamaqui Leonardo, Castro Anna Celia, and Chang Ha-Joon (Cheltenham, UK: Edward Elgar, 2000).

14. Sabel and Zeitlin, "Stories, Strategies, Structures," 9.

15. Karl Polanyi, "The Economy as Instituted Process," in *Trade and Market in the Early Empires: Economies in History and Theory*, ed. Karl Polanyi, Conrad Arensberg, and Harry Pearson (1957; reprint, Chicago: Henry Regnery, 1971). Polanyi characterizes "embeddedness" in the following terms: "The human economy . . . is

embedded and enmeshed in institutions, economic and noneconomic. The inclusion of the noneconomic is vital. For religion or government may be as important to the structure and functioning of the economy as monetary institutions or the availability of tools and machines themselves that lighten the toil of labor" (250). This definition has been picked up and elaborated by several economic sociologists. See, in particular, Mark Granovetter, "Economic Action and Social Structure: The Problem of Embeddedness," *American Journal of Sociology* 91 (1985).

16. See, e.g., Douglass C. North, *Structure and Change in Economic History* (New York: W. W. Norton, 1981) 201–5.

17. For critiques of the new institutional economics and its approach to institutions, see Storper, "Conventions and the Genesis of Institutions," and Geoffrey M. Hodgson, "The Approach of Institutional Economics," *Journal of Economic Literature* 36, no. 1 (1998). Hodgson argues for a more evolutionary theory of institutional change that "would stress the evolution of institutions, in part from other institutions, rather than from a hypothetical, institution-free 'state of nature'" (184).

18. Sabel and Zeitlin, "Stories, Strategies, Structures," 5.

19. Karl Marx, *The Eighteenth Brumaire of Louis Bonaparte* (1869; reprint, New York: International Publishers, 1963), 15.

20. Ronald H. Coase, "The Nature of the Firm," *Economica* 4, no. 16 (1937). Chandler's work, of course, is very well known. See, e.g., Alfred D. Chandler Jr., *Scale and Scope: The Dynamics of Industrial Capitalism* (Cambridge, MA: Belknap Press of Harvard University Press, 1990) and *Visible Hand.*

21. For an elaboration of the constructivist political economy perspective and the notion of industrial orders, see Gary Herrigel, *Industrial Constructions: The Sources of German Industrial Power* (Cambridge: Cambridge University Press, 1996). Herrigel's notion of industrial order is similar to the industrial system concept used by Annalee Saxenian in her comparative study of Silicon Valley and Route 128. See Annalee Saxenian, *Regional Advantage: Culture and Competition in Silicon Valley and Route 128* (Cambridge, MA: Harvard University Press, 1994). For similar perspectives, see Sabel and Zeitlin, "Stories, Strategies, and Structures"; and Gerald Berk, *Alternative Tracks: The Constitution of American Industrial Order, 1865–1917* (Baltimore, MD: Johns Hopkins University Press, 1994).

22. See Herrigel, *Industrial Constructions.*

23. Quoted in Mary Carr Mayle, "CEO Discusses International Paper's Place in Global Markets," *Savannah Morning News*, April 9, 2004.

24. Figures and quote from Ray's speech cited in James M. Fallows, *The Water Lords* (New York: Grossman, 1971), 56–57, 61, 263.

25. Per capita consumption = (annual production plus imports less exports) / population. American Forest & Paper Association, *Paper, Paperboard, and Wood Pulp: 1995 Statistics* (Washington, DC, 1995), 2.

26. Witness the words of Marianne Parrs, vice president and chief financial officer at International Paper Company, in 1997: "One of the most deep-seated notions in the forest products industry—one that I think must change—is the basic idea that when a company makes more tons, it also makes more money. The evidence is very

clear that making more tons does not lead to making more money industry-wide." Quoted in Kirk J. Finchem, "Top Managers, Analysts Say Paper Industry Is Slow to Change," *Pulp & Paper* 71, no. 12 (1997): 61.

27. For discussions of these larger macroeconomic changes, see Michael J. Piore and Charles F. Sabel, *The Second Industrial Divide: Possibilities for Prosperity* (New York: Basic Books, 1984); Harvey, *The Condition of Postmodernity*, esp. part 2; and Robert Brenner, "The Economics of Global Turbulence," *New Left Review* 299 (May–June 1998).

28. With exports accounting for an increasing proportion of domestic production and because export demand is generally more volatile than domestic demand, capacity utilization, prices, and profits have all become increasingly sensitive to global markets. See Peter J. Ince, "Global Cycle Changes the Rules for U.S. Pulp and Paper," *North American Papermaker* 81, no. 12 (1999).

29. James K. Flicker, "Structural Impediments to Profitability, *Pulp & Paper* 72, no. 1 (1998): 142; Mark Diverio, "Poor Financial Returns Are a Call to Action for the Paper Industry," *Pulp & Paper* 73, no. 4 (1999).

30. "Economics of Paper," *Fortune* 16, no. 4 (1937).

31. Interview with author, August 20, 1997.

32. Interview with author, August 13, 1997.

33. U.S. Census Bureau, "Census of Manufactures," 1963–1992; U.S. Census Bureau, "Concentration Ratios in Manufacturing," *1997 Economic Census* (2001); U.S. Census Bureau, "Concentration Ratios 2002," *2002 Economic Census* (2006); U.S. Census Bureau, "Concentration Ratios in Manufacturing," *2007 Economic Census* (2011). Data for 2007, 2002, and 1997 are organized and classified by the North American Industry Classification System (NAICS). Data for 1992 and prior years are organized and classified by the Standard Industry Classification (SIC) system.

34. Flicker, "Structural Impediments"; Diverio, "Poor Financial Returns." See also Foad Tamaddon, "Overcapacity Abounds, But Mills Can Still Secure Funds," *Pulp and Paper* 71, no. 9 (1997). Tamaddon noted that the industry had failed to earn its cost of capital for eight out of the last ten years preceding 1997.

35. Russ Banham, "Through the Woods, a Trail of Consolidations," *Journal of Commerce* 12 (November 1998): 8A.

36. Quoted in Hue Ha, "Paper Industry Seen Needing Mergers, but Few Occurring So Far," Dow Jones Newswires, May 20, 1997.

37. Quoted in Greg McIvor, "A Compelling Case on Paper: In Such a Soft Market, the Arguments for Further Consolidation Are Persuasive," *Financial Times*, December 7, 1998.

38. See, e.g., Scott Morrison, "Paper Makers Press on with Consolidation Process," *Financial Times*, April 5, 2001; Charles E. Swann, "Paper Profits: Will Consolidation Pay Off?" *North American Papermaker* 82, no. 3 (2000): 44. Between 1980 and 2000, the North American market share controlled by the top ten companies increased from 38 percent to roughly 52 percent. See Megan Mele and Alan Rooks, "Playing to Win in a Global League," *North American Papermaker* 83, no. 4 (2001).

39. U.S. Census Bureau, "Census of Manufactures," 1963–1992; U.S. Census Bureau, "Concentration Ratios in Manufacturing," *1997 Economic Census* (2001); US Census Bureau, "Concentration Ratios 2002," *2002 Economic Census* (2006); US Census Bureau, "Concentration Ratios in Manufacturing," *2007 Economic Census* (2011). Data for 2007, 2002 and 1997 are organized and classified by the North American Industry Classification System (NAICS). Data for 1992 and prior years are organized and classified by the Standard Industry Classification (SIC) system.

40. American Forest & Paper Association, *50th Annual Survey of U.S. Paper, Paperboard, and Pulp Capacity* (2012). See also James McLaren et al., "More Discipline, Supply Management Critical to North American Stability," *Pulp & Paper* 75, no. 1 (2001): 42–46; and Swann, "Paper Profits."

41. At the turn of the twenty-first century, the top ten global producers included companies from the United States, Finland, Ireland, Canada, Japan, and Norway. International Paper led the pack accounting for roughly 5% of global paper and board markets in 2001. Altogether, the top ten global producers in 2001 accounted for slightly less than 30% of global markets, a substantial increase over the previous decade. See Mele and Rooks, "Playing to Win in a Global League." See also "Global Cycle Changes the Rules for U.S. Pulp and Paper," *North American Papermaker* 81, no. 12 (1999).

42. For a discussion of the declining competitiveness of the U.S. pulp and paper industry relative to new low-cost producers in Latin America and Asia, see David N. Wear, Jeffrey Prestemon, Robert Huggett and Douglas Carter, "Chapter 9: Timber Product Markets," in *The Southern Forest Futures Project: Technical Report*, Gen. Tech. Rep. SRS-178, ed. David N. Wear and John G. Greis (Asheville, NC: Southern Research Station, 2011), 15–17.

43. John Schmid, "China Becoming Mighty Oak of World's Paper Industry," *Los Angeles Times*, December 28, 2012. Schmid noted that China had tripled production during the decade up to 2012 and in 2009 overtook the United States to become the world's largest papermaker.

44. For discussion of subsidies to China's paper industry, see Usha C. V. Halley, *No Paper Tiger: Subsidies to China's Paper Industry from 2002–2009*, EPI Briefing Paper 264 (Washington, DC: Economic Policy Institute, 2010), 1–2.

45. Ibid., 3, 14–17.

46. Wear and Greis observed, "The divestiture of forest lands by the forest products industry from 1998 to 2010 is the most substantial transition in forest ownership of the last century." David N. Wear and John G. Greis, *The Southern Forest Futures Project: Summary Report*, Gen. Tech. Rep. SRS-168 (Asheville, NC: Southern Research Station, 2011), 60. During this period, leading forest products companies sold roughly two-thirds of their timberland holdings during the 2000s, primarily to large timber investment management organizations (TIMOs) and real estate investment trusts (REITs). Between 1998 and 2008, the amount of timberland owned by the forest products industry declined from 23.4 million acres to 7.5 million acres. Conversely, timberland ownership by TIMOs increased over the same time period from 2.1 million acres to 13.4 million acres, and ownership by REITs increased from

5.4 million acres to 6.5 million acres. See Brett J. Butler and David N. Wear, "Chapter 6: Forest Ownership Dynamics of Southern Forests," in *Southern Forest Futures Project: Technical Report*, 29, table 6.1. See also Wear and Greis, *Southern Forest Futures Project: Summary Report*, 60–62 (showing changing patterns of forest land ownership in the southern United States since the late 1990s).

47. William H. Nicholls, *Southern Tradition and Regional Progress* (Chapel Hill: University of North Carolina Press, 1960), 162–63.

48. As a professor of economics at Vanderbilt University, Nicholls spent much of his career studying the challenges of southern economic development, often comparing the southern experience to that of developing countries. His view of economic development shared a great deal with modernization theory, which was very much in favor among American social scientists during the post–World War II decades.

49. See, e.g., William H. Nicholls, "Southern Tradition and Regional Economic Progress," *Southern Economic Journal* 26, no. 3 (1960): 191–93; and *Southern Tradition and Regional Progress*, chaps. 3–6. See also William H. Nicholls, "The South as a Developing Area," *Journal of Politics* 26, no. 1 (1964).

50. Nicholls, "Southern Tradition and Regional Economic Progress," 191.

51. See Daniel Joseph Singal, *The War Within: From Victorian to Modernist Thought in the South, 1919–1945* (Chapel Hill: University of North Carolina Press, 1982).

52. William Faulkner, "The Bear," in *Go Down Moses* (1940; reprint, New York: Vintage, 1973), 193, 254.

53. See Twelve Southerners, *I'll Take My Stand: The South and the Agrarian Tradition* (1930; reprint, Baton Rouge: Louisiana State University Press, 1983). See also Singal, *War Within*, 198–231.

54. One of the leading Nashville Agrarians, John Crowe Ransom, put the matter this way: "The South must be industrialized—but to a certain extent only, in moderation . . . [I]t will be fatal if the South should conceive it as her duty to be regenerated and get her spirit reborn with a totally different orientation toward life." John Crowe Ransom, "Reconstructed but Unregenerate," in Twelve Southerners, *I'll Take My Stand*, 22.

55. "How may the little agrarian community resist the Chamber of Commerce of its county seat, which is always trying to import some foreign industry that cannot be assimilated to the life-pattern of the community?" Twelve Southerners, *I'll Take My Stand*, xlvii.

56. Cf. Nicholls, *Southern Tradition and Regional Progress.*

57. As James Scott puts it: "Formal order . . . is always and to some considerable degree parasitic on informal processes, which the formal scheme does not recognize, without which it could not exist, and which it alone cannot create or maintain" (*Seeing Like a State*, 310).

ESSAY ON SOURCES

Any regional industrial history inevitably draws on multiple sources, disciplines, and methodologies. Getting started requires familiarity not only with the distinctive history of the region but also the structures and strategies that mark the specific industry. Telling the story adequately requires that one give voice to the many different actors involved. Taking the environment seriously adds another level of complexity. All of this, moreover, must be situated within the context of national-level changes in politics, economics, law, and, over time, the broader dynamics of globalization. Such an interdisciplinary perspective inevitably carries with it risks of sloppy analysis and unwieldy narrative.

This book has sought to avoid these risks by taking the regional industrial system as its unit of analysis. By focusing on the problems that confronted pulp and paper firms as they moved into the South and attempted to construct a new industrial complex, the goal has been to provide analytical coherence without sacrificing narrative force. The result of such a focus, of course, means that the perspectives of some actors will sometimes prevail over those of others. In the story told here, industry managers and executives emerge as the dominant actors. They were often the prime movers, and much of the story is told from their perspective. But they were hardly the only actors involved, and in each of the chapters, the perspectives of other important actors are brought to bear, often in considerable detail.

Gathering sufficient evidence to tell such a story proved to be a somewhat daunting task. Rather than start with known collections or data sets, my point of departure was more ethnographic. Interviews with key actors—from high-level industry executives to small contract loggers, wood dealers, local politicians, and environmentalists—provided much of the empirical foundation for the project. During seven months of fieldwork in south Georgia in 1997 and several follow-up trips to the region, I interviewed more than one hundred people. All of the interviews were taped, and all were conducted under a confidentiality agreement. Given the sensitivities involved and the vulnerabilities that some of my sources felt, maintaining confidentiality was important to allow them to speak freely. All of the tapes from all of the interviews are in my possession, as are the signed confidentiality statements. If this book has any interest for other scholars, it will surely be in part because of the stories that people so generously shared with me. Aside from the obvious places where such stories explicitly inform the narrative, these conversations added substantially to my understanding of the industry and its relationship to the region.

Interviews were complemented with archival research and extensive work with time series data, government sources, trade journals, and legal documents.

Specifically, I conducted research at the Forest History Society in Durham, NC; the Georgia Historical Society in Savannah; the University of Georgia library in Athens; the Southern Historical Collection at the University of North Carolina at Chapel Hill; and the National Archives regional facility in East Point, Georgia. For industry trade journals and government documents, the forestry collection at the University of California, Berkeley, proved invaluable.

In terms of overall framing for the project, important works from several secondary literatures are discussed below. In addition, government and industry trade association statistics provided important sources of data for the charts and tables throughout the book. Specifically, data from the U.S. Census of Manufactures, the U.S. Forest Service, the American Forest & Paper Association and its predecessors, the American Paper Institute and the American Pulpwood Association, were used throughout. Newspaper editorials and articles—in particular, articles and editorials from the *Savannah Morning News*, *Savannah Evening Press*, and the *Atlanta Journal* (later renamed the *Atlanta Journal & Constitution*)—provided important sources on key developments and the general reception of the industry in the region. Although the story told here is quite different in orientation and emphasis, Jack P. Oden's "Development of the Southern Pulp and Paper Industry, 1900–1970" (PhD diss., Mississippi State University, 1973) provided an indispensable starting point for understanding the broad features of the historical development of the industry.

Chapter 1, which focuses on the industrialization of the southern forest, relied primarily on government documents, proceedings of various forestry conferences, trade journals, academic reports and articles, and various documents from archival collections. Interviews were used to develop and refine the overall scheme and fill in important details. Reports from the Southern Pine Association, the American Institute for Economic Research, and the U.S. Forest Service provided important background regarding concerns over a national timber famine and the growing interest of the pulp and paper industry in southern forests during the 1920s and 1930s. See, e.g., U.S. Department of Agriculture, *How the United States Can Meet Its Present and Future Pulpwood Requirements*, Dept. Bulletin no. 1241 (Washington, DC: Government Printing Office, 1924); Southern Pine Association, *Economic Conditions in Southern Pine Industry* (New Orleans, 1931); and American Institute for Economic Research, *A Report on the Future of the Paper Industry in the Southeastern United States and the Effects on Stumpage Values* (Cambridge, MA, 1938).

On the destruction of southern forests during the late nineteenth and early twentieth centuries, see Charles S. Sargent, *Report on the Forests of North America* (Washington, DC: Government Printing Office, 1884); U.S. Forest Service, *Timber Depletion, Lumber Prices, Lumber Exports, and Concentration of Timber Ownership: Report of Senate Resolution 311* (Washington, DC: Government Printing Office, 1920); Howard W. Odum, *Southern Regions of the United States* (Chapel Hill: University of North Carolina Press, 1936); Rupert B. Vance, *Human Geography of the South: A Study in Regional Resources and Human Adequacy*, 2nd ed. (Chapel Hill: University of North Carolina Press, 1935). Important commentary by professional foresters included R. D. Forbes, "The Passing of the Piney Woods," *American For-*

estry 29, no. 351 (1923); and Gifford Pinchot, "Southern Forest Products, and Forest Destruction and Conservation Since 1865," in *The South in the Building of the Nation*, vol. 6 (Richmond, VA: Southern Historical Publication Society, 1910).

With respect to efforts to rationalize and improve the conditions for investment in industrial forestry through forest surveys, fire control, and tax and credit reform, chapter 1 relied on a mix of government reports and commentary from professional foresters. Important sources on the forest survey effort included U.S. Department of Agriculture, *Forest Resources of South Georgia*, USDA Misc. Pub. no. 390 (Washington, DC: Government Printing Office, 1941); Inman F. Eldredge, *The Forest Survey in the South*, Occasional Paper no. 31 (New Orleans: Southern Forest Experiment Station, 1934); and E. L. Demmon, *Economics for Our Southern Forests*, Occasional Paper no. 59 (New Orleans: Southern Forest Experiment Station, 1937). On fire, see W. W. Ashe, "Forest Conditions in the Southern States and Recommended Forest Policy," *South Atlantic Quarterly* 22, no. 4 (1925); Southern Forestry Educational Project, *Second Annual Report* (Washington, DC: American Forestry Association, 1930); James E. Mixon, "Progress of Protection from Forest Fires in the South," *Journal of Forestry* 54, no. 10 (1956); H. H. Chapman, "Forest Fires and Forestry in the Southern States," *American Forestry* 18 (1912); G. Wahlenberg, *Fire in Longleaf Pine Forests*, Occasional Papers no. 40 (New Orleans: Southern Forest Experiment Station, 1935); and various reports in the *Proceedings of the First Annual Tall Timbers Fire Ecology Conference* (Tallahassee, FL, 1962). Stephen J. Pyne's remarkable work, *Fire in America: A Cultural History of Wildland and Rural Fire* (1982; reprint, Seattle: University of Washington Press, 1997), contains an important chapter on woods burning and fire in the southern forests.

On credit reform, Resources for the Future, *Forest Credit in the United States: A Survey of Needs and Facilities* (Washington, DC, 1958), is the place to start for understanding the challenges and opportunities associated with efforts to allow long-term loans on forest assets during the middle decades of the twentieth century. A series of essays in the journal *Forest Farmer* also provided important context and commentary on the availability of credit for nonindustrial private timberland owners. With respect to taxes, see Fred R. Fairchild, *Forest Taxation in the United States*, USDA Misc. Pub. no. 218 (Washington, DC: Government Printing Office, 1935); Ronald B. Craig, *Reversion of Forest Land for Taxes Increasing in the South*, Occasional Paper no. 32 (New Orleans: Southern Forest Experiment Station, 1934); Ronald B. Craig, "The Forest Tax Delinquency Problem in the South," *Southern Economic Journal* 6, no. 2 (1939); R. Clifford Hall, "The Rise of Realism in Forest Taxation," *Journal of Forestry* 36, no. 9 (1938); Charles W. Briggs and William K. Condrell, *Tax Treatment of Timber* (Washington, DC: Forest Industries Committee on Timber Valuation and Taxation, 1978); and Margaret P. Hamel, ed., *Forest Taxation: Adapting in an Era of Change* (Madison, WI: Forest Products Research Society, 1988).

The regeneration of southern forests that began in earnest during the interwar decades was chronicled in conference proceedings, government reports, and trade journals. Specifically, the *Proceedings of the Southern Forestry Congress* in 1916 and 1921 provided important early statements on the importance of forest regeneration. The report of the 1917 Southern Cut-Over Land Conference in New Orleans, *The Dawn of the New Constructive Era* (New Orleans, 1917), documented the problem of forest destruction and articulated a program for regional forest replanting. Government reports on regeneration included Philip C. Wakeley, *Artificial Reforestation in the Southern Pine Region*, USDA Technical Bulletin no. 492 (Washington, DC: Government Printing Office, 1935); and F. A. Silcox, *Report of the Chief of the Forest Service, 1937* (Washington, DC: Government Printing Office, 1937). The impact of the Soil Bank Program on southern forest regeneration was chronicled in periodicals such as *Forest Farmer* and the *Journal of Forestry*. Important historical surveys and data are available in the *Proceedings of the Southern Forest Nursery Association*.

The history of tree improvement efforts drew heavily on early statements by key players in the *Journal of Forestry*, including Scott Pauley, Lloyd Austin, Aldo Leopold, and Philip Wakeley. The short history by Bruce Zobel and Jerry Sprague, both associated with North Carolina State University and the cooperative tree improvement program housed there, provided an important "insider" perspective on tree improvement. See Bruce J. Zobel and Jerry R. Sprague, *A Forestry Revolution: The History of Tree Improvement in the Southern United States* (Durham, NC: Carolina Academic Press, 1993). See also Bruce J. Zobel, "Forest Tree Improvement—Past and Present," in *Advances in Forest Genetics*, ed. P. K. Khosla (New Delhi: Ambika Publications, 1981); and Bruce J. Zobel and John Talbert, *Applied Forest Tree Improvement* (New York: John Wiley & Sons, 1984). The annual *Proceedings of the Southern Forest Tree Improvement Conference* also provided valuable perspective on the overall effort. For a more general history of forest tree breeding in the United States, with particular attention to the South and Pacific Northwest, see William Boyd and Scott Prudham, "Manufacturing Green Gold: Industrial Tree Improvement and the Power of Heredity in the Postwar United States" in *Industrializing Organisms: Introducing Evolutionary History*, ed. S. R. Schrepfer and P. Scranton (New York: Routledge, 2004).

Several reports from the U.S. Forest Service that focused specifically on southern forests and forestry also provided important context. See especially, U.S. Forest Service, *The South's Fourth Forest: Alternatives for the Future*, Forest Resources Report no. 24 (Washington, DC: Government Printing Office, 1988); and H. R. Josephson, *A History of Forestry Research in the Southern United States*, USDA Misc. Pub. no. 1462 (Washington, DC: Government Printing Office, 1989). In addition, see Michael Williams, *Americans and Their Forests: A Historical Geography* (New York: Cambridge University Press, 1989); Henry Clepper, *Professional Forestry in the United States* (Baltimore, MD: Johns Hopkins University Press, 1971); William G. Robbins, *American Forestry: A History of National, State, and Private Cooperation* (Lincoln: University of Nebraska Press, 1985); and William G. Robbins, *Lumberjacks and Legislators: Political Economy of the U.S. Lumber Industry, 1890–1941* (College Station: Texas A&M University Press, 1982). For a general history of the southern longleaf forest, see Lawrence S. Early, *Looking for Longleaf: The Fall and Rise of an American Forest* (Chapel Hill: University of North Carolina Press, 2004).

Chapter 2 relied more explicitly on interviews and company documents in addition to other materials such as trade journals, congressional hearings, and government reports. Because the evidence needed to tell the story of southern logging and wood procurement was not readily available, interviews with loggers, wood dealers, and company procurement managers proved critical in shaping the overall analysis and adding depth to the story. Through my interviews and archival research, I was able to secure several important internal company documents and unpublished reports on how companies developed their wood procurement systems and managed their wood dealers and pulpwood producers. Chief among these were E. H. Wilson's unpublished report to Union Bag, "Savannah Woodland Operations and Contract System of Wood Procurement as It Affects the Farmers and Landowners" (1938), Myron Kahler's unpublished report from the Southern Forest Experiment Station, "Logging Methods, Costs, and Woodlands Procedure of the Pulp and Paper Industry in the South" (1936), and W. N. Haynes's report prepared for Union Bag's Savannah Woodlands Division, "A Time and Cost Study of Pulpwood Harvesting Operations in Southeast and Central Georgia" (1949). The multiple volumes of the Battelle Institute's Report on Southern Pulpwood Production, which were commissioned by the American Pulpwood Association on behalf of nineteen pulp and paper companies operating in the South and released between 1960 and 1963, provided invaluable detail on the problems confronting the wood dealer system in the 1950s and early 1960s and the considerable technological and financial challenges associated with efforts to mechanize logging operations while maintaining the flexibility of a system based on independent contracting.

Industry and trade association conference proceedings; journals such as *Forest Farmer*, the *Forest Products Journal*, the *Journal of Forestry*, and *Pulp & Paper*; and periodic reports and surveys of southern producers conducted by the American Pulpwood Association and the Southern Pulpwood Conservation Association were all useful in understanding industry perspectives on the wood procurement challenge. Congressional hearings and government reports also helped fill out the picture of how the pulp and paper companies constructed local pulpwood markets and worked to organize a logging system that would insulate them from many of the risks and liabilities associated with company employees. In particular, the 1941 U.S. House pulpwood investigation hearings, several House and Senate hearings throughout the 1940s and 1950s on the Fair Labor Standards Act and its application to small logging crews, and hearings during the 1970s and 1980s on the status of independent contractors in the logging industry all contained important testimony from various players on the nature of pulpwood markets and the procurement system. A series of technical reports issued by the Southern Forest Experiment Station starting in the 1930s were particularly helpful in understanding the general practices of most logging operations, wages, and working conditions. Reports by the U.S. Department of Labor and OSHA provided useful information on the conditions prevailing in the logging industry.

Judicial decisions also provided valuable detail on the relationships between companies, wood dealers, and pulpwood producers. Chief among these was Scott

Paper Co. v. Gulf Coast Pulpwood Assoc'n, 1973 WL 881 (S.D. Ala. Sept. 21, 1973). In addition, a series of federal cases from the 1940s on the applicability of the Fair Labor Standards Act to small logging operations and multiple state court decisions from across the region spanning the 1930s–60s dealing with the wood dealer system and its implications for taxes, workers compensation, and tort liability helped to fill out the picture regarding the nature of the various contractual relationships that constituted the wood dealer system.

As for the secondary literature, very little has been written about the southern pulpwood procurement system, but a few articles provided context and insight on logging and the wood dealer system in the South. See Warren A. Flick, "The Wood Dealer System in Mississippi: An Essay on Regional Economics and Culture," *Journal of Forest History* 29, no. 3 (1985); and John C. Bliss and Warren A. Flick, "With a Saw and a Truck: Alabama Pulpwood Producers," *Forest and Conservation History* 38, no. 2 (1994). James Pikl's "The Southern Woods Labor 'Shortage' of 1955," *Southern Economic Journal* 27, no. 1 (1960), was helpful in understanding the state of the southern pulpwood procurement system in the 1950s and the growing pressure to mechanize. Robert Outland's *Tapping the Pines: The Naval Stores Industry in the American South* (Baton Rouge: Louisiana State University Press, 2004) provides a detailed portrait of the naval stores industry and the particular labor regimes that prevailed in that industry.

Chapter 3, which is the least original in terms of evidence, draws primarily on trade journals and secondary sources. Useful background works on papermaking and the development of the industry in the United States included Dard Hunter, *Papermaking: The History and Technique of an Ancient Craft* (New York: Knopf, 1947); Louis Tillotson Stevenson, *The Background and Economics of American Papermaking* (New York: Harper & Brothers, 1940); David C. Smith, *History of Papermaking in the United States, 1691–1969* (New York: Lockwood, 1970); Judith A. McGaw, *Most Wonderful Machine: Mechanization and Social Change in Berkshire Paper Making, 1801–1885* (Princeton, NJ: Princeton University Press, 1987); James D. Studley, *United States Pulp and Paper Industry*, U.S. Department of Commerce, Trade Promotion Series no. 182 (Washington, DC: Government Printing Office, 1938); and Nicholas A. Basbanes, *On Paper: The Everything of Its Two-Thousand-Year History* (New York: Vintage, 2013).

For more specific information on industrial organization and the changing economics of the industry, several books and articles proved useful. See Avi Cohen, "Technological Change as Historical Process: The Case of the U.S. Pulp and Paper Industry, 1916–1940," *Journal of Economic History* 46, no. 3 (1984); Nancy Kane Ohanian, *The American Pulp and Paper Industry, 1900–1940: Mill Survival, Firm Structure, and Industry Relocation* (Westport: Greenwood, 1993); Naomi R. Lamoreaux, *The Great Merger Movement in American Business, 1895–1904* (Cambridge: Cambridge University Press, 1985); and Alfred Chandler, *Scale and Scope: The Dynamics of Industrial Capitalism* (Cambridge: Belknap Press of Harvard University Press, 1990). The 1937 series of articles on the paper industry in *Fortune* magazine provided an important picture of the economic challenges facing the industry during that period.

Not surprisingly, the issue of race relations and mill labor has attracted the attention of several scholars. Herbert R. Northup's "The Negro in the Paper Industry," in *Negro Employment in Southern Industry: A Study of Racial Policies in Five Industries* by Herbert R. Northup, Richard L. Rowan, Darold T. Barnum, and John C. Howard, Industrial Research Unit, Wharton School of Finance and Commerce (Philadelphia: University of Pennsylvania, 1970), is an important starting point for any effort to understand this particular aspect of the industry's history. Timothy J. Minchin's "Federal Policy and Racial Integration of Southern Industry, 1961–1980," *Journal of Policy History* 11(2) (1999), and *The Color of Work: The Struggle for Civil Rights in the Southern Paper Industry, 1945–1980* (Chapel Hill: University of North Carolina Press, 2001) provide first-rate accounts of race relations and the struggle to integrate the industry during the civil rights era. Other important works on race relations, organized labor, and southern industry include Ray Marshall, *The Negro and Organized Labor* (New York: John Wiley & Sons, 1965); and Herbert Hill, "Racism within Organized Labor: A Report of Five Years of the AFL-CIO, 1955–1960," *Journal of Negro Education* 30, no. 2 (1961). Adam Fairclough's *Race and Democracy: The Civil Rights Struggle in Louisiana, 1915–1972* (Athens: University of Georgia Press, 1995) provides a detailed account of the conflicts around efforts to integrate the Crown Zellerbach mill in Bogalusa, which became the subject of the landmark decision Local 189 v. United States, 416 F.2d 980 (5th Cir. 1969), applying Title VII of the Civil Rights Act to end the discriminatory seniority system at the mill and providing a template for efforts to prohibit such practices across the industry.

Several edited collections on southern labor history also provided useful context for chapter 3. See Gary M. Fink and Merle Reed, eds., *Race, Class, and Community in Southern Labor History* (Tuscaloosa: University of Alabama Press, 1994); Robert H. Zeiger, ed., *Organized Labor in the Twentieth Century South* (Knoxville: University of Tennessee Press, 1991); and Melvyn Stokes and Rick Halpern, eds., *Race and Class in the American South since 1890* (Oxford: Berg, 1994). Needless to say, there is a very large literature on race relations in the postbellum South. See, in particular, Joel Williamson, *The Crucible of Race: Black-White Relations in the American South since Emancipation* (New York: Oxford University Press, 1984); and C. Vann Woodward, *The Strange Career of Jim Crow*, 2nd rev. ed. (New York: Oxford University Press, 1966). On civil rights and its implications for economic development in the South, see James J. Heckman and Brook S. Payner, "Determining the Impact of Federal Antidiscrimination Policy on the Economic Status of Blacks: A Study of South Carolina," *American Economic Review* 79, no. 1 (1989); Gavin Wright, "The Civil Rights Revolution as Economic History," *Journal of Economic History* 59, no. 2 (1999); and Gavin Wright, *Sharing the Prize: The Economics of the Civil Rights Revolution in the American South* (Cambridge, MA: Harvard University Press, 2013).

Chapter 4, which deals with the air and water pollution problems associated with pulp and paper production, utilized interviews, congressional hearings, government reports, judicial decisions, trade journals, and secondary sources. In terms of early water pollution problems associated with pulp and paper mills, a series of nuisance cases in state and federal court from the 1930s through the 1950s provided valuable information on the scale of wastewater discharges from specific mills, the resulting damage to local environments, and the considerable challenges facing local landowners and others seeking to preserve their livelihoods in the face of industrialization.

Government reports and proceedings from the water pollution enforcement conferences convened in 1965 and 1969 on the Savannah River contained important information on the nature of the water pollution problem and the relative contribution of the pulp and paper industry. See, specifically, U.S. Department of Health, Education, and Welfare, *Pollution of Interstate Waters of the Mouth of the Savannah River, Ga.* (Washington, DC, 1964); U.S. Public Health Service, *Conference on the Pollution of the Interstate Waters of the Lower Savannah River and Its Tributaries, South Carolina–Georgia, February 2, 1965* (Washington, DC: U.S. Department of Health, Education, and Welfare, 1965); and Federal Water Pollution Control Administration, *Second Session of the Conference in the Matter of the Interstate Waters of the Lower Savannah River and its Estuaries, Tributaries, and Connecting Waters in the States of Georgia and South Carolina, October 1969* (Washington, DC, 1970). Various House and Senate hearings on water pollution during the 1960s contained factual information and testimony from industry representatives and other stakeholders on the need for stronger federal regulation. See especially *Water Pollution Control and Abatement (Part 1A—National Survey): Hearings Before the Subcomm. on Natural Res. and Power of the H. Comm. on Gov't Operations*, 88th Cong. 682 (1964); and *Water Pollution Control and Abatement (Part 7—Alabama, Georgia, Mississippi, and Tennessee): Hearings Before the Subcomm. on Natural Res. and Power of the House Comm. on Gov't Operations*, 88th Cong. 3645 (1965).

The 1971 book by James Fallows and his colleagues in the Ralph Nader "study group," on the Union Camp mill in Savannah, *The Water Lords* (New York: Grossman, 1971), offered a hard-hitting indictment of that company's environmental practices. Other useful secondary sources on water pollution included Paul Charles Milazzo, *Unlikely Environmentalists: Congress and Clean Water, 1945–1972* (Lawrence: University Press of Kansas, 2006); Peter C. Yeager, *The Limits of Law: The Public Regulation of Private Pollution* (New York: Cambridge University Press, 1991); A. Myrick Freeman III, "Water Pollution Policy," in *Public Policies for Environmental Protection*, ed. Paul R. Portney (Washington, DC: Resources for the Future, 1990); and William L. Andreen, "The Evolution of Water Pollution Control in the United States—State, Local, and Federal Efforts, 1789–1972: Part I," *Stanford Environmental Law Journal* 22, no. 145 (2003).

On air pollution, reports from the EPA and the series of hearings on federal air pollution legislation during the late 1960s provided context for the overall problem, as well as insights into the positions that state and industry leaders took with respect to a stronger federal role in controlling pollution. See, especially, U.S. EPA, *Kraft Pulping: Control of TRS Emissions from Existing Mills* (Washington, DC, 1979); *Air Pollution—1967 (Air Quality Act): Hearings before the Subcomm. on Air and Water Pollution of the S. Comm. on Pub. Works*, 90th Cong. 2362 (1967). The Georgia Conservancy's report, *Air Pollution in Savannah: A Report on the State of the Air in Chatham County, Georgia and What Can Be Done to Improve It* (Savannah, GA, 1979), and Dr. John Northup's unpublished history of Savannah's Citizens for Clean Air, "History of the Citizens for Clean Air" (November 9, 1990), helped illuminate the campaign to force the Union Camp mill to adopt additional air pollution control measures.

With respect to dioxin contamination from pulp and paper mills, important government reports included U.S. EPA, *National Dioxin Study—Report to Congress*, EPA/530-SW-87-025, Office of Solid Waste and Emergency Response (Washington, D.C., 1987); U.S. EPA, *Summary Report on U.S. EPA/Paper Industry Cooperative Dioxin Study: The 104-Mill Study* (Washington, DC, 1990); U.S. EPA, *Exposure and Human Health Reassessment of 2,3,7,8-Tetrachlorodibenzo-p-Di oxin (TCDD) and Related Compounds, Part III: Integrated Summary and Risk Characterization for 2,3,7,8-Tetrachlorodibenzo-p-Dioxin (TCDD) and Related Compounds* (Washington, DC, 2000); U.S. EPA, "Health Assessment Document for 2,3, 7,8-Tetrachlorodibenzo-p-dioxin (TCDD) and Related Compounds," External Review Draft, EPA/600/BP-92/001c (Washington, DC, 1994); and National Academy of Sciences, *Health Risks from Dioxins and Related Compounds: Evaluation of the EPA Reassessment* (Washington, DC, 2006). Hearings included *Dioxin Contamination of Food and Water: Hearings Before the Subcomm. on Health and Environment of the H. Comm. on Energy and Commerce*, 100th Cong. (1989), and *Dioxin Pollution in the Pigeon River, North Carolina and Tennessee: Hearings Before the Subcomm. on Water Res. of the H. Comm. on Pub. Works and Transp.*, 100th Cong. (1988). Relevant court cases included Carol Van Strum v. EPA, 680 F. Supp. 349 (D. Or. 1987); Leaf River Forest Products, Inc. v. Simmons, 697 So. 2d 1083 (Miss. 1996); Leaf River Forest Products, Inc. v. Ferguson, 662 So. 2d 648 (Miss. 1995); and J. A. Shults & Joan Shults v. Champion Int'l Corp., 821 F. Supp. 517 (E.D. Tenn. 1992).

Mark R. Powell's *Science at EPA: Information in the Regulatory Process* (Washington, DC: Resources for the Future, 1999) provided a very valuable treatment of the struggles over dioxin science at the EPA. Richard A. Bartlett's *Troubled Waters: Champion International and the Pigeon River Controversy* (Knoxville: University of Tennessee Press, 1995) chronicled the conflicts over pollution of the Pigeon River in North Carolina and Tennessee. Maureen Smith's *The U.S. Paper Industry and Sustainable Production: An Argument for Restructuring* (Cambridge, MA: MIT Press, 1997) contained a valuable discussion of the effort to move to chlorine-free bleaching and the broader push for more sustainable production systems in the pulp and paper industry.

More generally, useful secondary sources on American environmental law and policy included Richard N. L. Andrews, *Managing the Environment, Managing Ourselves: A History of American Environmental Policy* (New Haven, CT: Yale

University Press, 1999); Richard J. Lazarus, *The Making of Environmental Law* (Chicago: University of Chicago Press, 2004); Robert V. Percival, "Environmental Federalism: Historical Roots and Contemporary Models," *University of Maryland Law Review* 54, no. 1141 (1995).

In terms of the broader regional historical context for the book, anyone writing about virtually any aspect of the American South in the twentieth century must contend with the giants of southern historiography. Chief among them for this book have been C. Vann Woodward and George B. Tindall. Their companion volumes in LSU's History of the South series provided important context for this book as well as the kind of inspiration that comes from the finest historical writing. See C. Vann Woodward, *Origins of the New South, 1877–1913* (Baton Rouge: Louisiana State University Press, 1951) and George Brown Tindall, *The Emergence of the New South, 1913–1945* (Baton Rouge: Louisiana State University Press, 1967). See also Numan V. Bartley, *The New South, 1945–1980* (Baton Rouge: Louisiana State University Press, 1996).

Likewise, the remarkable work of sociologists Howard Odum and Rupert Vance proved invaluable—not only in the detailed information they compiled on regional trends but also in their efforts to build a new model of regional social science and a policy agenda founded on responsible management of natural resources. Key works include Howard Odum, *Southern Regions of the United States* (Chapel Hill: University of North Carolina Press, 1936); Rupert Vance, *Human Geography of the South: A Study in Regional Resources and Human Adequacy* (Chapel Hill: University of North Carolina Press, 1935); and Rupert Vance, *All These People: The Nation's Human Resources in the South* (Chapel Hill: University of North Carolina Press, 1945). In their efforts to develop a social science of regionalism and in their close attention to environmental degradation, its effects on southern life, and the vital need for conservation, they provided strong arguments in favor of a progressive future for the South that matched their diagnoses of the many problems confronting the region.

More recent work on southern history, particularly southern economic history, also provided important background for understanding and tracing the broader, structural trends in the region's economy, especially the impact of the New Deal on southern agriculture and regional labor markets. Gavin Wright's *Old South, New South: Revolutions in the Southern Economy since the Civil War* (New York: Basic Books, 1986) stands alone in this respect—a model of analytical force combined with clear writing. Other more specialized works included Bruce Schulman, *From Cotton Belt to Sun Belt: Federal Policy, Economic Development, and the Transformation of the South, 1938–1980* (New York: Oxford University Press, 1991); Gilbert C. Fite, *Cotton Fields No More: Southern Agriculture 1865–1980* (Lexington: University Press of Kentucky, 1984); Pete Daniel, "The Legal Basis of Agrarian Capitalism: The South since 1933," in *Race and Class in the American South Since 1890*, ed. Melvyn Stokes and Rick Halpern (Oxford: Berg, 1994); Harold Woodman, *New South, New Law: The Legal Foundations of Credit and Labor Relations in the Postbellum Agricultural South* (Baton Rouge: Louisiana State University Press, 1995); Jack Temple Kirby, *Rural World Lost: The American South, 1920–1960* (Baton

Rouge: Louisiana State University Press, 1987); and James C. Cobb, *The Selling of the South: The Southern Crusade for Industrial Development*, 2nd ed. (Urbana: University of Illinois Press, 1993).

Although the South has not received the attention it deserves from environmental historians—a gap that is starting to close—several key works of environmental history provided inspiration: William Cronon's *Nature's Metropolis: Chicago and the Great West* (New York: W. W. Norton, 1991); Richard White's, *The Organic Machine: The Remaking of the Columbia River* (New York: Hill & Wang, 1995); Donald Worster's, *Rivers of Empire: Water Aridity, and the Growth of the American West* (New York: Oxford, 1985); and Arthur McEvoy's *The Fisherman's Problem: Ecology and Law in the California Fisheries, 1850–1980* (New York: Cambridge University Press, 1986). Samuel Hay's masterpiece, *Conservation and the Gospel of Efficiency: The Progressive Conservation Movement, 1890–1920* (Cambridge: Harvard University Press, 1959), even though it was a generation before the emergence of the "field" of environmental history was another important guide. Scholarship in history of technology—a field long preoccupied with industrialization—has also begun to take the environment seriously. For an overview, see Jeffrey K. Stine and Joel Tarr, "At the Intersection of Histories: Technology and the Environment," *Technology and Culture* 39, no. 4 (1998). In addition to historical studies of pollution and the making of industrial environments, scholars working in this field have also begun to explore some of the ways nature actually gets incorporated into technological systems—a key theme for this book. See, for example, the essays in *Industrializing Organisms: Introducing Evolutionary History*, ed. Susan Schrepfer and Philip Scranton (New York: Routledge, 2004).

Several southern historians have also begun to explore in more direct ways the region's environmental histories. See Paul S. Sutter, "No More the Backward Region: Southern Environmental History Comes of Age" in *Environmental History and the American South: A Reader*, eds. Paul S. Sutter and Christopher J. Manganiello (Athens: University of Georgia Press, 2009); Mart A. Stewart, *"What Nature Suffers to Groe": Life, Labor, and Landscape on the Georgia Coast, 1680–1920* (Athens: University of Georgia Press, 1996); and Jack Temple Kirby, *Mockingbird Song: Ecological Landscapes of the South* (Chapel Hill: University of North Carolina Press, 2006). Their work builds on the rich tradition of southern historiography that attends to the many connections between people and the land, as well as a handful of earlier efforts directed explicitly at the environmental history of the region. See, specifically, Albert E. Cowdrey, *This Land, This South: An Environmental History*, rev. ed. (Lexington: University of Kentucky Press, 1996). Outside of environmental history, sociologist Robert Bullard put the issue of environmental justice and its relationship to the South's regional political economy squarely on the agenda with his landmark work *Dumping in Dixie: Race, Class, and Environmental Quality* (Boulder, CO: Westview, 1990).

Scholars of southern history and southern economic development have long emphasized the environmental costs of the region's postbellum economic development. C. Vann Woodward and George Tindall wrote eloquently about how the

South's dependent and underdeveloped status perpetuated a mutually reinforcing cycle of poverty and environmental degradation. Howard Odum and Rupert Vance, along with other scholars associated with the Institute for Research in Social Science at the University of North Carolina, described and analyzed how the destruction of the southern soil, the ravages of the boll weevil, and the pathologies of an extractive, resource-based economy reinforced the region's underdeveloped status and denied economic opportunity to so many who lived on the margins. And, of course, novelists such as William Faulkner and Thomas Wolfe offered powerful depictions of the waste and destruction visited on southern landscapes by industrialization.

This book was also deeply informed by work in agrarian political economy and, specifically, the so-called agrarian question literature, which emerged out of a series of European debates at the beginning of the twentieth century focusing on the political potential of the peasantry, the varying forms and effects of capitalist agriculture, and the role of agrarian class structures in industrialization. A key insight from this literature is that agrarian class structures and the incentives embedded within those structures have a profound effect not only on the development of capitalist agriculture but also on the nature and pace of industrialization. Such a perspective goes a long way toward explaining the stunted nature of postbellum economic development in the South and the transformative effect of New Deal agricultural and labor policies on the regional economy. It also resonates with the work of Odum and Vance, historians such as C. Vann Woodward and Gavin Wright, and the rich tradition of southern agricultural history. Key works in the agrarian political economy tradition include Karl Kautsky, *The Agrarian Question*, 2 vols., trans. Peter Burgess (1899; London: Zwan, 1988); Robert Brenner, "Agrarian Class Structure and Economic Development in Pre-Industrial Europe," *Past and Present* 70 (1976); and Robert Brenner, "The Social Basis of Economic Development," in *Analytical Marxism*, ed. John Roemer (Cambridge: Cambridge University Press, 1986).

Another important focus of the agrarian political economy literature has been the ways that agriculture's "difference," as manifested in its deep roots in land-based production and its irreducible biological qualities, shapes and informs efforts to industrialize agricultural systems. Work in this vein includes Susan Mann and James Dickinson, "Obstacles to the Development of a Capitalist Agriculture," *Journal of Peasant Studies* 5, no. 4 (1978); Susan Mann, *Agrarian Capitalism in Theory and Practice* (Chapel Hill: University of North Carolina Press, 1990); David Goodman, Bernardo Sorj, and John Wilkinson, *From Farming to Biotechnology: A Theory of Agro-Industrial Development* (Oxford: Basil Blackwell, 1987); and Michael Watts and David Goodman, "Agrarian Questions: Global Appetite, Local Metabolism: Nature, Culture, and Industry in *Fin de Siècle* Agro-Food Systems" in *Globalising Food: Agrarian Questions and Global Restructuring*, ed. David Goodman and Michael Watts (London: Routledge, 1997). By training attention to the distinctive obstacles (and opportunities) confronting efforts to industrialize biophysical systems, this literature proved highly relevant to understanding how the pulp and paper industry mobilized the productive capacities of southern land and labor.

As for the broad secondary literatures on American business and industrial history, the work of Alfred D. Chandler Jr. obviously looms large. See, specifically, *The Visible Hand: The Managerial Revolution in American Business* (Cambridge, MA: Belknap Press of Harvard University Press, 1977) and "Organizational Capabilities and the Economic History of the Industrial Enterprise," *Journal of Economic Perspectives* 6, no. 3 (1992). Equally important has been the alternative understanding of American industrialization, much of it focused on regionally specific trajectories of industrial development, advanced by Charles Sabel and others (including several of Sabel's students). Where Chandler emphasized throughput imperatives, coordination challenges, and economies of scale and speed, this literature focuses more on the contingent circumstances and thicker institutional environments within which particular forms of industrial organization emerge. See Michael J. Piore and Charles F. Sabel, *The Second Industrial Divide: Possibilities for Prosperity* (New York: Basic Books, 1984); Charles F. Sabel and Jonathan Zeitlin, "Stories, Strategies, and Structures: Rethinking Historical Alternatives to Mass Production," in *World of Possibilities: Flexibility and Mass Production in Western Industrialization*, ed. Charles F. Sabel and Jonathan Zeitlin (Cambridge: Cambridge University Press, 1997); Philip Scranton, *Endless Novelty: Specialty Production and American Industrialization, 1865–1925* (Princeton, NJ: Princeton University Press, 1997); Annalee Saxenian, *Regional Advantage: Culture and Competition in Silicon Valley and Route 128* (Cambridge, MA: Harvard University Press, 1994); Gerald Berk, *Alternative Tracks: The Constitution of American Industrial Order, 1865–1917* (Baltimore, MD: Johns Hopkins University Press, 1994); and Gary Herrigel, *Industrial Constructions: The Sources of German Industrial Power* (Cambridge: Cambridge University Press, 1996).

Although it might seem out of fashion today, this book was explicitly conceived and executed as a work of political economy. As understood here—and borrowing from Eric Wolf, *Europe and the People without History* (Berkeley: University of California Press, 1982)—political economy encompasses a broad field of inquiry concerned principally with the production and distribution of wealth, the role of classes and other groups in the genesis and distribution of wealth, and the role of the state in relation to the various classes and groups. In a world marked by academic specialization and disciplinary turf wars, such a perspective all too often gets lost in the shuffle. For many, it seems, political economy is nothing but an empty label, an outdated approach that has lost any claim to insight and explanation. But its original articulation is certainly no less relevant today than it was in the time of Adam Smith, Karl Marx, or Max Weber—all thinkers who explicitly conceived of their own work as political economy. In the "constructivist" approach taken here, attention to structure and process, actors and institutions, and history and geography defines the basic analytical framework. There is no master variable, no single "carrier" class or unconstrained logic of capital driving the story forward. At the same time, however, the story cannot be reduced to contingencies and particularisms. Structure matters, and the imperatives of capitalist development are very real indeed, even as they get worked out through vernacular landscapes and institutions. In the case of the southern pulp and paper industry, full appreciation for this dynamic process of industrialization requires attention to how various actors worked together and in conflict to make a highly competitive industrial system out of the distinctive historical-geographical milieu of the post–New Deal American South.

INDEX

Page numbers in *italics* indicate figures.